Patrick Moore's Practical Astronomy Series

For other titles published in the series, go to
www.springer.com/series/3192

Hunting
and
Imaging Comets

Martin Mobberley

 Springer

Martin Mobberley
Denmara
Cross Green
Cockfield
Bury St. Edmunds,
Suffolk IP30 0LQ
United Kingdom

ISSN 1431-9756

ISBN 978-1-4419-6904-0   e-ISBN 978-1-4419-6905-7
DOI 10.1007/978-1-4419-6905-7

Library of Congress Control Number: 2010937433

Printed on acid-free paper

Springer is part of Springer Science+Business Media (www.springer.com)

Preface

The death knell of the amateur comet discoverer has been prematurely sounded many times in the last few decades and yet, in the twenty-first century, a steady stream of comets found by amateur astronomers appears in the astronomical headlines. The qualities possessed by the visual observers of yore are still required, namely infinite patience and a love of the night sky, but the technology has changed as has the ability to take great images of the new discoveries. While remorseless robotic discovery machines scour the sky each night from cloud free sites, seemingly invincible, every fighter knows that every adversary has weaknesses that can be exploited. The machines cannot patrol in twilight and they do not look for fuzzy objects, but moving dots. In addition, even sites like New Mexico have cloudy nights and every professional observatory is dependent on endless funds to keep the facilities going and the machines maintained. Comets are unpredictable things too, which helps the amateur patroller. The hour after a CCD patrol has swept over a twentieth magnitude fuzzy star, it may choose to turn on and brighten by several magnitudes, placing it firmly within the amateur's grasp; alternatively a comet might brighten with a few days of full Moon when the professional detection software cannot work and so the robotic patrols are idle. There are opportunities for the determined amateur everywhere. The current multi-million dollar blitz on the sky is largely as a result of US funding to detect NEOs in the inner solar system, but how long will that funding continue in these post credit-crunch times, especially if there is confidence that almost every object more than a few hundred meters across has been found? Anyway, even if the NEO patrols do continue unabated there is no doubt that amateurs will relish the challenge even if the competition gets tougher.

Now, you may well have already decided that I am a fraud in this context! After all, you probably haven't heard of comet Mobberley. There is a good reason for this: there isn't one! In the early 1980s I spent several years and hundreds of hours sweeping for comets with a 36-cm f/5 Newtonian; I had quite a few near-misses, but they don't count of course. With Halley returning in 1985/1986 I resolved to concentrate more on comet photography (which I had dabbled with since 1982) and so that has been my main thrust since then. I was encouraged in this regard by one of the greats of British astronomy in the twentieth century, the late Harold Ridley (1919–1995). Harold was the UK's top comet photographer for many years and a true gentleman. He did not think much of reflectors and always took very long photographs using photographic plates or sheet film. His system comprised of a 7-inch (178-mm) f/7 triplet refractor mounted alongside a very long 6-inch Cooke refractor guidescope. Harold was the master of the offset guided long exposure in which he manually guided, with his eye literally glued to the eyepiece, for 20–40 minutes, while trailing the guide star along a micrometer cross wire, at a carefully calculated rate, to allow for the comet's motion. It was a badge of honor in those days to end up with a photograph in which the star trails were ruler straight and the comet frozen on the photograph. No-one ever produced straighter star trails than Harold Ridley! He was a perfectionist and learned his comet photography skills from the legendary wheelchair-bound Reggie Waterfield (1900–1986) and from Michael Hendrie.

I always recall an amusing tale Harold told me regarding a particularly clear night in late 1985 when he had meticulously offset guided a photograph of the brightening comet Halley and then retired with the sealed photographic plate holder into his darkroom to develop the photograph. The night had been very clear indeed and Harold was excited at what the plate might show. The developing and fixing took place smoothly in the very dim red glow of the darkroom light and then, with the photograph fixed he turned on the main light. To his horror he saw that the photographic plate was still lying on the worktop and therefore now bleached by the darkroom light! He had not developed the 4×5-inch glass photographic plate at all, but the 4×5-inch ground glass focusing screen from the dismantled plate holder, which felt the same in the dark and looked the same in the dim red light. "It took me a long while to get over that" he laughed......many, many, months later! If Harold had not been such a heavy smoker he might still be alive now. However, he died just before the

Great Comets Hyakutake and Hale-Bopp came along and a few friends have suggested he arranged their arrival from behind the scenes!

Memories of dedicated amateurs like Harold spur me on and so, although I no longer hunt for comets, I image them on a regular basis, still inspired by those photographic era stalwarts. The CCD toys we now have are twenty times more sensitive than any photographic plate and we are thoroughly spoiled with the options now available. It is an exciting time to be a comet imager and since the mid 1990s some incredible sights have come our way, not least Shoemaker-Levy 9 colliding with Jupiter, the splendid comets Hyakutake and Hale-Bopp and then, in 2007, comet C/2006 P1 (McNaught) and the exploding comet 17P/Holmes. Added to those amazing objects we have now seen fascinating space probe images of four cometary nuclei at close range.

Whether you are a comet hunter or a comet imager I hope this book has something to inspire you to become even more engrossed in a phenomenon that, more than any other, used to have our ancestors staring in awe, presumably open-mouthed, at the sight of a hairy "broom star" monster in the sky.

M. Mobberley   Cockfield, UK

Acknowledgements

No book of this type, containing so many astronomical images, would be financially viable without the generosity of astronomers and professional organizations that are happy for their fine images to be used for educational purposes. I am especially indebted to NASA in this regard whose images are generally copyright-free, thereby enhancing the quality of thousands of space and astronomy publications worldwide. I would also like to thank the following astronomers and scientific facilities whose images I have used and advice I have received while compiling this book: Charles Bell; Ed Beshore; Peter Birtwhistle; Reinder Bouma; John Broughton; Eric Bryssink; Bernhard Häusler; Michael Hendrie; Rik Hill; Karen Holland; Mike Holloway; Michael Jäger; Mark Kidger; Gary Kronk; Steve Larson; Terry Lovejoy; Dr Robert S. McMillan (on behalf of Spacewatch, the Lunar and Planetary Laboratory, the University of Arizona and NASA); Robert McNaught; Richard Miles; MIT Lincoln Laboratory, Lexington, Massachusetts; Gustavo Muler; Michael Oates; Prof. Greg Parker & Noel Carboni; Arnie Rosner; Juan Antonio Henríquez Santana; Ian Sharp; Clay Sherrod; Starlight Xpress; Roy Tucker.

Thanks are also due to all at Springer New York and SPi for guiding this book through the production process. Finally, I thank my father Denys for his constant support in all my observing and writing projects.

About the Author

Martin Mobberley is a well-known amateur astronomer from Suffolk, England, who joined the British Astronomical Association in 1969, aged eleven, initially as a visual observer. Since the early 1980s he has been a regular photographer and imager of comets, planets, asteroids, variable stars, novae, and supernovae. He served as one of the youngest presidents of the British Astronomical Association, from 1997 to 1999, and in 2000 he was presented with the association's Walter Goodacre Award. In 1997 the International Astronomical Union (IAU) named asteroid number 7239 as "Mobberley" in recognition of Martin's contribution to amateur astronomy. Martin is the sole author of seven previous practical astronomy books published by Springer as well as three children's "Space Exploration" books published by Top That Publishing. In addition he has authored hundreds of articles in the UK magazine *Astronomy Now* and numerous other astronomical publications, as well as appearing from time to time on Patrick Moore's long-running BBC TV program *The Sky at Night*.

Contents

CHAPTER ONE

Comets, Their Orbits, and Where They Hide!

Within the visible universe our solar system occupies a very local region of space. Sunlight reflected from the distant planet Neptune takes only 4 h to reach us here on Earth and yet we can also see quasars and gamma ray bursts whose light has taken up to 13 billion years to arrive. It may take our rockets a year or two to reach the planets but at least they can visit them within a time period that is short compared to our lifespan so we are seeing these objects in almost real time compared to everything else in the night sky. One added benefit of nearby objects is that they can drift against the background stars thereby adding an extra degree of reality and fascination to their study. Objects that move and change their appearance dramatically through a telescope are especially fascinating to us amateur astronomers. We all know that the Universe is a three dimensional place (and maybe even 11 dimensional if you are a string theorist!) but most of the time the objects in it look like they are fixed to a two dimensional star chart plastered on the sky a few miles above our heads. However, dramatic nearby phenomena are extra special and prove to us, if any proof where needed, that we are living in a solar system where absurdly large boulders orbit the Sun at incomprehensible speeds. Undoubtedly the most dramatic example of a rapidly changing spectacle in the sky is a total solar eclipse. Few who have seen one of these ultimate solar system alignments would dispute this; but unless you are on the narrow umbral track and have clear skies for those few minutes you will miss it!

A dramatic display of aurora or a true meteor storm, involving thousands of meteors per hour streaking through the sky, can have a similar effect on the observer too. There is one other category of object that can send shivers down the spine though and that is the appearance of a truly "Great" comet. During every century there are always a number of comets that, for a few days, weeks or even months can hold the general public spellbound as they gaze in amazement into the sky. The most awesome comets can have tails that span 60° or more, with heads bigger than the full Moon and they can perceptibly move against the starry backdrop of the constellations from night to night. Even for those people with no interest in astronomy, such a spectacle can hold their fascination as, suddenly, the Universe is three dimensional and they are standing on a small planet watching something enormous float through the inner solar system. Imagine if your name was attached to that comet, as its discoverer. Surely, it must be the biggest ego trip an astronomer can have!

The best comets can have tails that are millions, or even hundreds of millions, of kilometers long. They can, briefly, become the largest objects in the solar system, even if, in terms of mass they are relatively insignificant. The tails of comets are incredibly ghostly away from the head. It would not be a lie to say that the density of material in a comet's tail is, by normal standards, as tenuous as a laboratory vacuum; but when illuminated by the Sun, against a dark sky, it is remarkable how well that gossamer thin material reflects the light. All the gas and dust that forms those two distinct tails in a comet originates from a region called the nucleus. Cometary nuclei are tiny by solar system standards. Even the nuclei of "Great Comets" can range from just 5 km in diameter to a massive 60 km in the case of a comet like Hale–Bopp. OK, you would not want either of those two objects landing in your backyard, but by planetary standards those dimensions are very small.

Comets have always held a great fascination for amateur astronomers and potential naked eye comets are always highly anticipated. In the twenty-first century, technology has moved on and keen amateurs can do really useful scientific research where comets are concerned. While the visual astronomer is still able to observe comets through the eyepiece in the traditional way, the CCD imager can easily go five magnitudes deeper and bring out details using image processing that no dark adapted eye can ever see. In addition, the precise position of a comet, against the stellar background, can be measured in minutes, as opposed to the hours of work it took in the 1980s with photographic negatives and mechanical measuring engines. Amateurs can also now capture incredibly beautiful images of comets which would have been the envy of every professional astronomer only a few decades ago. Unfortunately, the CCD era and the age of the robotic telescope and automated motion checking software has meant that it is very tricky for amateurs to discover comets *visually* any more, but they do still discover comets as we shall see in the following chapters. In an era where amateur supernova discoveries occur on an almost weekly basis the most highly prized discovery of all is still that of a "Great Comet." Finding a comet destined to be a zero magnitude monster writes the discoverer's name across the sky and into the history books and, for a brief period, anyone on Earth gazing skyward will know the name of that discoverer.

Cometary Tails and Structure

In the cold outer solar system any comet will normally look pretty inactive. It might resemble something between a mountain sized boulder and a mountain sized snowball and the temperature will be too cold for almost anything to evaporate from the surface. Just imagine a very low density Mount Everest with the edges chiseled down, the vegetation missing and the Yeti removed and you are almost there! However, as the comet moves in towards the Sun it will heat up and two types of material will depart from the low gravity of the rotating boulder's surface. These types of material are gas and dust, as clearly shown in the magnificent photograph of comet 1P/Halley, taken by the comet imaging maestro Michael Jäger, in Fig. 1.1. I have already mentioned the term snowball and the term "dirty snowball" used to be a popular one for a cometary nucleus. However, recent space probe results (see Fig. 1.2a–e) suggest that all frozen ice on comets may be locked some distance underneath the surface, meaning that snowball term may soon become less popular. In addition cometary nuclei are surprisingly dark, only reflecting 3–4% of the light falling on them: most unlike a real snowball. We should learn even more in November 2014 when the Rosetta space probe's Philae lander is due to soft land on the nucleus of comet 67P/Churyumov–Gerasimenko, 9 months before it reaches perihelion, and monitor the activity as the nucleus heats up.

Gases detected spectroscopically in the tails of comets include carbon monoxide, carbon dioxide, methane and ammonia as well as some distinctly organic compounds such as hydrogen cyanide, methanol, ethanol, ethane and formaldehyde. It is quite common to read about water being

Fig. 1.1. Comet 1P/Halley exhibiting a classic narrow and straight gas tail as well as the broad fan of a dust tail in this splendid photograph by the comet imaging maestro Michael Jäger. The photograph was taken with a 20-cm Schmidt camera on April 17, 1986. Image: Michael Jäger.

produced in comets and as water immediately creates a mental connection with "life" the term is, perhaps, used far too often without explanation. In fact what is being detected is usually the Hydroxyl radical OH,

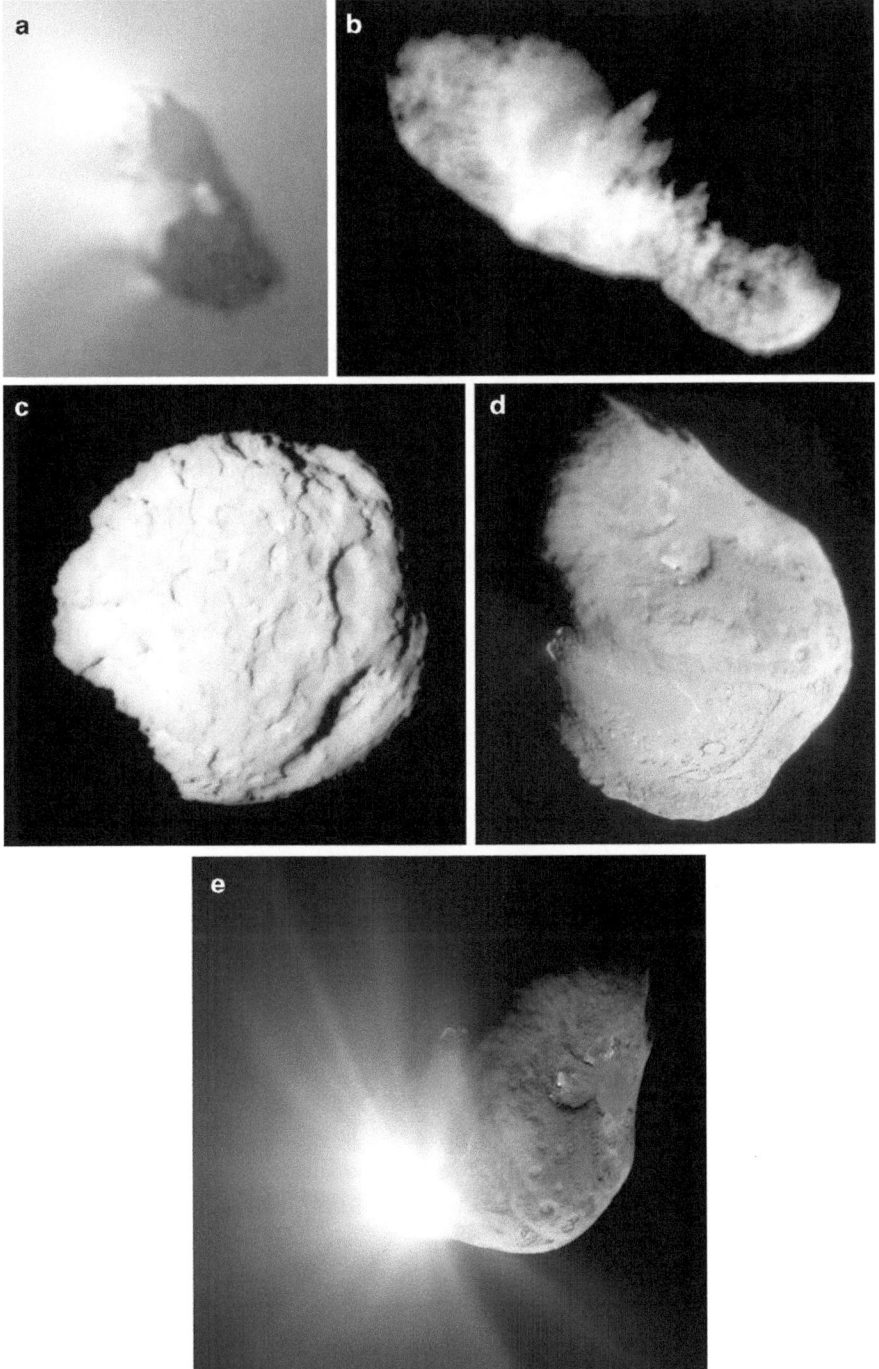

Fig. 1.2. (a) The nucleus of comet Halley as imaged by the space probe Giotto on March 14, 1986 from a distance of 600 km. The nucleus is 15 km long and the diameter varies from 7 to 10 km. Image: ESA. (b) The nucleus of comet 19P/Borrelly as imaged by the space probe Deep Space 1 on September 21, 2001.

Fig. 1.2. (continued) The nucleus has maximum dimensions of 8 × 4 × 4 km, a density of approximately 0.3 g/cc, a mass of 20 trillion kilograms and an albedo of only 3%. Image: NASA. (**c**) The 5.5 × 4.0 × 3.3 km diameter nucleus of comet 81P/Wild as imaged by the space probe Stardust on January 2, 2004. Particles from the comet's coma were returned in a canister which crash landed in Utah on January 15, 2006 and analysis of these revealed many organic compounds and crystalline silicates. The density of the nucleus is only 0.6 g/cc and it weighs in at around 23 trillion kilograms. Image: NASA. (**d**) The nucleus of comet 9P/Tempel imaged by the NASA Deep Impact probe on July 4, 2005. The nucleus measures 7.6 × 4.9 km and has a density of 0.6 g/cc. The mass is estimated as 75 trillion kilograms. The nucleus rotates every 40 h. Image: NASA. (**e**) The moment that a projectile from the NASA Deep Impact probe collided with the nucleus of comet 9P/Tempel. The time was 05:52 UT on July 4, 2005. The comet was 1 day before perihelion. The impacting projectile weighed 362 kg and collided with the nucleus at 10.3 km/s with an energy level roughly equivalent to 5 tons of high explosives being detonated. It is thought to have released 10,000 tons of loosely bound material. The crater formed by the impact was roughly 200 m across and 40 m deep and various compounds were detected spraying from the impact including silicates, carbonates, smectite, metal sulfides amorphous carbon and polycyclic aromatic hydrocarbons. Image: NASA.

which is one of the products of water dissociation. Water itself cannot be directly observed and, of course, water as we understand it, would have to be frozen solid beneath the top layer to be preserved for any significant duration. The same applies to the recent evidence that the Moon's southern pole contains "lake loads" of water, when all that has been observed is the OH radical.

Nevertheless, most professional astronomers agree that water is probably the dominant ice in comets and its sublimation is thought to control the evaporation of other volatile gases when the comet is closer to the Sun than 3 or 4 AU (AU = 149.598 million kilometers: the Earth–Sun distance). Further out than that 3–4 AU distance volatile compounds, such as CO, probably dominate cometary performance. The water production rate is estimated by detecting its photo dissociation products.

When the Giotto probe passed close to Halley in 1986 the ESA scientists interpreted the data as indicating that a total of three tons per second of matter was being ejected from the nucleus and emitted from about seven active sites on the comet. The spectral makeup of the material indicated that it had 80% water (OH) content; a 10% carbon monoxide content and 2–3% was an Ammonia/Methane mixture. The remaining 7 or 8% was made up from hydrocarbons and small amounts of iron and sodium.

Sometimes a cometary gas tail is described as an ion tail or a Type I tail or sometimes as a plasma tail. However, whatever the name it is formed by solar radiation ionizing the cometary coma particles so that the positively charged particles stream away from the Sun and give the coma and comet its own magnetosphere. The classic "bow shock" buffer forms against solar wind particles on the sunward side of the coma and solar magnetic field contours form around the comet and its tail. The ion tail in a bright comet will often look like a spring onion in this situation and over hours subtle details are occasionally seen spiraling around or moving down the tail axis. Sometimes the solar environment's magnetic field lines break the cometary barrier and temporarily disconnect the ion tail and it floats away, to be replaced by a new one.

The tails of comets always point away from the Sun so that after perihelion the comet moves tail first, which can seem a bit strange the first time you learn about it. When you see a picture of a comet with a long tail it is easy to imagine that it is plowing at high speed through the mythical ether and the tail is flowing behind it like the tresses of a modern day Rapunzel, seated on a motorbike! In fact the tails are pushed outward by the force of the solar wind. The ions in the tail are lightweight and so flow directly away from the direction of the Sun, along the so-called "extended radius vector."

However, the dust is heavier stuff and so although the dust tail (sometimes called a Type II tail) also streaks away from the Sun there is an amount of lag due to the masses of the dust particles; by the time the dust particles have moved a significant distance away from the cometary nucleus that nucleus has moved a significant amount in its orbit. So the dust tail often appears curved and at an angle to the gas tail. As the gas tail often contains exquisite fine detail it is good when the dust and gas tails are separated as the dust tail can hide the gas tail's beauty. Gas tails often record as a green/blue color on images because the coma contains cyanogens (CN) and diatomic carbon (C_2) which glow green–blue in sunlight due to fluorescence. In contrast dust tails simply reflect the sunlight and so appear yellowish.

Fig. 1.3. Comet C/2006 P1 (McNaught) and its amazing dust tail, photographed by Terry Lovejoy with a 23-mm lens on January 20, 2007. This is a 50° wide field and a 172-s exposure at f/3.2.

Exactly how the tail of a comet appears to the earthbound observer can be tricky to predict. After all, if a comet is heading inward toward the Earth the tail will be behind it and appear greatly foreshortened. If a comet has passed within the Earth's orbital radius of 1 AU (that value we saw earlier of 149.6 million kilometers or 93 million miles) and is returning from a close perihelion passage it may be heading out towards the Earth and so the end of the tail will be closer to us than the head. However, we will just see a two dimensional view which suggests the tail is simply at right angles to the head. Appearances can be very deceptive!

When healthy comets fly very close to the Sun the combination of the heat and the force of the solar wind can produce amazing and enormous "synchronic" band structures in the tail, unlike anything else. This was dramatically illustrated by comet West in 1976, comet Hale–Bopp in 1997 and, most dramatically of all, by the comet C/2006 P1 (McNaught) in January 2007 (see Fig. 1.3). The remarkable striations in comet McNaught's tail have to be related to gravity, solar wind and radiation forces acting on the dust. The time of particle release from the rotating nucleus as well as the dust grain sizes will define the patterns seen. The multiple tail effect caused by dust grains released at the same time, and then released again over many days, are called synchrones, whereas the more horizontal patterns caused by dust grains of similar mass are called syndynes. Comet McNaught easily displayed both patterns, even to the casual naked eye observer!

Anti-tails

A comet can even exhibit a tail pointing in two completely opposite directions because while the solar wind can force the traditional tail material out in a direction opposite to the Sun the flat plane of an active comet's orbit will often contain a considerable amount of dust especially close the head of the comet. Precisely how much dust you see ahead of the comet, or behind it, depends on a number of factors including the direction of rotation of the nucleus relative to the orbital plane, the activity levels on the comet and our proximity to the orbital plane of the comet. This final criterion is the most significant as if the Earth passes through the orbital plane of the comet the dust along the orbit is briefly viewed edge-on and so a narrow spike of material is seen. Whether the Earth actually does pass through the orbital plane of a very active comet close to perihelion, with the object in a dark sky, is just down to luck although the Earth must pass through the orbital plane of any comet every 6 months as both objects orbit the Sun and the Earth obviously orbits the Sun once per year. However, some comets do not lay down much dust in the orbital plane and if that orbital plane is at almost 90° to the Earth's orbit (the ecliptic plane) our passage through that plane can be very swift indeed: it can all be over in a day or two! So, being able to observe the comet at plane crossing time is crucial and many of the factors, such as altitude, cloud, solar elongation (and therefore twilight) are beyond the observer's control. The ideal situation for seeing an anti-tail is when a newly discovered and very dusty comet, with plenty of activity left on the surface of its nucleus, has an orbital plane with a very similar inclination to the ecliptic, namely an "i" close to 0 or close to 180°. One of the most perfect comets in this regard was discovered quite recently, in 1997 in fact. Astronomers at Taiwan's Lulin observatory in Nantou discovered a comet on July 11 of that year that was 6 months from perihelion. The prospects looked good and when the orbit was calculated the inclination was found to be 178.4°. The comet was orbiting within 2° of the ecliptic and precisely backwards (retrograde) with respect to the Earth's direction and the direction of all the major planets and main belt asteroids. Astronomers soon realized that this comet might have a permanent anti-tail close to perihelion and

Fig. 1.4. Comet C/2007 N3 (Lulin) showing its dust spike/ anti-tail feature to the *left* and a gas tail feature to the *right*. The field is 1° wide. Imaged remotely by Martin Mobberley using a Takahashi 0.25 m f/3.4 astrograph and SBIG ST10XME based at GRAS in New Mexico.

this turned out to be the case (see Fig. 1.4). For much of its immediate post perihelion phase C/2007 N3 (Lulin) exhibited two dust spikes either side of the head as well as an occasional, rather fainter, gas tail. For comets with less perfectly aligned orbits the key to determining whether an anti-tail will develop is the orbital parameter called the longitude of the ascending node. We will look at orbital elements in more detail very shortly.

Short and Long Periods

Comets and their tails come in all shapes and sizes as do their orbits, but historically they have been classified into two groups by the professionals, namely they are either short period or long period. The short period comets come with a "P/" before their name and the long period comets have a "C/" prefix. Some examples may be of help here and so here are five comet names: P/2009 U4 (McNaught); P/2009 T2 (La Sagra); C/2009

T3 (LINEAR); C/2009 S3 (Lemmon); 17P/Holmes. The alert reader will notice that the discoverer's name, whether the name of an individual or an organization, appears in parentheses for the long period C/ comets and the newly discovered short period P/ comets but that the style for the long established short period comet 17P/Holmes is rather different. In this latter case the comet has not just been recognized as one of short period, but has returned to perihelion two or more times and allotted a permanent number (17 for Holmes) in the list of periodic comets. These permanent numbers are given in chronological (historical) order in which the comets are confirmed as being a multiple return to perihelion object. The initial un-numbered P/ type designation, as with P/2009 T2 (La Sagra) for example, is immediately applied if the orbit is calculated to have an eccentricity low enough for it to return in less than 30 years. If the period of the newly discovered object seems to be more than 30 years it will have to return once more to get a P/ prefix and two or more times to be numbered.

At the time of writing there are 230 known periodic comets with a designated number and many more recent discoveries which will push that number above 250 very soon. In total some 3,700 comets are in the records, both historical and modern. However, this number needs to be treated with some caution as more than 1,700 of these are tiny cometary fragments discovered by the SOHO satellite and known as sungrazing comets. Most of these fragments probably originate from the break-ups of just a few big comets as they approached the Sun, as we shall see later. So, excluding them, we have a known comet population of around 2,000 of which more than 85% are long period. However, the true number of comets out there in the furthest reaches of our solar system must be truly enormous, running into countless millions.

You would not have to be a genius to work out that the famous comet Halley is numbered 1P, as it was the first comet to be numbered as a returning object. The other letters and numbers in a comet's designation, such as the U4 in P/2009 U4 (McNaught), denote the order that the comet was discovered during a specific half-month of the year. Thus McNaught was the fourth comet discovered in half-month "U" which represents the second half of October (the letters I and Z are not used in the system). Occasionally you see a comet with a really complex designation like P/2009 SK280 (Spacewatch–Hill), which means this is a short period comet which was originally classed as an asteroid. Sometimes the numbers are written in a subscript font. The S still represents the half month for this example, but the 280 means the alphabet has been used up 280 times before we got to the letter K on the 281st run through. Yes, there are that many asteroid discoveries; as many as 12,875 in the second half of October 2005 when the last asteroid labeled was 2005 UW512. Crazy! But, getting back to comets, the current system has been established since 1995 and works very well. However, historically, there was a different system and so all the comets discovered prior to 1995 have had to be adopted into the new system. The old system used Roman numerals (I, II, III, etc.) to number comets in order of discovery within the year and this was then replaced with a letter (a, b, c, etc.) when the date of the comet's perihelion was calculated. In the modern system "C" and "P" are not the only letters used; the letter "D" indicates a comet that has disintegrated or is "defunct" and the letter "X" denotes a historical comet for which no reliable orbit was calculated. Finally the rarely used "A" prefix indicates an object that was mistakenly identified as a comet, but is actually an asteroid (also known as a minor planet).

Maybe I just have a pitiful and puny lump of grey matter between my ears and my "far too close together for comfort" eyes but there are still aspects of this numbering system that confuse me, a confirmed comet fanatic. When I see a modern style cometary designation with "V1" or "X1" designated as its half-month time (November and December's first fortnight) and order of discovery (first in both cases) my brain does a double-take as it looks, at first glance like the good old Roman numeral designations have returned! Also, a comet with a half month discovery designation of "P" automatically triggers my brain into thinking that comet must be periodic, when it is actually just telling me that it was found within the first fortnight of August.

200 or 341 Years?

I have digressed considerably in the previous paragraph as what I was trying to convey to the reader was the fact that there are two basic groups of comets which can be hunted and discovered: the long period (C/) comets and the short period (P/) comets. But, I hear you cry, where is the dividing line between a short period and a long period comet? After all, comets orbit the Sun, so surely they are all periodic and so why make a distinction? Well, it is a very good question and I will try to answer it. The vast majority of short period comets orbit the Sun roughly in the plane of the ecliptic, just like the main belt asteroids, and so, to comet observers, there are plenty of "old friends" amongst these periodic comets. Typically, these comets orbit the Sun with periods between 5 and 20 years, although comet

Fig. 1.5. Comet 2P/Encke imaged on 2003 November 26.745 with a Celestron 14 at f/7.7. 14× 20 s using an SBIG ST9XE CCD. The field is 13′ wide. Image: Martin Mobberley.

2P/Encke orbits in only 3.3 years (see Fig. 1.5) and they are being herded into these orbits by the influence of Jupiter. In general, comet orbits are far more elliptical (less circular) than the orbits of the asteroids/minor planets between Mars and Jupiter. However, there are notable exceptions to the regular, well-behaved cometary visitors and comet 1P/Halley with its 76-year retrograde orbit inclined at 162° to the ecliptic is the most famous example. There are bright comets with even longer periods too, like 109P/Swift–Tuttle with a 130-year period and 35P/Herschel–Rigollet with a period of 155 years. Prior to 2002 an arbitrary cut-off point of 200 years was deemed to be the dividing point between short and long period comets simply because no comets with longer periods had ever been observed in the telescopic "orbit aware" (for want of a better term) era. Then, in 2002, two amateur astronomers, Kaoru Ikeya and Daqing Zhang, in Japan and China respectively, discovered a comet which the orbital calculations proved had previously been seen by Chinese astronomers in 1661, some 341 years earlier. So, the dividing line is now 341 years

Fig. 1.6. Comet 153P/Ikeya–Zhang: the longest orbital period short-period comet. This image was taken with an 80-s exposure, on 2002 March 26.822, using a 16-cm aperture f/3.3 Takahashi Epsilon astrograph and Starlight Xpress MX916 CCD. The field is 1° high. Image: Martin Mobberley.

because comet Ikeya–Zhang is now, officially, 153P/Ikeya–Zhang (see Fig. 1.6) and no other periodic comet recognized by the International Astronomical Union/Minor Planet Center has a longer period.

From an astronomical discoverer's point of view the ultimate dream find has to be hunting down a bright naked eye long period comet, but, even before the automated machines started searching for moving objects in the night sky, such discoveries were very rare. In recent memory the two zero magnitude comets that peaked in 1996 and 1997, namely C/1996 B2 (Hyakutake) and C/1995 O1 (Hale–Bopp) marked the absolute zenith of comet hunting success and, remarkably, they were discovered by human beings who were not employed as professional salaried astronomers. Comet Hale–Bopp probably has a period of around 2,500 years and Hyakutake around 100,000 years. The latter comet may have had a period of only 8,000 years prior to its 1996 return but the influence of the planets on its inward journey made the orbit considerably longer; it will not be back for a very long time. It is hardly surprising that the short period comets do not put on displays as spectacular as these two objects. Many of the really short period comets, originally from the outer solar system, but steered into shorter orbits by Jupiter, will have orbited the Sun hundreds of times and each time they warm up and lose gas and dust from their surfaces the comet gets weaker. Some astronomers place an active lifetime of a thousand orbits on these objects before they become dead. Taken to the extreme all comets will eventually become asteroids (a term interchangeable with the description "minor planet") or even disintegrate as more and more volatile material evaporates. So if you are looking for another zero magnitude comet, you need to search the whole night sky, as those undiscovered short period comets, mainly situated within about 25° of the ecliptic plane and herded by giant Jupiter, are unlikely to suddenly become naked eye objects. Having said this, in October 2007, 17P/Holmes outburst by half a million fold to become an easy naked eye object and 17P has an orbital period of 6.9 years and a 19° orbital inclination. The comet was discovered, by Edwin Holmes, during a similar outburst some 115 years earlier in 1892. In 2007 the outburst of 17P was first detected by the Spanish amateur Juan Antonio Henríquez Santana but the name of this established comet was not changed to 17P/Holmes–Santana. However, the case of 17P is pretty unique and I will have more to say about this remarkable object in Chap. 10.

The Ecliptic Plane

All eight major planets orbit the Sun within 7° of the Earth's orbital plane, also known as the ecliptic plane, and if we exclude Mercury that figure becomes 3°. Jupiter, the dominant planetary body of the solar system orbits the Sun in a plane tilted by slightly more than 1° to the ecliptic. Perhaps surprisingly the Sun itself rotates at an angle of 7° to the ecliptic. But, essentially, if you suspend yourself above the ecliptic the vast majority of objects are trundling around in an anti-clockwise direction, just like the good ship Earth.

Over aeons of time the primordial gas and dust in the Solar System either collapsed to form the Sun or coalesced into planets, asteroids and comets. The entire Solar System would once have been full of trillions of chunks of debris with regular massive impacts occurring on the major planets, but these days the only region of the Solar System that contains hundreds of thousands of mountain sized objects is the region between Mars and Jupiter, known as the asteroid belt. Comets are, in effect, similar to icy or fragile asteroids. A comet appears fuzzy on an image or a photograph, or in the telescope eyepiece, because material is sublimating and out gassing from the surface as the object heats up. Close up the comets look like low density asteroids with a few deposits on their surfaces, but from millions of miles away the nucleus (typically a few kilometers across) shrinks to a point and the tenuous gas and dust spreads over thousands or millions of kilometers, thereby reflecting far more sunlight than the nucleus itself. As a result comets can look infinitely bigger than one might expect from their tiny nuclei. As I mentioned earlier, against the blackness of space a huge cloud of dust and gas reflects sunlight very noticeably.

Use any decent planetarium software (my personal favorite is Project Pluto's Guide 8.0) and set the magnitude limits for comets and asteroids to fainter than magnitude 20 and you will see the ecliptic plane light up with asteroids and periodic comets. The comets that dominate such a plot will, not surprisingly, be the ones discovered in the past year by the professional patrols. Such comets are often captured when they are moving in to perihelion and well situated in the night sky; in other words, transiting at midnight, as seen from our Earth-based observing platform.

Magnitude Formulae

Typically, comets brighten highly exponentially, explosively sometimes, as they move closer to the Sun. An asteroid will also brighten exponentially but less dramatically. Essentially a dead cometary nucleus can be regarded as an asteroid, although the color and albedo may be quite different. For an inactive cometary nucleus the nuclear magnitude, $m2$, as it approaches the inner solar system, ignoring its phase (so it will be a "full" asteroid) will be given by $m2 = H_0 + 5 \log \Delta + 5 \log r$ where H_0 is the absolute magnitude, Δ is the Earth-nucleus distance and r is the Sun-nucleus distance. Distances are in Astronomical Units (AU), to reiterate, the average Earth–Sun distance of 149.598 million kilometers. The 5 log terms are simply a standard inverse square law. Halve the distance to the Earth and an object looks four times brighter if the solar distance is maintained. Halve the distance to the Sun and an object looks four times brighter if the Earth distance is maintained. Halve both at the same time (frequently the case for distant comets approaching Earth and Sun) and an object looks 16 times brighter. We can clearly see that there is a "double-whammy" effect here for objects a long way out. You only have to take an inactive asteroidal object twice the distance from Sun and Earth and it looks 16 times fainter because it is collecting a quarter of the sunlight and it has an angular area a quarter of the size when viewed from the Earth! The formula $m2 = H_0 + 5 \log \Delta + 5 \log r$ makes it look far less severe than this because this formula incorporates not only the inverse square law but also the astronomical magnitude law where a factor of 100× dimmer gives you a mere five magnitude drop. Nevertheless, take an asteroid twice the distance from Sun and Earth and it fades by three magnitudes. As a rough guide the value of H_0 for a 3-km diameter asteroid with an albedo of 0.25 will be about 15.0.

However, things are about to get far worse now because we are not dealing with asteroids in this book at all but active comets that fade far more quickly as they recede from the inner solar system. The standard magnitude law for an active comet is given by $m1 = H_0 + 5 \log \Delta + 10 \log r$, but the $10 \log r$ term is far more dramatic, reflecting the fact that if you halve a comet's distance from the Sun alone (with the Earth distance fixed) it will, typically, brighten 16 times: an inverse fourth power rule. Now we can see that if we double an active comet's distance from Sun and Earth simultaneously it will typically fade 64-fold, or 4.5 magnitudes. Painful! Now, the discerning reader will have spotted that the $10 \log r$ term does not correspond to any fundamental law of physics. It is simply a typical historical value for many comets which brighten to the tune of an inverse fourth power law as they move closer to the Sun. However, in practice, once a comet has passed through perihelion, and been well observed, that $10 \log r$ term may have to be modified to, say, $8 \log r$ for a rather more modest cometary fading. Indeed, the $10 \log r$ term is a bit too fixed and exponential for many comet experts and is often substituted by the term $2.5n \log r$. The rate of brightening coefficient n has to be set to 4 to give $10 \log r$, but many observers feel a more modest value of 3, giving $7.5 \log r$, is better. Historically the $10 \log r$ term comes from Vsekhsvyatskij, whereas a value nearer $8 \log r$ was favored by Bobrovnikoff. Of course, comet nuclei kilometers across have a thermal lag and often peak a few weeks after perihelion. These simple formulae are not sophisticated enough to incorporate the thermal lag factor and so, quite often, different formulae are applied pre and post perihelion.

In addition, the absolute magnitude H_0 needs to be determined to fit the light curve. At 1 AU from Sun and Earth the comet will have a magnitude equal to the value H_0, but it may not come that close to either body. An average comet may have an H_0 of, say, six or seven, but every comet is different and the behavior of some cannot be modeled by the basic magnitude formula, regardless of any thermal inertia. There is an additional factor that should be mentioned here which is called forward scattering. Sometimes when a comet passes almost between us and the Sun, so that its elongation from the Sun is small, the geometry and properties of the dust particles in the coma enhance the brightness by a phenomenon known as forward scatter. The phase of the comet might be small (as it is mainly being illuminated from behind) but light actually ends up being scattered *through* the coma and on towards us. If the geometry and coma dust properties conspire an enhancement of several magnitudes is possible when the comet is less than 30° from the Sun.

The reason I am laboring these points about comet magnitudes is simply to explain that while the aphelion and perihelion distances of a short period comet in the ecliptic may not seem that extreme, halving the distance a comet is from the Sun can increase its brightness by three magnitudes or more. In addition, when moving in towards perihelion, the comet invariably moves closer to the Earth, thereby adding another magnitude or two. Bearing this in mind it is easy to see that a comet seemingly happily moving in the zone between Jupiter and Mars can easily brighten by a 100-fold on its way to perihelion. For faint comets that may be, say, 22nd magnitude at aphelion, that rise to 17th magnitude brings it well within the range of the professional CCD (Charge-Coupled Device) patrols and it sees a comet rise from being just a fraction of a percent as bright as the light polluted sky background to it sticking well above the noise.

As most amateur astronomers will be well aware the ecliptic plane, as seen in our night sky, performs a sine wave shape as it travels through right ascension and declination. This sine wave peaks at +23.5° declination near the Gemini/Taurus border and bottoms out near the Sagittarius/Ophiuchus border at −23.5°. This, of course, is a consequence of the Earths axial tilt relative to the ecliptic plane. The ecliptic crosses the celestial equator at 0° declination in the constellations of Virgo and Pisces. The ecliptic is actually defined as the path that the Sun travels against the constellations during the year, even if, as the sky will be bright blue, the constellations will be totally invisible. Take a few deep CCD images in the ecliptic plane on any clear night and there will be faint asteroids in there somewhere. More than 400,000 known asteroids or minor planets live in that 360° line around the sky and so that averages out at more than 1,000 per degree of ecliptic longitude! Of course, the ecliptic plane is not infinitesimally thin, its main concentration spans many degrees of ecliptic latitude too; nevertheless, I think you can see how packed with moving objects the asteroid belt is. Many of these 400,000 asteroids will spend most of their lives below twentieth magnitude, but undoubtedly many more await discovery. Professional surveys easily detect asteroids that are just a kilometer or so in diameter but surely there must be much smaller objects out there as yet undiscovered which will take the tally into the millions at some point. Amongst all those moving dots are comets too, many masquerading as asteroids when first discovered, and some will only reveal themselves when close to perihelion, despite remaining dormant for much of their lives. There are currently more than a hundred known periodic comets within 15° of the ecliptic plane.

Orbital Elements

I have always felt (although some will disagree) that to appreciate the characteristics of any comet it is vital to be able to visualize its orbit in three dimensions. The orbital elements of a comet define its path through the solar system with respect to the ecliptic plane and with respect to the so-called "spring" or "vernal" equinox point on the ecliptic (see Fig. 1.7).

These orbital elements are defined as follows:

T = the date of perihelion, when the comet is closest to the Sun. For asteroids the date of perihelion is rarely used, partly because the brightness of an asteroid is not so dependent on its proximity to the Sun and also because asteroids do not have the sort of extremes of aphelion and perihelion that bright comets do. So, for asteroids the perihelion date T is replaced by the term M, where M = the "mean anomaly," and essentially gives you an angle for how far an asteroid was along its orbit at the quoted "Epoch time."

q = the perihelion distance in AU.

e = the eccentricity of the orbit. A value of 0 means the orbit is a perfect circle. A value of 1 means the orbit is a parabola (i.e., long period comets). Between 0 and 1 the orbit is an ellipse. A value greater than 1 signifies hyperbolic: more on this later.

Ω, or simply "node" = the longitude of the ascending node. This is the angle, in the plane of the ecliptic from the spring equinox, to the point where the comet ascends through the ecliptic. It is measured in the direction of the Earth's motion.

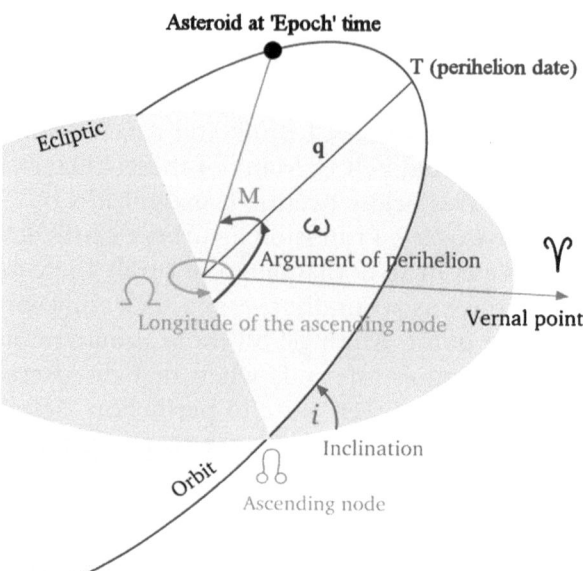

Fig. 1.7. The orbital elements that define the orbit of a comet or an asteroid. The argument of perihelion and the perihelion date are specified for comets, whereas the Mean Anomaly (M) and Epoch time are used for asteroids. Otherwise the elements are the same.

ω, or simply "peri" = the argument of perihelion. This is an angle in the plane of the comet's orbit, measured from the point where the comet ascends through the ecliptic to the perihelion point and measured in the direction that the comet is moving.

i = the inclination angle that the asteroid's orbital plane makes with the ecliptic. If the comet is moving around the Sun in the same direction as the Earth (direct), i.e., anti-clockwise, this will be between 0° and 90°. If the comet is moving clockwise around the Sun (retrograde), the inclination will be between 90° and 180°.

No other parameters are required to define a cometary orbit, although you will often see the letters a, n and P in the orbital elements too. The first of these letters, a, represents the semi-major axis of the ellipse that defines the orbit of the comet. In other words, take the long axis of the ellipse, from perihelion to aphelion, and halve it. The letter n indicates the mean daily motion of the comet in degrees per day and P gives the orbital period in years.

Two additional symbols appear when the orbital elements are used to produce a day by day ephemeris for the position off the comet. These are

Δ (or R) which represents the Earth to Comet distance and r which represents the Sun to Comet distance, typically in Astronomical Units.

It is worth mentioning here that there is a subtle difference between the terms Equinox and Epoch when defining an object's position in space. The Equinox defines the coordinate system, which, in practice, is the grid of right ascension and declination used on star charts. The J2000.00 Equinox is, not surprisingly, the one that is currently in use as it is defined for the year 2000.00. The term Epoch does not relate to the defining of the coordinate grid system but to the orbit of the celestial object itself. Thus an orbit will be calculated for a specific Epoch, such as, say 2010 January 3.9444, even though the position of the comet will be calculated with respect to the 2000.00 Equinox, so you can easily find it with a standard star chart (paper or software based).

Anti-tail Prospects

Earlier in this chapter I mentioned that the likelihood of an active comet developing an anti-tail could be deduced from studying the parameter called the longitude of the ascending node (indicated by the symbol Ω); but as I had not explained the full meaning of orbital elements I decided to delay any further clarification of this issue until now. Looking down on the orbit of the Earth from above the ecliptic plane we can see that our planet moves around the Sun in an anti-clockwise direction. At the point in the orbit where our longitude measured in the ecliptic plane, from the vernal equinox, is the same as the longitude of the ascending or descending node of the comet we are then effectively looking through the edge-on plane of the comet. The longitude of the descending node is, fairly obviously, 180° different from that of the ascending node. But how can we calculate when these situations occur and when the possibility of seeing a cometary anti-tail is therefore at a maximum? Well, an approximate way is by acquiring a nautical almanac which gives a table of the Sun's ecliptic longitude for each day of the year. You do not have to be a genius to realize that this will increase by roughly a degree per day because there are 360° in a circle and 365.24 days in a year. When the Sun's ecliptic longitude equals or is 180° different from the longitude of the comet's ascending node the Earth is close to the plane of the orbit of the comet and an anti-tail is more likely to be visible. Of course, if the comet is like C/2007 N3 (Lulin) and has an orbit almost in the ecliptic plane, we are always close to the plane of the cometary orbit. But such instances are exceedingly rare.

Constructing a Cardboard Model of a Cometary Orbit

When I first encountered cometary orbital elements as an 11-year old I found it very hard to visualize them in three dimensions. My solution was to make a three dimensional model out of cardboard and, even in this era of 3D graphics I still think it is an entertaining project! To construct any cardboard model of a comet's orbit the starting point is obtaining two sheets of cardboard! One represents the Earth's orbital plane; the other represents the comet's orbital plane. For our model let us pick a bright comet, say, C/2001 Q4 (NEAT). Its orbital elements, to create accuracy appropriate for a cardboard model, are as follows:

T = May 16, 2004; q = 0.96; e = 1.0; peri = 1°; node = 210°; i = 100°

We will choose a scale of 100 mm to 1 AU for our model.

The first step to take is to mark the card representing the ecliptic plane to show where the Earth is at each month of the year. We also mark the Spring Equinox and Ω, the longitude of the comet's ascending node: 210° for this comet. It is vital to remember that the line from the Sun, indicating the Spring Equinox is approximately 20 days (or 20°) past September 1. This is because the Spring Equinox position is the *Sun's apparent position against the constellations as viewed by the Earth on about March 21* and not the Earth's position in spring. The line from the ascending node to the Sun and out to the opposite side of the Earth's orbit represents the intersection of the two orbital planes, i.e., the slot through which the cardboard plane of the comet's orbit will slide through the ecliptic (see Fig. 1.8a).

The second step is to mark the cometary plane card with the position of the Sun and the angle (a tiny 1° for this comet) and line indicating the argument of perihelion. It is vital to remember that this angle is measured from the point where the comet ascends through the ecliptic and is measured in the direction that the comet is moving. If this angle is more than 180° then perihelion occurs below the ecliptic. The comet at perihelion can also be marked at the end of the line. In this case, q = 0.96, so q is 96 mm from the Sun.

Next, a cardboard shape or triangle, set to indicate the angle of inclination, can be glued to support the comet's orbit plane. The inclination for C/2001 Q4 is 100°, so an 80° triangle will suffice (180–100°), but we must

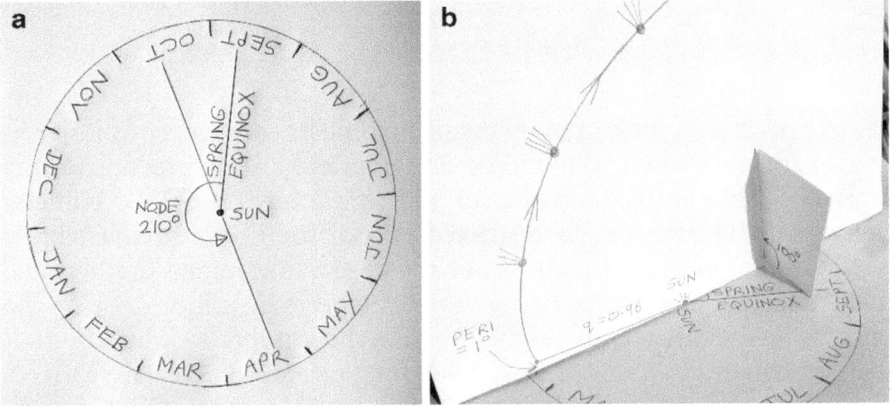

Fig. 1.8. (**a**) The first step to making a cardboard comet orbit model is marking up the piece of card representing the ecliptic plane with a pen. (**b**) The completed cardboard model consists of an ecliptic plane card, a comet orbital plane card and a small cardboard insert defining the angle of inclination between the planes. The actual position of the comet in its orbit can be marked using the ephemeris Earth and Sun distances and two sets of compasses.

remember to put it on the correct side so that the comet will be orbiting in a direct manner, i.e., more like the Earth than opposite the Earth.

So how do we plot the comet's physical position in space at different dates? From the ephemeris, the Comet–Sun distance, r, and Earth–Sun distance Δ, for a series of dates, typically weeks apart, can be employed to draw more triangles to triangulate the comet's position in its plane. For example, Q4 is at perihelion on May 16 when q=r=0.96 which we have already marked. About a month later, on June 14, r=1.09 and Δ=1.18. After a further month, on July 14, r=1.39 and Δ=1.80. If we construct a triangle for June 16 with side lengths 109, 118 and 100 mm (Earth–Sun) distance and another with side lengths 139, 180 and 100 mm (Earth–Sun) distance, we can use the triangle apex to pinpoint the comet's position on the comet plane, by gluing the triangle at the Earth's position for those dates. Alternatively, if you do not want the hassle of making lots of tiny cardboard triangles, just using two sets of compasses with the first compass anchor point digging into the Sun and the second compass anchor point digging into the Earth (at a set date) enables the point where the compass needle tips meet on the cardboard comet plane to represent the comet's position (see Fig. 1.8b).

Short Period Orbits

Let us now have a look in more detail at some of the orbits of those 250 or so periodic comets. While they only represent a tiny fraction of the total of 3,700 comets so far discovered (if we include SOHO comets) they tend to live in a fairly restricted part of the sky (near the ecliptic plane) and at any one time most of the observable comets are periodic ones. The apparent contradiction here (the periodic minority being the observable majority) is, of course, down to the fact that over hundreds of years of comet observing almost 2,000 long period comets have arrived, performed and departed. The overwhelming majority of periodic comets have orbital planes tilted within 20° of the ecliptic plane. However, with so many periodic comets now known there are dozens that lie outside that 20° range. A typical short period comet, if there is such a thing, might have a perihelion distance of 1.5 AU, an aphelion distance closer to 7 AU and an orbital eccentricity of 0.6, as well as an orbital period between 5 and 9 years. Many short period comets have parameters well outside these limits but most can be thought of as spending the majority of their lives in the inner solar system (see Fig. 1.9).

When a cometary orbit is inclined at between 90 and 180° to the ecliptic the comet is, effectively, going backwards around the Sun. Angles of inclination "i" greater than 180° are not needed because the longitude of the ascending node parameter, combined with the inclination takes care of every orbital plane possibility. From above the solar system the Sun and planets rotate anti-clockwise, as do most other objects, but not every periodic comet conforms to this rule. As we saw earlier comet 1P/Halley itself, the first to be numbered, has an orbital inclination of 162°, in other words its orbital plane is tilted by 18° to the ecliptic plane, but it moves backwards around the solar system every 76 years. As I mentioned earlier the technical term for such an orbit is *retrograde* and other famous clockwise periodic comets include the progenitor of the Leonid meteor shower, Tempel–Tuttle (also with an inclination of 162°, but with an orbit of 33 years) and Swift–Tuttle (inclination 113°, orbital period 130 years). Of course, by normal periodic standards these are unusually long periodic orbits. In comparison 153P/Ikeya–Zhang rotates the correct way around the Sun, inclined at a mere 28°, every 340 years or so. Of course, orbital inclinations close to 90° are harder to visualize as either clockwise or anti-clockwise orbits as they are nearer to moving in orbital planes at right angles to the solar system. Thus the orbits of 12P/Pons–Brooks, 35P/Herschel Rigollet, 96P/Machholz and 109P/Swift–Tuttle with inclinations of 74, 64, 60 and 113° respectively are well outside the ecliptic except at the two nodes where their orbits intersect that plane. However,

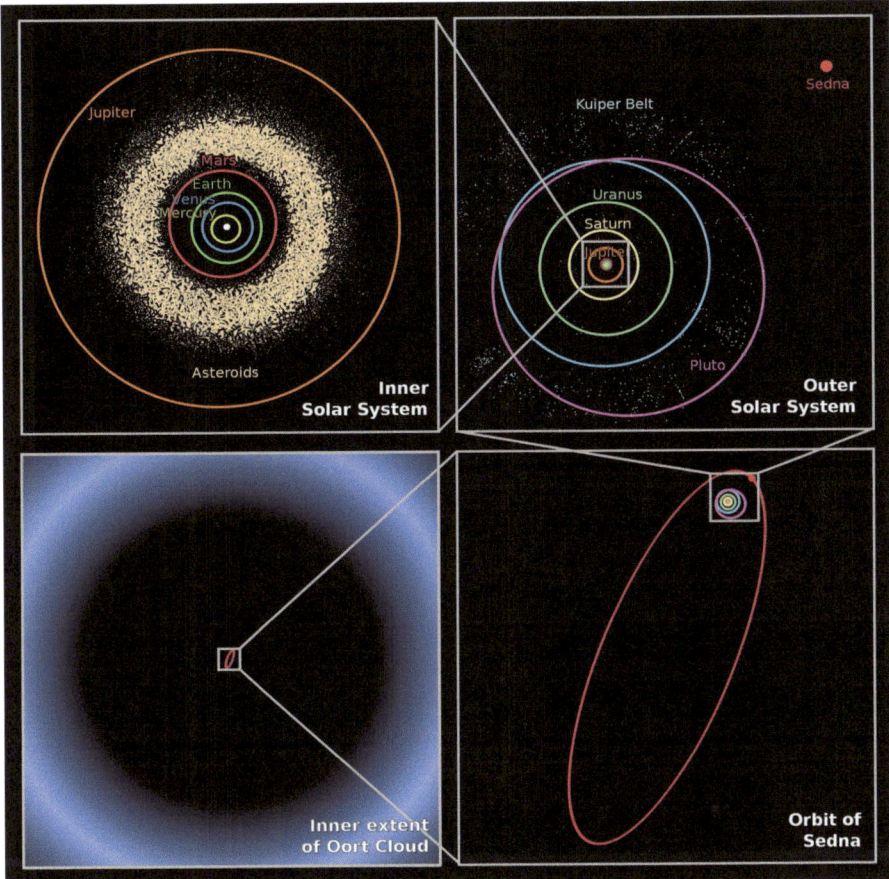

Fig. 1.9. This diagram by NASA indicates the scale of the solar system, the sort of distance long period comets move out to at aphelion and the size of the Oort cloud. Moving clockwise from *top left* the *first box* shows the orbits of the three inner planets, the asteroid belt and the orbit of Jupiter. This entire region fits into the *square box* on the *top right diagram* which shows orbits out as far as Neptune and Pluto, a similar distance from the Sun as the aphelion point of the orbit of comet Halley. This entire region fits into the *square box* in the *lower right diagram* which shows the orbit of the dwarf planet Sedna, which, with a period of about 11,000 years and an aphelion distance of almost 1,000 AU, travels some three times further from the Sun than comet Hale–Bopp at its furthest point (Sedna, close to perihelion, is also shown as a *red dot* in the *top right diagram*). The *final diagram* shows the orbit of Sedna relative to the inner rim of the Oort cloud. Image courtesy of NASA/JPL-Caltech/R. Hurt.

the reader of this book is probably more interested in the recent periodic comet discoveries and where they were made and so let us digress for a moment to just study the periodic comets discovered in the year 2009, the last full year prior to this book being written.

In 2009 there were 15 new periodic comet discoveries. These were made by Boattini (2), Gibbs (2), Yang-Gao (1), Hill (4 including one shared with Spacewatch), McNaught (3), La Sagra (2) and LINEAR (1). The orbital inclinations of all these new discoveries ranged from 5 to 41° and the orbital periods from 5 to 22 years. In terms of averages, the average inclination was 17° and the average orbital period was 12 years. We should not be too surprised that the average inclination of these discoveries was as high as 17° as obviously the ecliptic plane itself has been scoured by photographic and CCD patrols for years and so those comets which live close to the ecliptic plane have had plenty of opportunity to be discovered before. The comets with orbital inclinations of more than 10° are less likely to have been bagged already. Nevertheless, regions close to the ecliptic are still fertile discovery areas and consume less hunting time than attempting to patrol the entire night sky. From a human discoverer's viewpoint there must also be some comfort in knowing that if you discover a faint short period comet, at least it will be observed on many future returns, whereas a faint long period comet will come, go, and likely be forgotten very quickly.

Long Period Orbits

Long period comets, when first discovered on their way into the inner solar system, hold out the prospect of potentially being naked eye objects or even those zero, first and second magnitude examples that are eventually classed as "Great Comets." Of course, the vast bulk of long period comets become nothing of the sort and may not ever come within amateur visual range, even when using a telescope. Long period comets have orbits that more resemble parabolas, not ellipses. Put another way, on the scale of the inner solar system their orbits do not appear to re-converge, their minor axes simply widen as the comet moves further from the Sun. An elliptical orbit with an eccentricity value of 0 is a circle with the Sun at the centre or focus. As the eccentricity increases above 0 the circle becomes more elongated and develops a minor axis and a major axis and two "foci." The Sun lives at one focus. As we have already seen the longest orbital period short period comet, 153P/Ikeya–Zhang, has a period of roughly 341 years and it also has an eccentricity close to 0.9900, whereas comet C/1996 B2 (Hyakutake) has a period of 100,000 years and an eccentricity close to 0.9999. In general the longer the orbital period of a long period comet the closer the eccentricity gets to 1.00000, or a perfect parabola. Long period comets travel very far from the Sun before they reach aphelion. One way of visualizing this is to think of long period comets originating from an aphelion point at least three times further than the eighth planet Neptune, which orbits the Sun at a distance of 30 AU, just inside the aphelion distance of the famous comet Halley. In the case of comet C/1996 B2 (Hyakutake) this aphelion distance may be as large as 4,000 AU which is approximately half a trillion kilometers or two-thirds of a light month! These very long period comets are thought to originate in the Oort cloud, light months away from Earth, where the gravitational influence of our Sun is only slightly greater than that of the nearest stars Alpha and Proxima Centauri, 4.3 light years away. Tiny variations in the very weak gravitational contours in interstellar space (caused by distant stars drifting) may dislodge cometary material so it falls towards the inner solar system over thousands of years, thereby producing a new long period comet. If a big chunk of icy material ends up in the inner solar system and passes closer to the Sun than the Earth's orbital radius, a truly spectacular and memorable comet can result. If the comet passes close to the Earth as well then things can hardly get better for comet lovers.

Some long period comets of recent times have probably been on their first trip to the inner solar system, while others may have been around the circuit many times; but, unlike the short period comets of the inner solar system, they will probably not have orbited the Sun hundreds of times. If they had they would almost certainly have run out of gas and dust (as long period comets seem to last for less orbital revolutions) and would quite possibly have had their orbits modified by Jupiter's influence too. If a comet is on that first trip to the inner solar system it may well carry an excess of volatile material on its surface which will make it appear abnormally bright as it passes within a few AU of the Sun. Do NOT be fooled by the hype this may generate! What am I talking about here? Well, I am old enough to remember the press frenzy that surrounded the comet Kohoutek in 1973. It was spotted, on photographs, by Lubos Kohoutek at an unusually large distance (for that time) of almost 5 AU from the Sun and was calculated as coming to a very small perihelion distance of 0.14 AU on December 28. Applying the usual $H + 5 \log \text{delta}$ plus $10 \log r$ formula, with H set to 5.0, would have been ambitious enough for a comet that far out at discovery, but the formula 2.5 + 5 log Delta + 15 log r was used and some even went totally berserk and chose to use −1.0 + 5log Delta + 20 log r, leading to insane values around magnitude −18 at perihelion. The newspapers screamed that the comet of the century was on the way and it would be visible in broad daylight. However, they had not taken into account that the comet might be on its first entry to the solar system, in which case it would initially look abnormally bright as it started to warm up. This proved to be precisely the case. Of course, the press pack then swung the other way and declared Kohoutek a "flop," claiming "the boffins had got it wrong." Well, it was not a flop at all as it did become a very nice comet in binoculars and small telescopes, but nowhere near the daylight spectacle the public had been warned about. Remarkably a similar level of excitement surrounded comet Austin 1989c1 in 1989/90. Even the well respected U.S. publication *Sky & Telescope* had a banner across the front screaming "Monster Comet is Coming!" Like Kohoutek, comet Austin was a splendid object for amateur astronomers, but not the negative magnitude spectacle the press had hyped, and, once again, it was probably a comet on its first visit to the inner solar system.

I mentioned the Oort cloud some time ago and it is crucial to my next point regarding the differences between short period ecliptic-based comets and the true long period comets. The Oort cloud is not a disc of matter, like the asteroid belt, but a gigantic sphere surrounding the outer solar system. Some professional astronomers think there may be a trillion comets in this spherical shell! So, when tiny gravitational disturbances make long period Oort cloud comets "fall" into the inner solar system they can fall in from any angle and therefore be discovered in any position in the night sky. You have to cover the entire celestial sphere if you want to have a comprehensive search for incoming long period comets. As short period comets generally have much more modest orbital inclinations and, by definition, shorter orbits, astronomers think the origin of these short period comets may be the transneptunian Kuiper belt/scattered disk/Centaur regions of the solar system. In other words, regions of the outer solar system roughly four to eight light hours from the Sun. It is easy to see that there can be no strict definition though. Simply looking at the retrograde orbits of Halley, Tempel–Tuttle and Swift–Tuttle tells us that an orbital period under 200 or 300 years is no guarantee of a well behaved comet orbiting near the ecliptic plane. Also, it is vital to remember that the Solar System contains four very big planets and the biggest, Jupiter, can drastically alter cometary orbits, even to the extent of causing them to collide with the Jovian atmosphere, as we saw in 1994 and 2009.

The Kreutz Family

Some of the most spectacular long period comets ever witnessed have been part of the so-called Kreutz sungrazer family. The Great Comets belonging to this family are characterized by having a healthy absolute magnitude (in other words they are very active or have big nuclei rich in material which will evaporate when heated) and they pass extremely close to the solar surface at perihelion. Remember that comet magnitude law of $H_0 + 5 \log \Delta + 10 \log r$, and the fact that the 10 log r term is an inverse fourth power law. Well, imagine what happens if you have a perihelion value r between 0.005 and 0.009 AU. Compared to a comet that reaches perihelion at 1 AU from the Sun such comets will, theoretically at least, be 20 magnitudes brighter! In practice, such sun grazing comets of the past have reached negative magnitudes while visible in strong twilight and with the Sun less than 10° below the horizon.

Not all sungrazing comets belong to the Kreutz group, but many of the most spectacular ones of the last few hundred years do. The family is named after Heinrich Carl Friedrich Kreutz (1854–1907) the German astronomer who was the editor of the *Astronomische Nachrichten* magazine. Kreutz realised that the orbits of the great comets of 1843, 1880, and September 1882 (he apparently did not know about the comet X/1882 K1 seen at the solar eclipse of May 1882) were all very similar and could have a common origin and so the group is named after him, although Kirkwood had come to a similar conclusion slightly earlier. Since the time of Kreutz the orbit master Brian Marsden and a number of other twentieth and twenty-first century astronomers have tackled the Kreutz comets orbit calculations using modern computing power. Astronomers now believe that a very large sungrazing comet, the Great Comet of February 1106, broke up and the two spectacular comets known as the Great Comet of 1882 (C/1882 R1) and C/1965 S1 (Ikeya–Seki) are derived from that break up. It is thought that other impressive comets, namely C/1945 X1 (du Toit), C/1963 R1 (Pereyra) and C/1970 K1 (White–Ortiz–Bolelli) can also be traced back to that Great Comet of 1106 breaking up. But we are not finished yet! It turns out that another large sungrazing comet, possibly one that arrived as much as 50 years prior to that of 1106, may have been involved, also broke up, and spawned the comets C/1843 D1, C/1880 C1 and C/1887 B1. So we now have two progenitor comets, both responsible for spectacular sungrazers and with similar orbital elements. Not surprisingly astronomers think that both these parent comets may have split from a massive grandparent comet, possibly in the third to fifth centuries, that is between 1,800 and 1,600 years ago. The favorite candidates for this monster comet, which may have had a nucleus 100 km across, are the comets of 214, 426 and 467 AD. When will the next spectacular Kreutz sungrazer come along? Well, it is impossible to say, but over the last two centuries a Kreutz group comet has appeared every 20 years, on average, and it is quite a while since the last one, so there could be another one along any time soon. Many of the faint cometary fragments discovered in the SOHO images have turned out to be tiny Kreutz sungrazers and these have added more data about the orbits of these enigmatic objects. I will have more to say about the Kreutz comets at various stages in this book but I will conclude this section by listing the orbital elements of Kreutz sungrazer comet C/1965 S1 (Ikeya–Seki), the brightest comet of the twentieth century. Apart from the perihelion date, which is, of course, unique to this particular large member of the Kreutz group, the other orbital elements are typical and are shown below.

C/1965 S1 (Ikeya–Seki) Orbital Elements:
Date of perihelion, *T*: 1965 Oct. 21.1837 TT
Perihelion distance, *q*: 0.0077860 AU
Argument of perihelion, ω: 69.0486
Longitude of ascending node, Ω: 346.9947
Inclination, *i*: 141.8641
Eccentricity, *e*: 0.9999150 AU
Semi-major axis, *a*: 91.6 AU
Period, *P*: 876.70 years

Beyond Parabolic: Hyperbolic!

Look at the near-parabolic orbit of a long period comet in plan view and, as I mentioned earlier, the ends of the curve, in the inner solar system at least, do not seem to converge. Indeed, with a semi-major axis that is light days or light weeks in length you have to go an order of magnitude beyond Neptune before the orbit will start to converge, signifying the comet is half way to aphelion. Strictly speaking a comet with an orbital eccentricity of 1.0, in other words a perfect parabola, will never return to the solar system, although for the overwhelming majority of very long period comets the eccentricity will be somewhere between 0.99 and 0.9999. When only limited data is available about a bright historic comet's orbit, or about a brand new long period comet the value of e is always set to 1.0000 by default.

Barring encounters with a large planet like Jupiter, an object is simply falling in from deep space, going into a slingshot around the Sun and then having the kinetic energy it gained on the downward slope, slowly drained away by the Sun's gravity on its way back to the Oort cloud. There is no friction to speak of on its journey through the inner solar system and the only non-gravitational forces are those caused by action and reaction forces when icy volatiles sublimate on its surface and a coma and tail develop. Nevertheless, there are a few comets which genuinely appear to be heading out of the solar system entirely, as their orbits have an eccentricity greater than 1.0 and so are, technically, hyperbolic. Perhaps the most famous comet in this regard was comet C/1980 E1 (Bowell) which Brian Marsden of the Minor Planet Center once described as "a comet in a million." At the time of its discovery it was remarkable in many ways. It was less than 2° from Jupiter, its orbit was less than 2° from the ecliptic plane (by far the lowest on record in 1980 for a long period comet) and after it ventured within 3.36 AU of the Sun it left the solar system with a hyperbolic orbit having a record eccentricity of $e = 1.0574$. The planet Jupiter was responsible for increasing the eccentricity to this record value and the fact that comet Bowell orbited so close to the ecliptic plane would have helped here. 1980 E1 is not the only hyperbolic comet though, although its eccentricity is remarkable. Other comets with unusually high e values include C/1997 P2 (Spacewatch) with an eccentricity of 1.0279, C/2008 J4 (McNaught) with an eccentricity of 1.0272 and C/1999 U2 (SOHO), with an eccentricity of 1.024.

Of course, apart from Jupiter bending a comet's path to become hyperbolic it is always possible that a genuine stray object from outside our solar system has crossed interstellar space and entered our solar system without falling in from the Oort cloud. Some astronomers regard this kind of comet as being extremely rare (say, one comet in every few thousand long period comets) but at the present time it is simply not mathematically possible to state categorically whether a comet with a hyperbolic orbit has come from outside our solar system.

CHAPTER TWO

Great Comet Discoveries Throughout History

This book is mainly aimed at modern hunting and imaging methods, but as so many great and mouthwatering comets have come and gone over the centuries it would be rather disappointing if I did not at least devote a chapter to them, if only so that the reader can hope a few similar examples are headed our way. The term "great" is used frequently in the twenty-first century and more often than not it means "fantastic," but in the context of comets the term has a slightly different historic meaning. From the late eighteenth to the early twentieth century a bright naked eye comet for whom there was no obvious single discoverer, was invariably named "Great," with a capital "G." There are more than 20 examples for which the official name included the word "Great," such as, the Great September Comet of 1882, now named C/1882 R1. The word "Great" in this sense obviously means "truly awesome," "magnificent," or "huge." In the days before rapid communication was possible and comet discoveries were always made visually it was often impossible to say who actually discovered a bright naked eye comet. Quite often such comets were discovered simultaneously by numerous people who just happened to spot a comet close to perihelion, sporting a long tail, in the twilight. In the centuries before aircraft con trails, nothing else could look like a comet! However, these days the descriptive term "Great" is not just reserved for comets for which there was no discoverer; in general a comet can be described thus

M. Mobberley, *Hunting and Imaging Comets*, Patrick Moore's Practical Astronomy Series, DOI 10.1007/978-1-4419-6905-7_2, © Springer Science+Business Media, LLC 2011

if it peaked in brightness at first magnitude or brighter, although a few second magnitude comets seen in a dark sky, in the centuries when there was no light pollution, also qualify.

Great Comet Criteria

What actually makes a comet truly spectacular though? Well, I touched on this briefly in the previous chapter and there are three main criteria. Firstly, a comet that has a substantial absolute magnitude, H_0, will stand a good chance of being labeled "Great." Remember that magnitude law, $H_0 + 5 \log \Delta + 10 \log r$? Well, if a comet is placed 1 AU from the Sun and 1 AU from the Earth the $5 \log \Delta$ and $10 \log r$ terms become zero and the comet will have a magnitude equal to its absolute magnitude. From this fact alone we can see that absolute magnitudes lower than four or so guarantee a naked eye object if the comet approaches Earth and Sun within 1 AU. Comet Hale–Bopp featured the most outstanding absolute magnitude of any comet in recent years. Its absolute magnitude was between −0.5 and −1.0 and its nucleus may have been as large as 60 km in diameter, which is truly massive for a comet. There have been comets with an even more impressive absolute magnitude though. Comet Sarabat of 1729 had an absolute magnitude of about −3 and Tycho's comet of 1577 is thought to have had an H_0 of −1.8. Only 50 years ago, in 1962, comet Humason was estimated to have an absolute magnitude of +1.3, although it never came into the inner solar system. By comparison, even the famous comet Halley only has an absolute magnitude of about 4.5 and a typical new comet will have an H_0 value between 6.0 and 7.0.

The second criteria, not surprisingly, is the comet's perihelion distance q. As a typical comet will brighten to an inverse fourth power law, with respect to solar distance, the perihelion distance is crucial to a good performance. Of course, as I mentioned in Chap. 1, for the so-called sungrazers the perihelion distance is so tiny that highly negative magnitudes can result, assuming the comet is large enough to avoid total disintegration. There has never been an official definition of what constitutes a sungrazing comet, although some astronomers loosely use a perihelion distance of ten solar radii or less (7 million kilometers or 0.047 AU). An even stricter criterion is two solar radii or less (1.4 million kilometers or 0.01 AU) from the solar center, in other words one solar radius from the solar surface. The problem with such comets is that they are invariably at their best when seen in twilight. For a comet to be really stunning against a dark sky it needs to be elongated from the Sun by at least 30°, so that the Sun is, say, 15° below the horizon when the comet is 15° above it. In addition, the elongation needs to be perpendicular to the horizontal horizon from the observer's latitude, which is rarely the case. So, in practice, when a comet approaches the Sun closer than about 0.6 AU (Venus orbits the Sun at roughly 0.72 AU) twilight can start to be a serious obstacle.

The third criterion is simply how close the comet approaches the Earth. If the orbital elements of the comet conspire to allow the two orbits to almost intersect each other then this is obviously good news, but it will be sheer luck if the Earth is in the right part of its orbit at the right time. In the worst case the comet may peak in brightness on the opposite side of the Sun, but with many comets there are two potential points where the comet will ascend or descend through the ecliptic plane (the ascending or descending nodes) and offer a reasonable view. Of course, if a comet's orbit passes extremely close to the Earth a potentially spectacular sight may be seen. The closest cometary fly-bys out to the 15 million kilometers distances of C/1996 B2 (Hyakutake) and C/1961 T1 (Seki) are shown in Table 2.1.

It may, at this point, be worth mentioning some future close approaches that are coming up. None are likely to qualify as a Great Comet, but close Earth fly-bys are always of interest to comet observers. Of course a close fly-by of a long period comet is impossible to predict in advance and the unpredictability of such events is one of the most fascinating aspects of comet observing. The short period comet 45P/Honda–Mrkos–Pajdusakova is a comet that orbits the Sun every five and a quarter years and has made many close approaches in the past and is scheduled to repeat this performance in the future. Its orbit comes close to the Earth at two points, in February and August. On August 15, 2011 comet 45P should pass us by at a distance of nine million kilometers and on February 11, 2017 it should come within 13 million kilometers. However, 45P is a relatively small comet and so do not expect anything spectacular. A relatively recently discovered comet that can perform the same trick is P/2006 U1 (LINEAR) which will ascend to having a numbered status at some point. This comet, with a 4.6-year period, can fly past the Earth in late October/early November and also in April. Its next close passage should be on November 2nd/3rd, 2029 when the closest approach will be just within seven million kilometers. On October 30, 2061 it is scheduled to pass only three million kilometers from Earth just a few months after 1P/Halley returns to perihelion.

Table 2.1. Comets that have passed closer to the Earth than C/1996 B2 (Hyakutake) and C/1961 T1 (Seki), compiled from various sources including the Minor Planets Center data and Gary Kronk's Cometography project.

Distance (km × 10⁶)	Date (UT)	Designation and name
*1.40	1491 Feb 20	C/1491 B1
2.26	1770 Jul 1	D/1770 L1 (Lexell)
3.43	1366 Oct 2	55P/1366 U1 (Tempel–Tuttle)
4.67	1983 May 11	C/1983 H1 (IRAS–Araki–Alcock)
5.00	837 Apr 10	1P/837 F1 (Halley)
5.48	1805 Dec 9	3D/1805 V1 (Biela)
*5.60	1014 Feb 25	C/1014 C1
5.83	1743 Feb 8	C/1743 C1
5.89	1927 Jun 26	7P/Pons–Winnecke
6.54	1702 Apr 20	C/1702 H1
*6.80	1351 Nov 28	C/1351 W1
*6.80	1132 Oct 7	C/1132 T1
*7.30	1345 Jul 31	C/1345 O1
*8.40	1499 Aug 16	C/1499 Q1
9.23	1930 May 31	73P/1930 J1 (Schwassmann–Wachmann)
9.39	1983 Jun 12	C/1983 J1 (Sugano–Saigusa–Fujikawa)
*9.60	1080 Aug 5	C/1080 P1
*9.60	1699 Oct 27	55P/Tempel–Tuttle
10.20	1760 Jan 8	C/1760 A1 (Great comet)
*10.40	1472 Jan 22	C/1471 Y1
*11.00	400 Mar 31	C/400 F1
11.77	2006 May 12	73P/Schwassmann–Wachmann

(continued)

Table 2.1. (continued)

Distance (km × 10⁶)	Date (UT)	Designation and name
*12.40	1639 Oct 26	C/1639 U1
*12.50	1556 Mar 12	C/1556 D1
12.55	1853 Apr 29	C/1853 G1 (Schweizer)
13.15	1797 Aug 16	C/1797 P1 (Bouvard–Herschel)
13.22	374 Apr 1	1P/374 E1 (Halley)
13.43	607 Apr 19	1P/607 H1 (Halley)
13.97	1763 Sep 23	C/1763 S1 (Messier)
*14.00	568 Sep 25	C/568 O1
14.42	1864 Aug 8	C/1864 N1 (Tempel)
14.69	1862 Jul 4	C/1862 N1 (Schmidt)
*15.20	868 Jan 25	C/868 B1
15.23	1996 Mar 25	C/1996 B2 (Hyakutake)
15.24	1961 Nov 15	C/1961 T1 (Seki)

Comets indicated by an asterisk are ones where there is considerable uncertainty in the proximity of the fly-by distance as the historical records are very patchy

1P/Halley

I cannot possibly start our look at specific "Great Comets" any other way than by looking at the first comet in the numbered list. The famous comet 1P/Halley (Figs. 2.1a, b) has been entertaining and even terrifying stargazers for millennia but some returns have been rather disappointing, solely because of where the Earth was in its orbit when Halley returned. Even so, when a comet can reach magnitude −3 and only slump to +3 at a really bad return it deserves to be classed as "Great." Halley is extra special because it is a very active comet that passes quite close to the Sun at perihelion (0.59 AU) and can come dramatically close to the Earth too and, above and beyond this, it returns every 76 years or so. No other comet with such a healthy set of characteristics returns so regularly. All the other truly "Great" comets have periods of thousands or tens of thousands of years, or longer. If you make

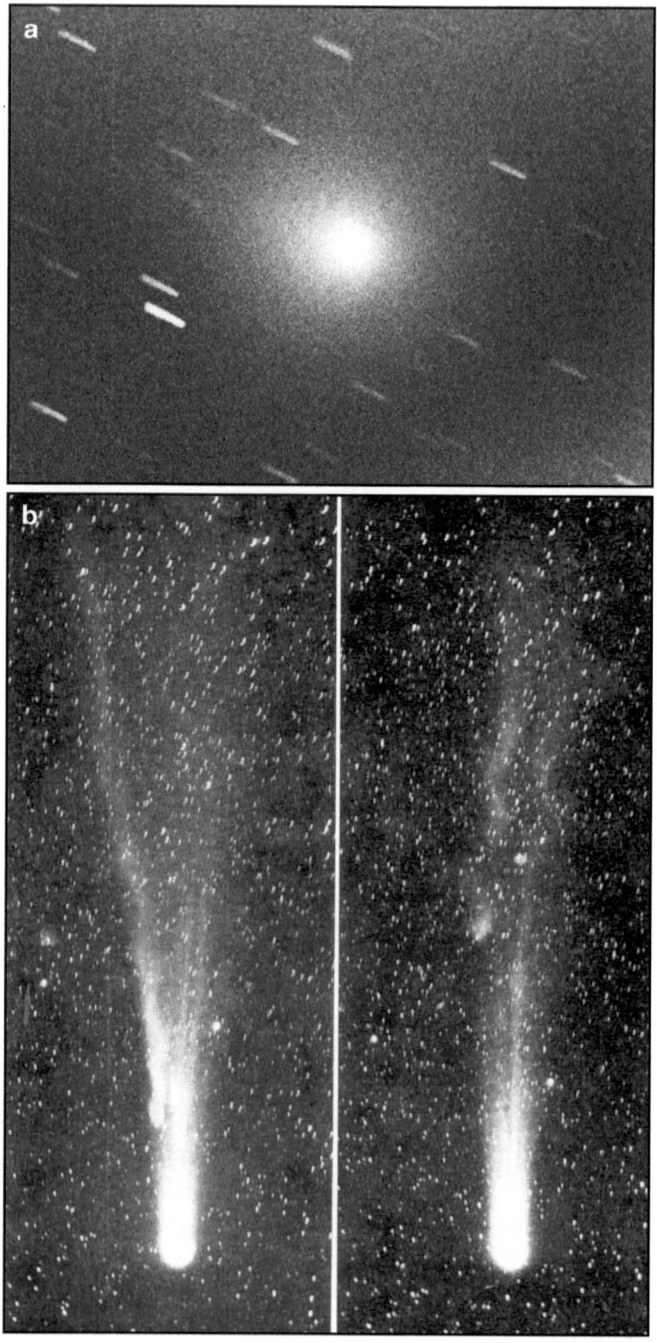

Fig. 2.1. (a) Comet Halley photographed from the UK on December 7, 1985 in an 18-min exposure with 1,000 ISO film using a 14-in. (36-cm) Newtonian. The motion of the comet was compensated for while manually guiding. Image by Martin Mobberley.

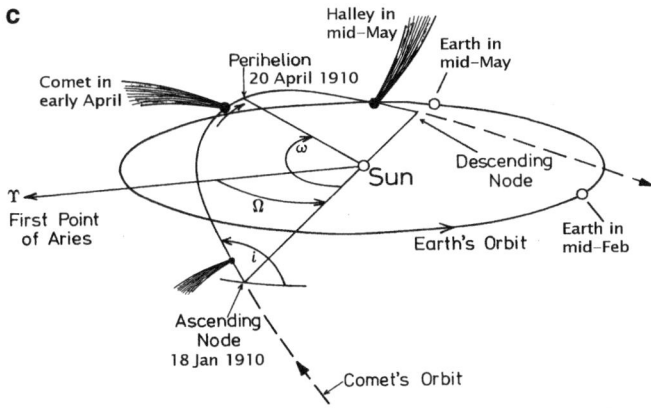

Fig. 2.1. (continued) (**b**) Comet Halley photographed from the Lick Observatory on its 1910 return. The two images were taken on June 6 and 7, and show a tail disconnection event. (**c**) The orbital circumstances of the 1910 return of Halley's comet.

a list of all the comets that have been magnitude 2.0 or brighter in the last 2,400 years or so you will come up with approximately 70 comet returns, or roughly one spectacular comet every 35 years. However, 22 of those 70 comets will be Halley. What a performance! An entertaining rhyme summarising Halley's characteristics, supposedly written by the tenth Astronomer Royal, Harold Spencer Jones (1890–1960), reads as follows:

Of all the comets in the sky,
There's none like Comet Halley.
We see it with the naked eye,
And periodically.
The first to see it was not he,
But yet we call it Halley.
The notion that it would return,
Was his originally

Over time orbits evolve, due to the gravitational influence of the planets, especially giant Jupiter, and also due to the non-gravitational factors, namely comets venting material. So, although "76 years" is often quoted as the orbital period for Halley, in practice anywhere between 74 and 79 years is possible. The precise calendar date that the comet returns to perihelion, its closest point to the Sun, determines how close it will fly past the Earth and how awesome it will appear. Halley spends most of its life below the Earth's orbit. It rises through the Earths orbital plane 92 days before perihelion and 270 million kilometers from the Sun. It then cuts above and inside the Earth's orbit, reaches perihelion and swoops back down and out again (see Fig. 2.1c for the 1910 circumstances). Thirty days after perihelion Halley passes down through the Earth's orbital plane, 127 million kilometers from the Sun and plunges below and outside the Earth's orbit. For the absolute best situation we need perihelion to occur from late February to late April with an Earth close approach, post-perihelion, from early April to late May, with Halley plunging south. Pre-perihelion encounters with Halley, from late September to early October, can still be impressive (perihelion dates from late October to early November) but the comet is further away and has not heated up as dramatically as after perihelion. The perihelion date in 1986 was February 9. If that date had been just 2 or 3 weeks later the return would have been spectacular in the extreme; as it turned out those 2 or 3 weeks made all the difference between an awesome comet and a naked eye smudge. Some of the most memorable historic returns of Halley are listed in Table 2.2 below, plus,

Table 2.2. A few selected returns of 1P/Halley from the last two millennia with perihelion dates, perigee dates of the closest Earth approaches, corresponding perigee distances to the Earth in AU (1 AU = 149.6 million kilometers) and the peak magnitude achieved or likely to be achieved.

Year	Perihelion	Perigee	Distance (AU)	Peak magnitude
141	Mar 22	Apr 22	0.17	−1
374	Feb 16	Apr 02	0.09	−1
607	Mar 15	Apr 19	0.09	−2
837	Feb 28	Apr 11	0.03	−3
1066	Mar 20	Apr 24	0.10	−1
1301	Oct 25	Sep 23	0.18	+1
1378	Nov 10	Oct 03	0.12	+1
1910	Apr 20	May 20	0.15	0
1986	Feb 09	April 11	0.42	+2.5
2061	July 28	July 29	0.48	+3??
2134	March 27	May 07	0.09	−1??

Dates up to and including 1378 AD are from the Julian Calendar. The next two returns have also been included

I have added the return of 1986, and the next two, in 2061 and 2134. It should be borne in mind that the dates up to and including 1378 are in the Julian calendar and by that return the calendar was so out of step that the November 10, 1378 perihelion date is equivalent to November 18 in today's Gregorian calendar.

Halley next comes to perihelion in July 2061. However, it will be the most dismal return imaginable with Halley unlikely to do better than third magnitude and being very poorly placed in the sky. However, if current predictions are correct, the 2134 return will be a splendid one with the comet even closer to the Earth than it was 100 years ago in 1910. Indeed, the 2134 return is very similar to that of the 1066 return when the appearance of Halley was depicted on the Bayeux tapestry and linked to the Battle of Hastings and the downfall of King Harold.

There have been many superb comets over the centuries and, excluding the numerous returns of 1P/Halley, roughly 50 spectacular comets have been recorded as easy naked eye objects, with long tails, in the last 2,000 years. However, much caution is necessary because prior to the era of orbit computation and the invention of the telescope only vague descriptions are usually recorded. Sometimes the movement of a comet between constellations can be deduced but at others a few references just record something like "a broom star was visible in the East." That is not much to go on!

I have already mentioned the Kreutz sungrazing family of comets which was responsible for several spectacular examples in the nineteenth century. One possible candidate for the grandparent comet of this group may have been observed by Aristotle and by Ephorus in 371 BC. Ephorus claimed to have seen this comet break in two, but this was not confirmed by other observers. Nevertheless, as we have already seen, the comets of 214, 426 and 467 AD are considered as serious progenitors of the Kreutz comets seen in the nineteenth and twentieth centuries. I mentioned the Great Comet of 1106 AD (X/1106 X1) in its Kreutz parent connection a while back and it is perhaps worth starting at the description of this comet in a brief look at the very best "Great Comets" of antiquity.

X/1106 C1: The Great Comet of 1106

This monster appears to have been observed throughout Europe (including England and Wales) as well as in China, Japan and Korea in February and March of 1106. Reports that it split in two are, of course, of crucial interest to orbital experts studying the Kreutz comet sungrazing family. A Welsh manuscript of 1106 (*Brut y Tywysogion*) famously recorded the comet as follows:

> In that year there was seen a star wonderful to behold, throwing out behind it a beam of light of the thickness of a pillar in size and of exceeding brightness, foreboding what would come to pass in the future: for Henry, emperor of Rome, after mighty victories and a most pious life in Christ, went to his rest. And his son, after winning the seat of the empire of Rome, was made emperor.

In addition the *Peterborough Chronicle* recorded:

> In the first week of Lent, on the Friday, 16 February, in the evening, there appeared an unusual star, and for a long time after that it was seen shining a while every evening. This star appeared in the south-west; it seemed small and dark. The ray that shone from it, however, was very bright, and seemed to be like an immense beam shining north east; and one evening it appeared as if this beam were forking into many rays toward the star from an opposite direction.

C/1577 V1: The Great Comet of 1577

This famous historic comet passed close by the Earth and was observed by the Danish astronomer Tycho Brahe from November 1577 to January 1578. It seems incredible but it was around this time that those who studied phenomena in the sky, started to grasp that new objects were actually above the atmosphere rather than meteorological phenomena! We have certainly come a long way in the past 433 years. Sketches made by Tycho show the comet passing close to Venus and his numerous observations, when compared with those of several other European observers, led him to conclude that the comet was several times further away than our Moon.

C/1680 V1: Kirch's Comet of 1680

A century after Tycho's observations of 1577, comet observations and orbit calculations were well advanced and the world was now in the era of Sir Isaac Newton. Sometimes the comet of 1680 is called Newton's comet because its motion through the sky was used by Newton to verify Kepler's laws. However, this Great Comet was discovered by Kirch and holds the distinction of being the first comet to be discovered by a telescope. Eusebio Francisco Kino (1645–1711), also charted the comet's course from Cadiz, before departing on a gruelling trip to Mexico. Kirch's comet was a sungrazer, but not a member of the Kreutz family. It probably had a perihelion distance of only 0.006 AU around December 16, 1680 and would have passed within 0.4 AU of Earth close to November 30. The comet was still visible in mid March 1681 but now, 330 years later, it will be at least 250 AU from the Earth.

C/1743 X1: de Chéseaux' Comet of 1744

The Great Comet of 1744 (see Fig. 2.2) is surely one of the best documented of the truly awesome comets of recent centuries. It has many alternate names as it was independently discovered by Jan de Munck in late November 1943, Dirk Klinkenberg in mid-December and

Fig. 2.2. The Great Comet de Chéseaux–Klinkenberg, on March 9, 1744 showing six tails rising above the dawn horizon. From *The World of Comets* (London, 1877) by Amedee Guillemin.

Jean-Philippe de Chéseaux a few days later. The latter observer recorded it as a tailless nebulous star of the third magnitude, with a 5-arc-min coma. It peaked in brightness in February and March of 1744. Sometimes the object is described as comet Klinkenberg–Chéseaux and sometimes as just Comet de Chéseaux. I have never heard it described as Comet Munck, which is a bit of a shame for the long departed Mr. de Munck! Undoubtedly it is most associated with Jean-Philippe de Chéseaux simply because he was the best known comet observer of the three. Regardless of the discoverer though this was a spectacular object. Some sources estimate the comet's perihelion magnitude as high as magnitude −7, although more modest estimates center on a value of −3 or −4. Like comet Hale–Bopp 253 years later it had an abnormally healthy absolute magnitude, probably slightly fainter than zero. Unlike Hale–Bopp the comet had a very small perihelion distance of only 0.22 AU. Compared to Hale–Bopp's 0.91 AU perihelion distance, this would, all other parameters being equal, give de Chéseaux' comet a six magnitude brightness advantage! The comet arrived at perihelion on March 1 by which time it was bright enough to be observed in daylight with the naked eye. After perihelion a spectacular multiple tail developed which rose well above the morning horizon, in a dark sky while the head was still well below that same horizon. The tail formed a giant fan comprised of six multiple tails. While astronomers were familiar with comets having two tails (a straight gas or ion tail and a curved dust tail), one with six was something completely different. It is likely that the large nucleus of the comet had several specific dust sources on its surface which, coupled with the rotation of the nucleus, would have been responsible for its appearance. Nowadays these sort of features are more familiar to us from the pictures of Comet West in 1976 and Comet McNaught in 2007, although Hale–Bopp's tail also featured "synchronic bands" which are a related phenomena that I mentioned in Chap. 1. In the second week of March 1744 de Chéseaux lost track of his comet as it headed south but southern hemisphere observers were able to follow the comet into late April and tail lengths of up to 90° were reported in late March. Mysteriously some of the Chinese records of this comet describe atmospheric sounds when the object was at its peak. In an era when there were no planes or cars how can this be explained? Maybe the sounds were distant animals howling in terror at the sight of such an awesome spectacle? However, some observers of impressive aurorae have occasionally reported sounds and so one can speculate about a cometary interaction with the Earth's magnetosphere. Interestingly, the teenage Charles Messier saw de Chéseaux' comet and went on to become one of the greatest comet discovers and deep sky object cataloguers of that era, so it is tempting to think that the comet played a pivotal role in his formative years.

C/1811 F1: The Great Comet of 1811

When Hale–Bopp was discovered in July 1995, and astronomers realised that it was such a huge object, the Great Comet of 1811 (see Fig. 2.3) was quickly unearthed as the most similar example seen in recent history. Amateurs were soon scouring the literature to try to unearth what we might expect from such an enormous object when it reached perihelion. Admittedly, the comet C/1729 P1 (Sarabat) was an even bigger comet with a nucleus that may have been 100 km across, but that comet, although it reached third magnitude never came within 4 AU of the Sun! As well as C/1995 O1 (Hale–Bopp) and C/1811 F1 having absolute magnitudes close to zero the orbital elements of the two comets have a few notable similarities too. C/1811 F1 had a perihelion distance of 1.04 AU, compared to Hale–Bopp's 0.91 AU. The eccentricity of both comets' orbits was 0.9951 AU implying a similar orbital period of, probably, 2,500–3,000 years. In addition both comets had orbits that were inclined at almost a right angle to the ecliptic, namely 106.9° for C/1811 F1 and 89.4° for Hale–Bopp. C/1811 F1 was discovered on March 25, 1811 by Honoré Flaugergues at 2.7 AU from the sun in the old constellation of Argo Navis. It was also found by the prolific comet discoverer Jean-Louis Pons on April 11. After 2 months of solid mental graft (there were no electronic

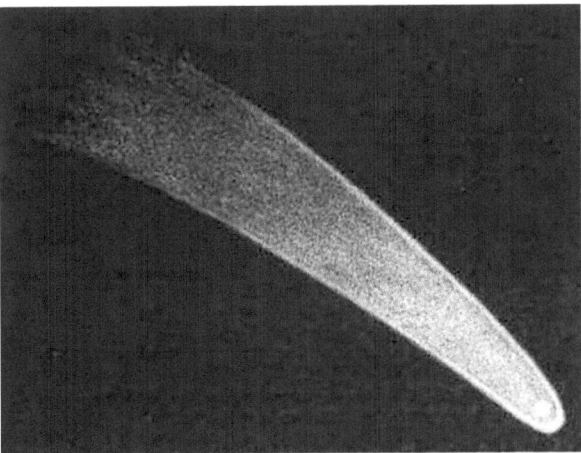

Fig. 2.3. The Great Comet of 1811, drawn by William Henry Smyth, as depicted in Guillemin's *The Heavens* (London, 1886).

calculators or computers in 1811!) an orbit was computed by Johann Karl Burckhardt which led the famous astronomer Heinrich Olbers to predict that this comet would become very bright indeed. The date of perihelion for C/1811 F1 was September 12 and by early September the comet was an easy naked eye object in Ursa Major. Although it was never going to develop a sky spanning tail, with its perihelion distance of 1.04 AU and its distance from the Earth, it was a splendid object with a coma far larger than the apparent size of the Moon in the night sky. After perihelion, in October, the great William Herschel observed the tail spanning an impressive 25°. C/1811 F1 was followed by telescopic observers for almost a year after perihelion with Vincent Wisniewski estimating it as being between magnitude 11 and 12 on August 12, 1812. Wisniewski was observing from Novocherkassk, a town in Rostov Oblast, Russia. C/1811 F1 is sometimes referred to as Napoleon's Comet as some saw it as a portent to his invasion of Russia and the 1812 War. However, on a lighter note 1811 was noted as a year famous for its fine wines and wine merchants made a healthy income selling "Comet Wine" in the coming years!

C/1843 D1: The Great March Comet of 1843

The Great Comet of 1843 (see Fig. 2.4) marks the first nineteenth century return of a big fragment from the Kreutz sungrazer family that I mentioned earlier. It was discovered on February 5 of that year and its orbit suggests it is an offspring not of the Great Comet of 1106 itself, but of that Great Comet's poorly observed other half, whose precise identity is uncertain. Only 3 weeks after it was first spotted it reached perihelion at 0.005527 AU or 827,000 km from the center of the Sun. Bearing in mind the mean solar radius of 696,000 km, that made it a true sun grazer! Precisely what its magnitude was at perihelion is unsure but it must have been highly negative as there are reports of it being visible when only a degree from the Sun! On March 6 it passed its closest point to the Earth and was at its most impressive on the following day, but was best seen in the southern hemisphere after perihelion. The tail of the comet was extraordinarily long, being more than 2 AU or 300 million kilometers in length. Never before had a comet been observed to pass so close to the solar surface and it was remarkable that it survived its encounter. The last known observation of it was on April 19, 1843.

COMET OF MARCH 1843.

Seen from Aldridge Lodge V.D. Land

Fig. 2.4. The Great Comet of 1843, viewed from Australia, as depicted in an old book by Mary M. Allport (1806–1895).

C/1858 L1 (Donati)

On June 2, 1858 the Italian comet hunter Donati swept up "a very faint small nebulosity of uniform brightness" (probably of eighth magnitude) which would turn out to be his most spectacular comet find and one

Fig. 2.5. Donati's Comet from the book "Bilderatlas der Sternen-welt," published in 1888. Drawing by E. Weiß.

of the most beautiful comets ever seen. Donati's comet (see Fig. 2.5) would become a naked eye object from early August until November and be truly spectacular in early October of that year. Donati had all the ingredients that are needed for a spectacular comet. Firstly, it was intrinsically bright with an absolute magnitude of approximately 3.3. Secondly, it was passing close to the Sun with a perihelion distance of 54 million miles (86 million kilometers). Thirdly it was passing close to the Earth, at a distance of 51 million miles (82 million kilometers) 10 days after perihelion. Finally, it was sufficiently elongated from the Sun, for several months (September to November) to be an obvious naked eye feature. On October 5 1858 5 days after perihelion, Donati's Comet passed in front of Arcturus in the evening sky, as seen from the northern hemisphere. The comet appeared almost as soon as the Sun had set and remained visible until the comet itself set (the head about 9 pm from the UK and the end of the tails many hours later). Its wonderfully broad, sweeping dust tail grew to 60° in length (roughly 50 million miles) on the 10th of October and two thin tails, one a gas tail, the other (possibly) a synchronic feature were also visible. A fortnight later the *Illustrated London News* of October 23 contained a Comet Supplement showing paintings from a variety of observers, including a rendering from the Cambridge University Observatory made on October 11.

By the start of October the comet was now attracting so much attention that observations of it were regularly being reported in almost every edition of national and local newspapers in the UK. For example, the following details in a local Suffolk newspaper were copied from a letter in a national newspaper, written by no less a person than J.R. Hind, the UK astronomer and asteroid discoverer:

> The comet will arrive at its least distance from the Earth about midnight on the 10th of October, when we shall be separated from it by rather over 51,000,000 miles. Its maximum brilliancy will be attained the day previous, when the intensity of light will be twice as strong as at the present time. It is, therefore, obvious that during the absence of moonlight in the evening hours for the next ten days or upwards the comet will form a splendid object in the Western heavens. On the evening of October 5th, the nucleus will make a near approach to Arcturus, the principal star in the constellation of Bootes. At six p.m. their distance will be little more than one-third of a degree. It is not probable that the comet will be visible in this country after the end of the third week in October, unless a few daylight observations be subsequently procured. In a somewhat hazy sky last evening the apparent length of the tail was about twelve degrees, corresponding to a real length of 16,000,000 miles. As usual in great comets, the tail is very visibly curved in the opposite direction to that of the motion of the nucleus.

After November 1858, Donati's comet faded but it had been a major attraction in the night skies for 3 months and for those observers old enough to remember the Great Comet of 1811 some 47 years earlier it was the only comet since 1811 to rival that monster. Donati had peaked at between magnitude 0 and +1, slightly fainter than the earlier comet, but both had been visible in a dark sky for several months and Donati had, arguably, been the most beautiful object of the two. With an aphelion distance of some 300 AU and a period of 2,000 years it will be a long time before we see Donati's comet again.

C/1861 J1: The Great Comet of 1861

The Great Comet of 1861 (see Fig. 2.6), like its predecessors in 1858 and 1811, was a naked eye object for several months, despite its relatively modest perihelion distance of 0.822 AU. It was discovered by John Tebbutt of New South Wales, Australia on May 13, a full 4 weeks before the comet reached perihelion on June 11 and it had an orbital inclination of 85°, in other words the plane of its orbit was pretty much a right angle to the ecliptic. The comet first became visible in the far northern hemisphere

Fig. 2.6. Tebbutt's Great Comet of 1861, from the book "Bilder-atlas der Sternenwelt," published in 1888. Drawing by E. Weiß.

around June 29 and 30 when it passed only 0.13 AU from the Earth itself. The comet was moving rapidly north through Auriga, usually regarded as a northern hemisphere winter constellation, but because it was now circumpolar from UK type latitudes it could be well observed above the northern horizon despite the summer twilight. Of course it was totally invisible to its Australian discoverer at this time. Also, a day or so earlier (June 28/29) the comet passed within 12° of the Sun and passed through the plane of the ecliptic. The Earth was thus only 20 million kilometers downstream of the comet's giant tail and we were looking almost along the tail towards the head and its starlike nucleus. Tail lengths of 90° were reported streaming from the zero magnitude comet's head at this time. Tebbutt's comet remained a naked eye object until mid August and keen comet observers followed it in telescopes until May the following year, when it was in Cepheus, only 10° south of Polaris, the pole star. The orbital elements of C/1861 J1 indicate an orbital period of around 400 years. In 1995 the Japanese orbit experts Ichiro Hasegawa and Syuichi Nakano, suggested that the Great Comet of 1861 was the same comet as the modest C/1500 H1. If they are correct the reader may wonder why there was not a Great Comet recorded in 1500 AD. Well, the orbital circumstances of C/1861 J1 were especially fortunate in that year with the comet rising up through the ecliptic plane just as the Earth passed by. If the Earth had been on the opposite side of the Sun things would have turned out very differently!

C/1882 R1: The Great September Comet of 1882

Thirty-nine years after the appearance of the comet C/1843 D1 another spectacular member of the Kreutz sungrazing family turned up. This time the object was a direct descendant of the comet seen in 1106 AD, some 776 years earlier and, in effect, a cousin of the Great Comet of 1843. C/1882 R1, the Great Comet of 1882 (see Fig. 2.7) was a naked eye object at discovery, which occurred during the earliest days of September from the southern hemisphere. W.H. Finlay, the chief assistant at the Royal Observatory in Cape Town, South Africa, recorded it as third magnitude on September 7 and with a 1° tail. On September 17, the day of perihelion, the comet passed only 0.00775 AU from the solar center and so was less than half a million kilometers from the actual surface of the Sun. Estimating the magnitude of comets when they pass this close to the blinding solar surface is, at best, a wild guess, usually vaguely based on their brightness with respect to the planet Venus and the popular $10 \log r$ magnitude law. Nevertheless, it is highly likely that C/1882 R1 brightened way beyond magnitude -10 at perihelion. When a comet passes this close to the Sun, within one solar radius of the surface itself, there is obviously a good chance that it will cross the solar disk as viewed from the Earth. There is also a very good chance that the intense solar heating, even for a huge chunk of icy material 10 km across, will weaken the structure until it breaks apart. Both of these events occurred with the Great September Comet of 1882. At Cape Town, Finlay observed the bright dot of the comet through a smoked glass filter until it disappeared at the solar limb, but it could not be seen against the solar disk itself. After perihelion the comet was an impressive sight in the night sky, but by the end of the month Finlay and the great observer Edward Emerson Barnard noticed the coma was elongated: a sure sign that the nucleus was breaking up. At first two nuclei were visible, then four, and even a fifth. Orbital elements for the four brightest chunks, labeled as C/1882 R1-A, B, C and D were eventually computed and will presumably return

1882 Nov 7.

Fig. 2.7. A photograph of the Great Comet of 1882, as seen from South Africa, taken on November 7, 1882 at the South African Astronomical Observatory by Sir David Gill (1843–1914).

to perihelion around 2660 AD, give or take a decade or two. The comet was still a naked eye object in February 1883 and was followed with telescopes until June 1883.

1P/Halley and the Daylight Comet C/1910 A1

I have already mentioned the recurring appearance of comet Halley as a great comet throughout the centuries and do not wish to repeat ground I have already covered. However, between 1882 and 1927 there were, arguably, only two truly "Great" comets and they both appeared in 1910: Halley, and the so-called "Daylight Comet" which tried, but failed, to upstage it. So, despite mentioning Halley already I really cannot restrain myself from covering its 1910 return in detail and mentioning the Daylight Comet at the same time.

On January 13, 1910 workers at the Premier diamond mine in Cullinan South Africa saw a bright comet in the dawn sky. The local Johannesburg newspaper (*The Leader*) reported this as a sighting of comet Halley. However, this was not Halley at all but a completely new and impertinent comet trying to steal the show! It would, in future years, simply be referred to as the Great January Comet, or the Daylight Comet of 1910 (see Fig. 2.8). The Daylight Comet was briefly spectacular, but it arrived unannounced and peaked in a twilight sky. It reached perihelion just 4 days after discovery, on January 17, 1910 at only 0.129 AU from the Sun.

Fig. 2.8. The Daylight Comet (C/1910 A1) sketched by W.B. Gibbs on January 27, 1910 from Biskra, North Africa. From the Journal of the British Astronomical Association, Volume 20.

Although there were a few reports around perihelion of the comet being seen just a few degrees from the Sun, in daylight, and brighter than Venus, most observers (including those in the UK who knew where to look) reported a peak in brightness between first and third magnitude in the weeks after perihelion and with a tail between a paltry 1° and a stunning 50° in length! The longest tail lengths were generally reported towards the end of January with experienced observers reporting the comet had faded to third magnitude by that time. February saw a dramatic fade in the Daylight Comet's brightness, from around fifth magnitude on February 10th to ninth magnitude by mid-month. The Daylight Comet had certainly been spectacular but only for the briefest of periods in the second half of January.

Max Wolf at Heidelberg recovered comet Halley on the night of September 11/12, 1909. Its return had been greatly anticipated and several observatories had tried, but failed, to beat Wolf to the first photograph on its return (see the amusing cartoon of Fig. 2.9).The perihelion date was soon calculated as April 20, 1910. The comet would come closest to the Earth on May 20, a month after its return to perihelion. It would be an excellent return, even if it would not be record breaking. If the comet had returned slightly later the apparition would rapidly have become an average one, but as it turned out Halley would pass within 23 million kilometers of the Earth on May 20, and be an awesome sight that month. Halley was a steady ninth magnitude object by the start of February and in the first week of that month its brightness pulled ahead of the rapidly fading upstart "Daylight Comet." All of the top astronomers of the day were now concentrating on comet Halley including legendary characters like Georges van Biesbroeck (Uccle, Belgium), Knox-Shaw (Helwan, Egypt), Millosevich (Italy), and Innes (Transvaal Observatory, Johannesburg) as well as Max Wolf at Heidelberg and E.E. Barnard at Yerkes.

By early March this most famous comet of them all was heading rapidly into the evening twilight as eighth magnitude fuzz with Earth and comet on opposite sides of the Sun and moving in opposite directions. The keenest observers recovered a much brighter Halley in the morning sky in the second week of April 1910 with estimates of its brightness ranging from fifth to fourth magnitude. Undoubtedly, from mid-April onwards, more people were observing Halley than at any time in its history. The comet passed perihelion, well above the Earth's orbital plane, on April 20 but was already moving rapidly down towards the Earth's orbit. It was now a magnitude 3.0 object with a tail 2° in length. The comet had entered the critical immediate post-perihelion phase when the weeks of maximum heating from the Sun were finally soaking into the icy nucleus creating the most activity. In addition the comet was now rapidly

Fig. 2.9. A humorous 1910 cartoon by W. Heath Robinson (1872–1944) depicting astronomers at Greenwich Observatory deploying all manner of techniques to be the first to spot the returning comet Halley!.

drawing closer to the Earth. The result of these two factors conspiring would be dramatic. By the first week in May comet Halley reached magnitude 2.0 with a tail some 18° in length; an impossible object to miss in a dark sky. Halley was well placed for northern hemisphere observers, despite the encroaching summer twilight. It was descending towards the Earth from above, keeping its declination nice and healthy, peaking at +20° north on May 19, just 1 day before its closest approach. Although Halley had moved westward from the evening twilight sky to the morning twilight sky, travelling through Pisces in March 1910 and passing less than 6° north of the Sun on the 27th, the gradual coming together of the Earth and the comet near to Halley's descending node would reverse this apparent westward trend around April 25. The comet would briefly clip the Pisces/Pegasus border before hurtling back east, through Pisces. On May 8, 1910 a classic photograph of the head of comet Halley was taken by the Observatory of the Carnegie Institute at Washington (see Fig. 2.10).

As May progressed, with the comet past perihelion and heading towards a rendezvous with the Earth, the motion of the comet eastward increased dramatically and the head rapidly started to depart the morning sky and approach the morning twilight once more.

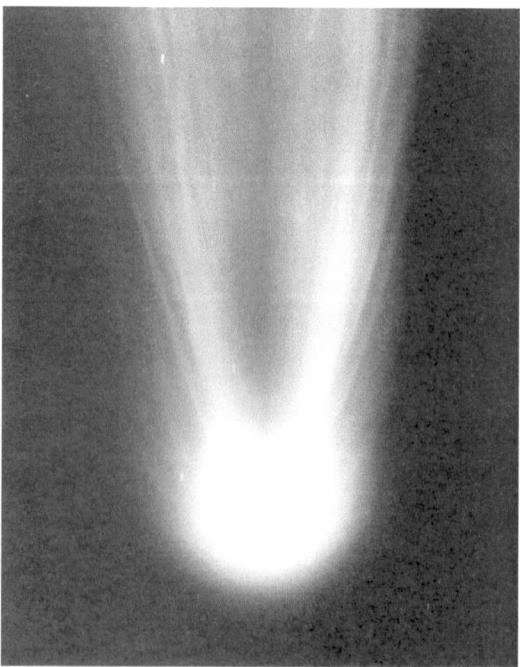

Fig. 2.10. A photograph of the head of comet Halley taken on May 8, 1910. Image courtesy of the Carnegie Institute of Washington Observatories.

However, with the comet now so close to the Earth its head would not be lost in-between the morning and evening sky for long, it was moving too fast through the sky (up to 15° per day, due east!) But there was to be an extra twist to this comet-earth flyby. As the comet plunged towards the Earth's orbital plane (the ecliptic) and its closest point to us, it would be in a perfect straight line between the Sun and Earth; so perfect that the comet's nucleus would appear to pass right across the face of the Sun for 25 min around 04:00 GMT on May 19 and, in theory, Halley's long gas tail would indeed sweep across the Earth itself on that day. Remarkably, the comet was observed in strong twilight low in the May 18th morning sky although there do not appear to be any cast iron observations of it by expert observers in the equally strong twilight of the May 19th evening sky. The straight gas tail and curved dust tail were visible in both the morning and evening skies for much of mid-May. Of course, many leading professional astronomers with suitably sun-filtered equipment were monitoring the Sun to see if the head of Halley could be seen crossing its brilliant face. However, observers at more than a dozen of the world's finest observatories across the globe failed to witness any transit of the nucleus across the Sun's blindingly bright disc. By far the most amazing feature of Comet Halley at the 1910 apparition was the sheer length of its tail, or tails. Not until 86 years later and the close flyby of Comet Hyakutake in late March of 1996 would a similar phenomena be observed. The tails of Great Comets can easily stretch for tens of millions of kilometers. When one passes within, say, 25 million kilometers of the Earth such that the tail stretches right past us, and is in a dark sky, the sight can be extraordinary. The Halley tail lengths reported by the most reputable observers in 1910 started to become impressive around the New Moon period of May 9 when lengths of 25° were reported. On the 13th Max Wolf reported a tail length of 52° with a width of 3° at a time when naked eye estimates of the comet's brightness were between magnitude 0 and 2. By the 18th E.E. Barnard reported a tail spanning a collosal 107°, just prior to the solar transit period and the tail engulfing the Earth. After May 20 with the comet's head once again visible in the evening sky the reported tail lengths became even more surreal, despite the full phase of the Moon. Experienced observers like Knox-Shaw were reporting a staggering 140° of tail up to 15° wide at the widest point. But only 5 days later Barnard reported a tail length of only 25° which the comet maintained until the end of the month with its magnitude dropping from zero to three. By late May Halley was fading noticeably as, in addition to being well past perihelion, its distance from Earth was increasing rapidly. Halley remained a naked eye object until mid June but, after that, observers followed it with telescopes and by using photography. By mid-August 1910 Halley was, once more, disappearing into the twilight and when, eventually, it re-emerged in November 1910 it was no brighter than twelfth magnitude; but the photographers persevered as none of them would be around for the 1985/1986 return. The last known successful observation of Halley was on June 16, 1911 when a 40-min exposure by Curtis using the 36-in. Crossley reflector at Lick Observatory recorded "a small hazy patch of utmost faintness."

C/1927 X1 (Skjellerup–Maristany)

After the two superb comets of 1910 the year 1911 saw two good comets, 1911 IV Beljawsky (C/1911 S3) and 1911 V Brooks (C/1911 O1); 1914 produced Delavan's comet (C/1913 Y1) but the next 17 years were unusually barren ones for spectacular naked eye comets; certainly nothing warranting the term "Great" appeared on the scene. But the drought was ended in 1927 with the arrival of Comet Skjellerup–Maristany 1927 IX (see Fig. 2.11) now designated as (C/1927 X1).

Rather confusingly the comet was first discovered by O'Connell on November 28, 1927 followed by a Melbourne night watchman 2 days later and finally, at third magnitude and with a 3° tail, by Skjellerup (or some even say it was first noticed by his cat!) on December 4 with a 7.6-cm refractor. However, Skjellerup was the first to be officially credited with the discovery, followed by co-discoverer Maristany on December 6, by which time the comet was an easy second magnitude object. Perihelion was on December 18 at a solar distance of 0.176 AU and the comet was a predominantly daylight object, indeed it was photographed from Lowell observatory in bright twilight at a reported magnitude of −6. From December 29 to January 3, 1928 the comet displayed a tail of some 35° in length.

C/1947 X1 and 1948 XI

In the immediate post war period two Great Comets appeared in the Southern skies. Comet 1947 XII (C/1947 X1) reached perihelion on December 2 at a solar distance of 0.11 AU. Like so many bright comets its small perihelion distance was the deciding factor in its brilliance. The comet was discovered 6 days after perihelion in bright twilight by a

Fig. 2.11. Comet Skjellerup–Maristany sketched by R.A. McIntosh, Posonby, Auckland, New Zealand with a 14-in. reflector 30x. Sketch from the Journal of the British Astronomical Association.

number of observers and became known as the "Southern Comet." It was unusual in having an orange head and it also featured a 25° long tail; the magnitude peaked at around magnitude −3 around discovery time but it had faded below naked eye visibility by Christmas 1947. Like a number of comets which pass very close to the Sun, the nucleus split into two components.

Comet 1948 XI (C/1948 V1), the "Eclipse Comet" was an interesting discovery; in fact, it was discovered twice and there was some confusion about the discovery. The comet was discovered during the November 1, 1948 Nairobi Total Solar Eclipse by a Dr Atkinson; hardly a difficult discovery as the comet was at least magnitude −1 and possibly much brighter.

Perihelion had already occurred some 5 days earlier on Oct 27, with a solar distance, q of 0.14 AU. After the eclipse the comet was not seen again until its second discovery on November 7 by Union astronomer Dr W.H. van den Bos. It was then an easy object in morning twilight with a 30° tail. By November 10, the comet had faded to second magnitude but it still had a tail some 15° in length.

C/1956 R1 (Arend–Roland) and C/1957 P1 (Mrkos)

Comet Arend–Roland (see Fig. 2.12) was discovered on November 8, 1956 at Uccle Observatory in Brussels at a distance of 2.84 AU from the Sun and 1.88 AU from the Earth. It was near opposition in Triangulum and, at 11th magnitude, already observable using amateur equipment. Calculations showing that perihelion would occur on April 8, 1957 with a q of 0.316 AU and an approach to the Earth within 0.6 AU some 10 days later augured well, especially as the absolute magnitude was in the same class as Halley (4.5–5.0). In addition, the comet was moving rapidly north after perihelion, peaking at more than 60° in declination in early May at an elongation of greater than 50° from the Sun. The comet lived up to expectations reaching zero magnitude at perihelion and was a naked eye object from late April to Mid May. Tail lengths of up to 20° were reported and the classic anti-tail, most associated with Arend–Roland, was the most remarkable feature. With the Earth passing through the plane of the comet's orbit (the longitude of the ascending node was 215°) on April 26, 18 days after perihelion and with the Earth and Sun distances little more than 0.6 AU, a good anti-tail was always on the cards. The comet had another role to play too as it peaked just as the first BBC *Sky at Night* TV program was broadcast by the well known astronomy author and personality Sir Patrick Moore. The first program, by sheer good luck, was ushered in by a naked eye comet and photographs of Arend–Roland where shown on the program.

Only 2½ months after Arend–Roland dropped below naked eye visibility Comet Mrkos (see Fig. 2.13) was discovered on August 2, at second magnitude, the day after its perihelion passage at a solar distance of 0.355 AU (a similar q to Arend–Roland). Mrkos was independently discovered on August 3 by Clive Hare at Tamworth UK, a 15-year-old BAA member, and by W. Clark at Hull, UK on August 4. Whereas Arend–Roland had been easily observable from the UK as a first magnitude object at more than 35° elongation from the Sun, Mrkos only achieved this happy elongation

Fig. 2.12. Comet Arend–Roland, photographed by Reginald Lawson Waterfield (1900–1986) from Ascot, England on 1957 April 28.8. This was a 30-min exposure on a Kodak Oa-E plate with a 60-mm f/3.5 Aldis lens. Photograph from the British Astronomical Association archives.

when it was fourth magnitude. Nevertheless, it was a superb bonus to the observing year! Not since 1910 had two comets of such quality appeared in our skies in a single year. Antonin Mrkos (1918–1996) was a formidable comet discoverer who swept up 13 comets during the twentieth century. Eleven of these were visual binocular discoveries as part of the Skalnate-Pleso patrol from 1948 to 1959; the other two, in 1984 and 1991 were photographic discoveries made at Klet Observatory.

Fig. 2.13. Comet Mrkos, photographed by Alan McClure (1929–2005) from California on 1957 August 20.2: a 25-min exposure. Photograph from the British Astronomical Association archives.

C/1965 S1 (Ikeya–Seki)

On September 18, 1965 in the dawn sky, the Japanese hunters Ikeya and Seki discovered a comet that was set to become (technically) the brightest of the twentieth century: this was Comet Ikeya–Seki 1965 VIII (see Fig. 2.14) which is now designated C/1965 S1. Seki discovered six comets from 1961 to 1970 and his young rival Ikeya discovered five from 1963 to 1967 and a sixth (the longest period periodic comet Ikeya–Zhang) in 2002. In later years Seki would become a prolific comet and asteroid astrometrist, comet recovery specialist and an asteroid discoverer. At the time of discovery the comet was an eighth magnitude object 1 arc-min in diameter but it brightened steadily and by the end of September it was a naked eye object with a 20 arc-min tail visible in large binoculars.

Fig. 2.14. Comet Ikeya–Seki, photographed by Alan McClure (1929–2005) from Mount Pinos California on October 31, 1965 at 12:52 UT. This was a 12-s exposure with a 35-mm camera and a "high speed film + lens." Photograph from the British Astronomical Association archives.

In the weeks following discovery the orbit was derived and its similarity to that of Comet 1882 II (C/1882 R1) was soon spotted. Yes, this was another member of the Kreutz family with typical perihelia of 0.0078 AU, arguments of perihelion around 69°, ascending node longitudes of 347°

and inclinations of 142°. In addition, their orbital periods are roughly in the 700–800 year range. With such a small perihelion distance, the comet was set to pass within 470,000 km of the solar surface on October 21. Ikeya–Seki, despite its arrival time some 83 years later, was literally a brother of the Great Comet C/1882 R1 and both were children of the Great Comet of 1106, X/1106 C1.

By mid October the new Kreutz comet was second magnitude and with a 10° tail visible in bright twilight, but it was, understandably, becoming more and more difficult to observe as it sank deeper into the twilight sky. Remarkably, the comet was observed, even on the day of perihelion, when it was only 10 arc-min from the solar limb, both by observers at the Tokyo Observatory Mount Norikura Coronograph station and by other naked eye observers. They glimpsed it as the Sun's red disk appeared above the horizon, a highly dangerous and inadvisable method! The experienced US comet observer John Bortle was reported as describing the comet as looking "like a dagger sticking out of the Sun" and magnitude estimates ranged from −10 to −16. Certainly, this was the brightest Great Comet of the century so far, and there has been nothing brighter since. But of course few observers would have seen it under these extreme conditions. Like so many sungrazers (including its twin in 1882) the comet started to disintegrate at perihelion, but Ikeya–Seki still put on a superb show even after its close brush with the Sun. By the end of October 1965, Ikeya–Seki was still approximately magnitude −2 and displayed an unusually bright, 25° dust tail; some observers could trace the tail as far out as 45°. The tail also featured the synchronic band striata later seen in Comet West in 1976 and Hale Bopp in 1997. From November onwards the nucleus was seen to consist of at least two, and possibly three, components. By the second week of December the comet had dropped below naked eye visibility but it had put on a superb show. Undoubtedly, Ikeya–Seki had staked its claim as one of the greatest comets of the twentieth century.

Comet C/1969 Y1 (Bennett)

On December 28, 1969 the comet hunter Jack Bennett, of South Africa, swept up an eighth magnitude fuzzy patch close to the Magellanic Clouds (see Fig. 2.15). Calculations of the orbit soon revealed that it would reach perihelion on March 20, 1970 at a very favorable 0.54 AU from the Sun. Five days after perihelion Comet Bennett (C/1969 Y1) would pass within 0.69 AU of the Earth. The third favorable parameter was Bennett's absolute

Fig. 2.15. Comet Bennett photographed by Mike Hendrie on 1970 April 4.2. A 40-min exposure on a 103a-O plate using a 150-mm f/4.5 Cooke triplet at Reggie Waterfield's relocated Observatory at Woolston near Wincanton, Somerset. Photograph from the British Astronomical Association archives.

magnitude of 4.0. Most Great Comets have a favorable perihelion distance, a few pass close to the Earth and a few have outstanding absolute magnitudes; these are the three big parameters that influence the brightness of a comet and all three were very favorable in Comet Bennett. The beautiful Comet Donati of 1858 was similarly endowed. Bennett peaked at magnitude 1.0 at the time of perihelion and in the subsequent weeks the comet was a superb northern hemisphere object at more than 40° elongation from the Sun, and in a dark sky. Indeed, from April 10 until the last observation in September 1970, the comet was a circumpolar object from the UK, steadily getting closer to the pole as the weeks went by. Comet Bennett was a truly superb object in the morning sky in the first 2 weeks of April 1970, being around second magnitude with tail lengths of up to 20° being reported. It was well observed by amateur astronomers and it must certainly rank amongst the very best comets of the twentieth century. Visual observations of the inner coma at high power revealed a wealth of detail. The comet expert David Seargent used the term "orange pin wheels" to describe the high power view and a dark "shadow of the nucleus" feature (a dark band stretching behind the dead center of the coma) was noted, possibly the best example of this since the beautiful comet Donati of 1858.

C/1975 V1 (West)

Comet West (formerly designated 1976 VI) was discovered by Richard West of the European Southern Observatory on November 5, 1975. The discovery was made on survey plates exposed 6 weeks earlier. A 15th magnitude suspect was spotted which would arrive at a small perihelion distance (0.197 AU), in a few months time, on February 25, 1976 in fact. It all sounded worryingly familiar because a similar comet discovery had occurred 2 years earlier in 1973, when a comet called Kohoutek had been discovered, predicted to be the greatest comet ever seen, and then did not live up to the frenzied media hype! Understandably, after Kohoutek, predictions of the comet's likely brightness were tempered with caution. However, when the first amateur visual magnitude estimates were received in December 1975, the comet was actually two magnitudes *brighter* than predicted. By mid-February 1975 the comet was a third magnitude naked eye evening object with a distinctly yellowish cast and a tail visible in binoculars. In the second half of February the magnitude accelerated dramatically and the comet was at least magnitude −2 at perihelion, but only observable in strong twilight. After perihelion, West was a spectacular morning object with a magnitude of −1 and a superb multiple dust tail 30° long , not dissimilar to the awesome De Chésaux comet of 1744; the comet that may have inspired the teenage Charles Messier to start hunting for comets. In the USA, John Bortle observed the nucleus as two distinct points on March 8 and within days professional observatories were photographing a quadruple nucleus. The awesome dust tail of West and its synchronic band striata and multi-tail appearance were relatively short-lived phenomena. I have, over the years, spoken to a number of observers on the subject of Great Comets and the effects on "Greatness" of the invariably cloudy British weather during their appearances. Undoubtedly the full glory of West (see Fig. 2.16) was lost to most UK observers, due to cloud. However, the accounts of the late Paul Doherty and of Guy Hurst were memorable and almost identical. Both recall getting up on a single morning in the early UK hours of March 1976, armed with binoculars (to find the comet!), leaving their respective houses, walking round to the

Fig. 2.16. Comet West as photographed from the Lick Observatory in 1976 March.

Eastern side, and then gaping in awe at the remarkable sight. Both told me they felt rather silly having brought their binoculars with them! The Great Comet West faded below naked eye visibility at the start of April but was still an easy binocular object, even in June and July.

C/1996 B2 (Hyakutake)

Comet Hyakutake (see Fig. 2.17) is, without doubt, the greatest cometary spectacle this author has ever seen! It also brings back many memories for me. At the February 1996 European Astrofest meeting in London I was to give a talk entitled "Hale–Bopp: Comet of the Century?" which

Fig. 2.17. Comet C/1996 B2 (Hyakutake) photographed with a Canon 85 mm f/1.2 lens set to f/1.8 with Fuji Super G 800. Three-minute exposure on the night of March 24/25, 1996 from altitude at Tenerife. Photograph by Martin Mobberley.

I duly delivered. That comet had been discovered 7 months earlier but still had another 14 months until it reached perihelion. Little did I know at that time, but even as I was giving the talk, the elements for the newly discovered Comet Hyakutake 1996/B2 were breaking. The next day we all learned that there was now a two horse race for the potential "Comet of the Century." My talk was out of date the day after it had been delivered! The Japanese amateur Yuji Hyakutake had discovered his first, much fainter comet (C/1995 Y1) using Fujinon 25×150 binoculars, on Dec 26, 1995. A month later he discovered his second, C/1996 B2. This comet was destined to be mouth-wateringly awesome for those with clear skies. Whereas Hale–Bopp's strength would, in the following year, prove to be its sheer size and activity levels, Hyakutake was a better-than-average comet coming extremely close to the Earth. The orbital elements showed that Hyakutake would pass a mere 0.1 AU from the Earth on March 25.3 UT The initial post-discovery photographic magnitudes looked, if anything, a little depressing and the visual magnitude estimates shortly afterwards made the comet far brighter, leading many to overestimate the perihelion prospects in store! However, as it would turn out the perihelion perform-ance would be irrelevant, at least compared to its late March passage. A close Earth fly-by is over in a few days and so the prospects for Hyakutake were critically weather dependent and the weather in the UK throughout February was incomprehensively dismal, with even a glimpse of the Sun being a truly rare daytime event. In a moment of unusual sanity for me, I decided to flee to Tenerife with two astro-colleagues (Nick James and Glyn Marsh) and experienced a comet observer's view of a lifetime. For UK observers, the weather broke in the days following closest approach, but by then the Moon was becoming a problem. The four nights I spent on Ten-erife were undoubtedly tiring and, of course, neither I nor my colleagues had any idea how well, or badly, our photographs would turn out until we returned home. But, despite the freezing cold (we were poorly prepared for 9 h at altitude) and the fatigue, the experience was an unforgettable one. On a personal level, I can only compare the memory of Hyakutake on March 23.0, 24.0, 25.0 and 26.0 with the total solar eclipses I have seen and with the 1999 Leonid Storm from the Sinai desert. Everything else is in a lower category. Without doubt the strongest recollection, as if from a sur-real dream, is of the night of March 23/24. The Mayor of Puerto de Gui-mar on Tenerife ordered the city lights turned off for 90 min at 0000 UT for the Hyakutake beach-side Star Party. With Nick James, Glyn Marsh and Mark Kidger (resident professional astronomer) I walked along the beach where thousands of people were assembled, just staring up at the sky, as the well-broken cloud started to really break up even more. I lost count of how many people I handed my 11×80 binoculars to; at one point I lost track of them but they eventually came back to me. Several thousand people staring skyward at Hyakutake was totally unreal. At altitude the view was even more impressive; to see a tail disconnection in a comet with the naked eye and to see it subtly change during the night through bin-oculars was awesome! My observing log records direct vision tail lengths of 20° and averted vision tail lengths of 40–50° with my comment "the head is just slightly fainter than Vega when defocusing my eyes." So where does Hyakutake stand in the league table of recent Great Comets. Accord-ing to the veteran comet observer John Bortle around closest approach

Table 2.3. Historical rivals of Hyakutake with *Ho* at least 5.5 and perigee distances no greater than 0.15 AU.

Comet	Perigee date	Distance (AU)	Best magnitude in a dark sky	Comments
1P/Halley	May 20, 1910	0.15	0–1	Halley. Tail up to 130°
C/1861 J1 (Tebbutt)	June 30, 1861	0.13	0	90°
C/1556 D1	Mar 12, 1556	0.08	–2	Reported in Germany/Asia
C/1471 Y1	Jan 22, 1472	0.07	–3	36° tail
1P/Halley	Oct 3, 1378	0.12	1	Halley
C/1132 T1	Oct 6, 1132	0.05	–1	Chinese/Japanese/Korean
1P/Halley	April 24, 1066	0.10	0–1	Halley
1P/Halley	April 11, 837	0.03	–3	Halley April 13. Tail >130°
1P/Halley	April 19, 607	0.09	–2	Halley
C/568 O1	Sept 25, 568	0.09	0	Chinese
C/400 F1	March 31, 400	0.07	0	Chinese/Korean
C/390 Q1	Aug 18, 390	0.10	–1	Chinese: Tail >70°
1P/Halley	April 2, 374	0.09	–1	Halley
C/-146 P1	Aug 3, –146	0.14	1?	Chinese

Hyakutake's coma was "as bright as any comet visible in a fully dark sky since the Great September Comet of 1882" and featured "the second longest credible tail length on record, exceeded only by 1P/Halley at the 1910 apparition." Going back in history, what other comets, with a respectable H_o (5.5 for Hyakutake) have passed within 0.15 AU from the Earth? There appear to be 14 cometary apparition rivals to Hyakutake in this respect going back to 147 BC (see Table 2.3) and six of these cases are returns of comet 1P/Halley.

In 1910 comet 1P/Halley, 86 years prior to Hyakutake, was a similar object if 50% further away, and one has to go back to 1556 to find a similarly active comet which came closer than Hyakutake. A reasonable generalization might be that Hyakutake-like cometary encounters come along about every 100 years or so.

On my fourth night in Tenerife in March 1996, with the comet having passed the Earth and moved north into Ursa Minor I really felt like I was standing on a rock in orbit around the Sun having witnessed a Great Comet fly past on the previous nights; a perception I've never felt with any other comet. To Nick James, Hyakutake's head and tail looked like someone had pierced a hole in the sphere of the heavens (the head), letting a ray of light (the tail) shine through. What on Earth would the ancients have made of such a sight?

A fascinating video by Canadian amateur astronomer Peter Ceravolo and his colleagues at *Cyanogen Productions* was made of comet Hyakutake's flyby. This was the result of taking 900 high quality film photographs from the Arizona desert, scanning all the images, and then combining them into an MPEG movie. The result showed Hyakutake's tail changing with time as it flapped in the solar wind and as dust production turned on and off: a remarkable piece of work! After perigee, Hyakutake moved towards its perihelion at a very respectable 0.23 AU on May 1. It was still a superb object, especially well seen from the UK, in the crystal clear skies of April 17, but it had faded to second magnitude by then and would never rival its perigee appearance.

C/1995 O1 (Hale–Bopp)

Two amateurs, now immortalized, discovered the monstrous comet C/1995 O1 Hale–Bopp (see Fig. 2.18) that would be remembered by all amateur astronomers who were active in the late 1990s. At the time of the discovery, July 23, 1995 the comet was a magnitude 10.5 smudge very close to the ninth magnitude globular cluster M 70. Alan Hale in

Fig. 2.18. Comet C/1995 O1 (Hale–Bopp) photographed with a 530-mm focal length Takahashi E-160 astrograph (160-mm aperture f/3.3) and Kodak Ektar Pro-Gold 400 film on March 30, 1997; a 10-min exposure. Photograph by Martin Mobberley.

Cloudcroft, New Mexico (a veteran comet observer) and Thomas Bopp in Stanfield, Arizona, were both observing this deep sky object with large 0.41 and 0.44 m reflectors, respectively, when they spotted the new object. This discovery, and that of C/1996 B2 Hyakutake, came at an interesting period in comet discovery history when professional sky patrols were at a minimum. The Shoemakers (and Levy) had been forced to wind down their patrols with the 0.45-m Palomar Schmidt only a year or so after the discovery of Shoemaker–Levy 9 .Other Near Earth Asteroid Surveys such as NEAT and LINEAR were not fully operational, although Space-watch was operating. So the skies were relatively free from software con-trolled discovery machines and the likes of Bradfield, Machholz, Levy and Brewington were free to trawl in more comets in the last few years, before the machines took over! However, even these famous names were not destined to discover the two Great Comets of the 1990s, and neither were the well known numerous Japanese discoverers like Takamizawa, Kiuchi, Kushida and Nakamura (to name but four!). As soon as Hale–Bopp was discovered it was obvious that the comet, like Kohoutek, was moving painfully slowly against the background stars; almost too slowly to believe in fact! The period around July 20 was starting to look like a guaranteed time for astronomical excitement. In 1969, man had landed on the Moon on this day; 25 years later, Shoemaker–Levy 9 fragments were bruising Jupiter and a year after that, Hale–Bopp was discovered. After a few days, a highly preliminary set of orbital elements were issued, suggesting perihelion might be on January 26, 1997 with a q of 0.8 AU (that would have been very impressive!); soon after the elements were revised to make T and q April 1 and 0.91 AU, respectively.

This enormous comet had been discovered visually, at magnitude 10.5, at a staggering 6.7 AU from the Sun! Only comet C/1976 D2 (Schuster) and C/1980 E1 (Bowell) had been discovered at similar distances from the Sun at that time. Bearing in mind that Kohoutek had been 16th magnitude when only 5 AU from the Sun (25% closer) and Halley was 24th magnitude when recovered 10.7 AU from the Sun (60% further) this, surely, was a true behemoth! Predictably, after Kohoutek and Austin, no-one believed that the magnitude would hold up; the initial magnitude law was given as $-1.5 + 5\log \text{Delta} + 10\log r$, implying an absolute magnitude (H_o) some 250× that of Halley! Early CCD images by Warren Offutt in Cloudcroft, New Mexico, and others, showed a hint of a spiral coma similar to that frequently displayed by 29P/Schwassmann–Wachmann after its numerous outbursts. Massive cometary outbursts of up to six magnitudes are not that rare; Halley, Humason, Nagata, Tago-Sato-Kosaka and Tuttle–Giacobini–Kresak are all comets seen to have undergone dramatic outbursts and we will learn more about those events later in this book. So maybe this apparent monster with a negative H_o had been discovered during an outburst? The alternative was that it was a truly monstrous comet in the same class as Comet Sarabat of 1729 ($H_o = -3$), De Ché-saux comet of 1744 ($H_o = 0.5$), the Great Comet of 1811 ($H_o = 0$), or even the Great September Comet of 1882 ($H_o = 0$). The Great Comet of 1811 had the most similar orbital and activity circumstances to Hale–Bopp. Over the next year, Hale–Bopp continued to brighten steadily (although there were a few hiccups) and having brightened to eighth magnitude by the time the comet emerged from the dawn sky in March 1996, few observers doubted that this would become another Great Comet. How remarkable that, since West in 1976, we waited 20 years and then had two zero magnitude comets in the space of 1 year. Only 1910 and 1957 had been similarly blessed this century. Just as Hyakutake was a very rare object in terms of its perigee distance, Hale–Bopp was just as rare in terms of its size. It was never closer to Earth than 1.35 AU but exceeded zero magnitude and was slightly brighter than Hyakutake at its peak. Although the size of Hale–Bopp's coma and tail could not match that of Hyakutake at its closest to the Earth (it appeared, to me about a third the size) it was an easy naked eye comet with a tail throughout February, March and April 1997. Thus, even allowing for the usually cloudy UK weather, members of the general public could not fail to spot it at some point during that period. In contrast, they did not stand a chance with the close approach of Hyakutake in the cloudy skies around March 25, 1996. In addition, although Hyakutake's nuclear region was a splendid sight in amateur telescopes; Hale–Bopp's was utterly staggering. Personally, I had always regarded nineteenth century sketches of the nuclear regions of Swift–Tuttle, Donati and other comets with considerable skepticisms until I saw Hale–Bopp through my 36-cm Newtonian in February 1997. Clearly one was seeing a pattern of material being laid down by a rotating nucleus when one observed Hale–Bopp at moderate powers. Even a year before perihelion, amateurs had been recording multiple jets of material emanating from the nucleus of Hale–Bopp. If Hyakutake and Halley's (1910) tails were the tails of the century, Hale–Bopp's inner coma must have been the cometary telescopic view of the century (with Shoemaker–Levy 9's bruises on Jupiter a close second). As with Hyakutake's nuclear regions, the pioneering UK amateur and CCD camera expert Terry Platt obtained some high resolution images of the Hale–Bopp inner coma which he combined to make a stunning MPEG movie showing the near nuclear region rotating and laying down a new "spiral arm" of material.

A comparison of Hale–Bopp and Hyakutake's magnitude laws, derived by Jonathan Shanklin of the BAA, is interesting.

$$\text{For Hyakutake} \quad m = 5.31 + 5 \log \text{Delta} + 7.69 \log r$$
$$(q = 0.23 \text{ AU}; \text{ perigee} = 0.1 \text{ AU})$$

$$\text{For Hale – Bopp} \quad m = -0.65 + 5 \log \text{Delta} + 7.49 \log r$$
$$(q = 0.91 \text{ AU}; \text{ perigee} = 1.35 \text{ AU})$$

As can be seen, Hale–Bopp's absolute magnitude would make it six magnitudes, or 250×, brighter than Hyakutake, all other things being equal. Hyakutake's passage within 0.1 AU of Earth, some 13.5× closer than Hale–Bopp gave it an advantage of nearly 200×, thus both comets had similar peak magnitudes of about 0. Of course, Hyakutake's close passage gave it a tail length which Hale–Bopp could not hope to match, but Hale–Bopp was a naked eye object for far longer. It is interesting to speculate on what might have happened if Hale–Bopp had arrived at perihelion

4 months earlier than April 1, 1997 specifically, December 1, 1996. Over such a long period orbit 4 months is a trivial adjustment! In that case the six magnitude H_o advantage of Hale–Bopp over Hyakutake would have been displayed to its full and awesome potential as Hale–Bopp would have crossed through the ecliptic plane around January 3, 1997 a month past perihelion and 0.1 AU *behind* the Earth by then: the same distance as Hyakutake flew past in the previous year. A magnitude −5 or −6 spectacle would then have appeared in the sky with the head alone being maybe 5° across. I am dribbling over the keyboard as I think of such a sight! Sadly, we missed that spectacle by 4 months, but I do not think we should register an official complaint.

So, back to reality and in the UK the period around Easter 1997 produced clear night after clear night and this proved somewhat exhausting for comet photographers. My best Hale–Bopp photographs were obtained around this time. I can confirm that it is possible to get a bit tired of a Great Comet when you have spent day after day photographing, developing and printing images. The trouble is, you feel terribly guilty if you do not go out when it is clear, whether you are sick of the comet or not! By late May, Hale–Bopp had disappeared forever from UK eyes and the comet is not predicted to return for another 2,500 years.

C/2006 P1 (McNaught)

Nine years after the spectacle of Hale–Bopp, it was business as usual on the night of August 7, 2006 when the world's greatest comet discoverer, Rob McNaught, made an apparently routine comet discovery using the Uppsala Schmidt telescope in New South Wales, Australia. This famous instrument and its equally well-known discoverer are based at Siding Spring Observatory, near Coonabarabran. The discovery itself might have been routine, but Comet C/2006 P1 (McNaught) would turn out to be anything but. It will be remembered as one of the most awesome cometary spectacles of the last 100 years. On the day of discovery, the new comet McNaught was in Ophiuchus (R.A. 16 h 40 m, Dec −18) and gave no indication that it would prove to be one of the most spectacular comets of the last 100 years. It was a small, 17th magnitude fuzz, moving slowly westward at 40 arc-sec per hour against the background stars. At that time it was 370 million kilometers from the Earth and 460 million kilometers from the Sun, and at 116° elongation it was well away from the solar glare. By August 10, a total of 23 measurements enabled a preliminary parabolic orbit to be calculated and things started to look quite promising.

Of particular interest was the predicted solar distance at perihelion, on January 17, of only 0.18 AU from the Sun, in other words about 27 million kilometers. Later refinements to the orbit modified these values slightly to 0.17 AU (25 million kilometers) and a slightly earlier perihelion date of January 12.

All other things being equal a typical comet passing only 0.17 AU from the Sun, like McNaught, will be 1,200× (7.7 magnitudes) brighter than one passing a mere 1 AU from the Sun. However, as we have seen, some caution is always needed in these situations as various comets have disappointed us in the past with Comet Kohoutek in 1973 and comet Austin in 1990 both failing to live up to the media hype, even though they were still very nice comets from a keen amateur astronomers' perspective. Kohoutek in particular was a very bright but much-maligned comet, simply because the media hype was so extraordinary. The problem is that the 10 log r "solar heating" part of the cometary magnitude law is only a generalization for the "typical" comet; and there is no such thing as a "typical" comet!

It had been obvious since August that the comet would probably enter a long period of invisibility from mid November and throughout December 2006, as its angular elongation from the Sun would prevent it being seen in a dark sky, but its distance from perihelion would be too far for it to be a naked eye object. This was a tempting challenge for CCD and digital SLR equipped observers who could image the comet against the twilight glare using short exposures. In addition the comet was moving steadily north, while the Sun was moving south towards its winter solstice low point. For this reason, at perihelion, northern hemisphere observers would, briefly, have the best view, before the comet plunged south once more.

With the comet peaking at its highest declination of −7° on Jan 3rd/4th, some 15° further north than the Sun, the first 12 days of January would see the comet as the sole property of northern hemisphere observers, as the comet would rise just before, and set just after the Sun. However, December had been one of the cloudiest in memory for many UK and European amateur astronomers; would anyone actually see the comet in the critical twilight period? As it turned out UK observers (including this author) were, for the most part, very lucky. A cloudy high pressure moved away from our skies in the second week of January to be replaced by wet and very windy conditions, but with large periods of crystal clear skies in-between, especially on the mornings of January 8 and 9, and the critical last evenings of January 10 and 11, prior to the comet heading south. The comet was now on the Aquila/Sagittarius border but, in strong twilight, these constellation boundaries were somewhat academic! On these four consecutive days I was able to view the comet when the head was barely 1° above the horizon and the Sun as little as 5° below it! For me, despite a quarter-century as a dedicated comet photographer and imager, this was a unique experience, as was a tripod mounted photographic system with exposures of only 2 s maximum! In the evening sky Venus (and Mercury) provided a useful magnitude reference when estimating the brightness of the comet's head. The comet was clearly almost as bright as Venus on the 10th and 11th with a magnitude of at least −3 (see Fig. 2.19).

Typically comets peak in activity a week or two after perihelion, when the ball of ice and rock has had time to really soak in the solar energy: McNaught was no exception. As this extraordinary comet plunged south, and its elongation from the Sun increased, moving it out of twilight and into a darkening evening sky, cometary activity increased. Even more fortunate was the fact that the closest approach to the Earth (120 million kilometers) occurred on January 15, 3 days after perihelion, and New Moon occurred on January 19, thus ensuring no significant moonlight interference for several days. During this time the comet's elongation from the Sun increased rapidly, reaching 20° by the 20th. The comet was, at last, visible in a dark sky, with its head fractionally above the horizon, from the country where it was discovered some 5 months earlier, namely from Australia. The spectacle seen in southern hemisphere evening skies around this time was truly awesome. Not since Comet West in March 1976 had such a sight been seen from any place on Earth. 2006 P1 was producing vast amounts of dust and a classic curved dust tail, easily visible even to casual onlookers in the evening sky. From January 17 to 25 the sight was remarkable, peaking around January 20 with the comet some 8 days past perihelion and in the southern constellation of Microscopium. As well as a main tail with a height of 10° above the comet's coma, fainter multiple tails, the remnants of previous days emissions, stretched horizontally (and vertically) across the horizon for at least 50°, equivalent to a 150 million kilometers swath in space. An image of this spectacle, taken by Terry Lovejoy, appeared in Chap. 1. As the comet's orbital plane was at 78° to the ecliptic and well away from the Earth there would be no chance of a spectacular meteor storm as the Earth ploughed through the debris. Oh well, you can't have everything!

Even to non astronomers looking at the southern hemisphere early evening sky in mid to late January 2007 it was as if dozens of comets were trailing behind the main one. Including the faintest visible extent of the tails, the cometary spectacle was covering a swath of sky 50° wide by 25° in height! The popular technical term for the intricate details in the tail(s) of

Fig. 2.19. Comet C/2006 P1 (McNaught) photographed with a Canon 300D DSLR and a 300-mm f/5.6 focal length mirror lens; a 2-s exposure on a fixed tripod at ISO 400 in very strong twilight. Field is 3.4° wide. Photograph by Martin Mobberley.

comet McNaught was striae. A comet's dust tail, driven by the solar wind will initially point directly away from the Sun and be driven outside the comet's orbit. Unlike gas particles the dust particles have significant mass and so tend to lag behind the position of the comet's head, producing the traditional curved dust tail. The remarkable striations in McNaught's tail have to be related to gravity, solar wind and radiation forces acting on the dust. As we saw in Chap. 1 and I will repeat it here, the time of particle release from the rotating nucleus as well as the dust grain sizes will define the patterns seen. The multiple tail effect caused by dust grains released at the same time, and then released again over many days, are called "synchrones," whereas the more horizontal patterns caused by dust grains of similar mass are called "syndynes." McNaught easily displayed both patterns, even to the casual naked eye observer. These visible striae stretched so far across the sky that their ends could even be photographed by observers in the far northern hemisphere, despite the fact that the comet's head was (sadly) now 30° below the horizon by January 20! As January 2007 came to a close the awesome spectacle faded. Not only was McNaught moving rapidly away from Sun and Earth, the Moon was fattening up too. But for those who had seen C/2006 P1, from either hemisphere, it was not a comet they would ever forget, coming second in brightness to only C/1965 S1 (Ikeya–Seki) in the previous 60 years (see Table 2.4).

Fear, Awe and Nutters!

Before I close this chapter about Great Comets I would like to digress slightly from purely scientific matters to mention a few of the paranoid and even barking mad responses there have been to Great Comets. It is hardly surprising that the zero magnitude broom stars of the ancient past caused fear and awe in the centuries before their nature was understood. After all, life, for many, was painfully short and all manner of religious beliefs were engrained in the minds of the people. Surely something occurring in the sky had to be down to the wrath of the Gods? Of course, in ancient times many thought that anything in the sky was in the atmosphere and extremes of weather were responsible for crop failures, famine and death; but it is rather more surprising that even when the nature of comets was deduced, fear, awe and hysteria survived amongst some unusual individuals.

At the 1910 return of comet Halley the pioneering spectroscopic work undertaken at Lick observatory had revealed the presence of poisonous cyanogen (cyanide gas) and carbon monoxide in the comet's tail. For some members of the public for whom comets were still regarded as portents of doom, the idea of a giant tail stuffed to the brim with cyanide flapping around the Earth was enough to trigger a simultaneous panic attack and hissy fit. The peddlers of wonder tonics and elixirs saw a chance to make a quick buck and as Halley brightened, and the astronomers predicted that

Table 2.4. The 16 brightest comets over the 60 years from 1947 to 2007.

Comets in perihelion date order	Perihelion date	Perihelion distance (AU)	Peak magnitude	Outstanding characteristic
C/2006 P1 (McNaught)	Jan 12, 2007	0.17	−5.5	Small q
C/2002 V1 (NEAT)	Feb 18, 2003	0.10	0	Small q
C/1998 J1 (SOHO)	May 8, 1998	0.15	1	Small q
C/1995 O1 (Hale–Bopp)	Apr 1, 1997	0.91	−1	Massive H_o
C/1996 B2 (Hyakutake)	May 1, 1996	0.23	0	Small Δ
C/1983 H1 (IRAS–Araki–Alcock)	May 21, 1983	0.99	1.5	Small Δ
C/1975 V1-A (West)	Feb 25, 1976	0.20	−3	Small q
C/1973 E1 (Kohoutek)	Dec 28, 1973	0.14	0	Small q
C/1970 K1 (White–Ortiz–Bolelli)	May 14, 1970	0.0089	1	Sungrazer
C/1969 Y1 (Bennett)	Mar 20, 1970	0.54	0	Good q and H_o
C/1965 S1-A (Ikeya–Seki)	Oct 21, 1965	0.0078	−10	Sungrazer
C/1962 C1 (Seki–Lines)	Apr 1, 1962	0.031	0	Very small q

C/1957 P1 (Mrkos)	Aug 1, 1957	0.35	1	Good q and $H_\circ$
C/1956 R1 (Arend–Roland)	Apr 8, 1957	0.32	−0.5	Good q and $H_\circ$
C/1948 V1 (Eclipse Comet)	Oct 27, 1948	0.14	−1	Small q
C/1947 X1 (Southern Comet)	Dec 2, 1947	0.11	−3	Small q

The final column shows that comets become exceptionally bright either because they have a small q (perihelion distance), they have a massive "size"/activity level (that is their absolute magnitude $H_\circ$ is large) or they come close to the Earth (small Δ). Sometimes it is a combination of all three. In some cases the q is so small that the comet qualifies as a sungrazer, by any definition the Earth would pass through the tail of the comet, a variety of life saving options appeared including comet pills and cyanide-proof gas masks! The situation was not helped by newspapers advertising these wares and reporting cases of mass hysteria where city dwellers were plugging their doors and windows to block the poisonous vapors! By contrast, in other cities the residents were holding comet parties and in New York sales of telescopes were reported to be outstripping the supply. On April 23, 1910 the New York Times headline reported that "Women and foreigners attribute darkness over Chicago to comet; some become hysterical." In the UK King Edward VII died on May 6, 1910, and the comet was reported to be "acting strangely upon his death"! Back in the USA the Boston authorities announced that they would sound the city's fire alarms if the comet became visible and on May 19 the New York Times headline read:

> HALLEY'S COMET BRUSHES EARTH WITH ITS TAIL; 350 American astronomers keep vigil; Reactions of fear and prayer repeated; all night services held in many churches; 1881 dire prophecies recalled by comet scare.

Now you might think that 1910 would see an end to all this crazy stuff, but no! When astronomers calculated that the fragments of comet D/1993 F2 Shoemaker–Levy 9 would hit Jupiter in July 1994, a Sister Marie Gabriel (as she called herself) made a prediction that comet Halley (!) would hit Jupiter and that the resulting cosmic flash would stop all planes and traffic unless certain things were carried out "at top speed," including a visit to London by the Pope, the reinstatement of capital punishment and a decision by everyone to "become pure, holy, angelic saints" overnight! Well, the scientific results from the Shoemaker–Levy 9 impact were fascinating but, as we all know, there were no dire consequences! Did the crazy stuff end there though? Er, no! In March 1997, 39 members of the California based "Heavens Gate" religious UFO cult committed suicide due to comet Hale–Bopp reaching its peak performance in the night sky. The sect saw the appearance of Hale–Bopp as an omen that the Earth would imminently be "wiped clean, renewed, refurbished and rejuvenated" and the only chance of survival would be to leave Earth immediately. Fortunately the sect knew that a UFO was trailing Hale–Bopp and if they killed themselves now their souls would be beamed up into this rescue craft. Handy! The method of suicide was an overdose of barbiturates mixed with cyanide, arsenic, applesauce pudding and Vodka. Yummy! Plastic bags were placed over the cult members heads too just in case the meal of death did not work. As far as I can recall the Earth was not "wiped clean, renewed, refurbished and rejuvenated" in 1997 so it looks like they got it wrong.

Next time a zero magnitude comet comes around, keep an eye out for the crazy stuff!

CHAPTER THREE

Professional Twenty-First Century Comet Hunters

For centuries comets have been discovered by both amateur and professional hunters, but now that we are in the twenty-first century amateurs have never faced such stiff competition. In the nineteenth century era, prior to routine astrophotography, where visual patrols were the only form of discovery possible, amateurs and professionals had the same tools to accomplish the job: a modest aperture telescope and the sheer determination to spend hundreds of hours per year sweeping the skies and the regions just above the solar twilight glare, where small perihelion distance comets are first likely to brighten and move within detection range. In fact the only real difference between amateurs and professionals of the visual comet discovery era was that professionals were associated with an observatory appointment and a few successful amateur comet discoverers were actually promoted to a professional status once they had discovered their first few comets. In nineteenth century America cash prizes were awarded for comet discoveries and so even if you were an amateur comet hunter there was money to be made. Even during the era of photography amateur astronomers still discovered substantial numbers of comets as the dark adapted human eye and the formidable human brain can efficiently scan the skies in real time without any need to develop, fix and inspect the film hours or days later. However, since the 1990s the automated CCD patrols have seriously dented amateur comet hunting prospects and so it is worth

M. Mobberley, *Hunting and Imaging Comets*, Patrick Moore's Practical Astronomy Series, DOI 10.1007/978-1-4419-6905-7_3, © Springer Science+Business Media, LLC 2011

understanding just how these systems operate if we amateurs stand any chance of beating them to a comet.

The NEO Connection

Hunting for comets is a very similar activity to asteroid hunting and when some of these asteroids can pass close to the Earth, posing a serious threat to mankind, even politicians start to take notice. Yes, I know that is hard to believe but there are microscopic bits of brain even inside a politician's skull! In the late 1980s geologists established that the impact of a huge asteroid or cometary nucleus at the edge of Mexico's Yucatán Peninsula, centered on the town of Chicxulub, accelerated the extinction of the dinosaurs. Admittedly the era of these massive beasts was probably coming to an end anyway due to massive eruptions on the Earth, but the Yucatan impact, at the end of the Cretaceous Period, roughly 65 million years ago, probably put the final nail in the dinosaurs' coffin. Then, in July 1994, the scientific world watched in amazement as the multiple fragments of the comet Shoemaker–Levy 9 collided with Jupiter, leaving huge visible bruises in the atmosphere of the planet (see Fig. 3.1a): these bruises were bigger than the Earth! Remarkably the same thing would happen 15 years later when amateur astronomer Anthony Wesley spotted a bruise on Jupiter; this time the impacting object had not been spotted in advance (see Fig. 3.1b). Although countless millions of collisions have taken place in our solar system over eons and the risk to the civilized world of an asteroid or comet impact was already understood, the Shoemaker–Levy 9 impacts seemed to concentrate the minds of governments and astronomers into trying to detect as many near-Earth objects (NEOs) and potentially hazardous asteroids (PHAs) as possible. In addition, the relatively new and huge CCD detectors which had recently been developed meant that very deep images could be taken of sensibly large areas of the sky, each night, with an automated software controlled system managing the telescope, the camera and even the image checking. Measuring the positions of moving objects in the sky could be automated too, resulting in a very quick patrolling, discovery and reporting system. As a result of this surge to discover PHAs in the 1990s the number of asteroids (also known as minor planets) in the Minor Planet Center database has grown exponentially in recent years. There are now more than half a million minor planets in the database and more than a quarter of a million of these have precise orbital elements, resulting in the asteroid achieving a numbered status.

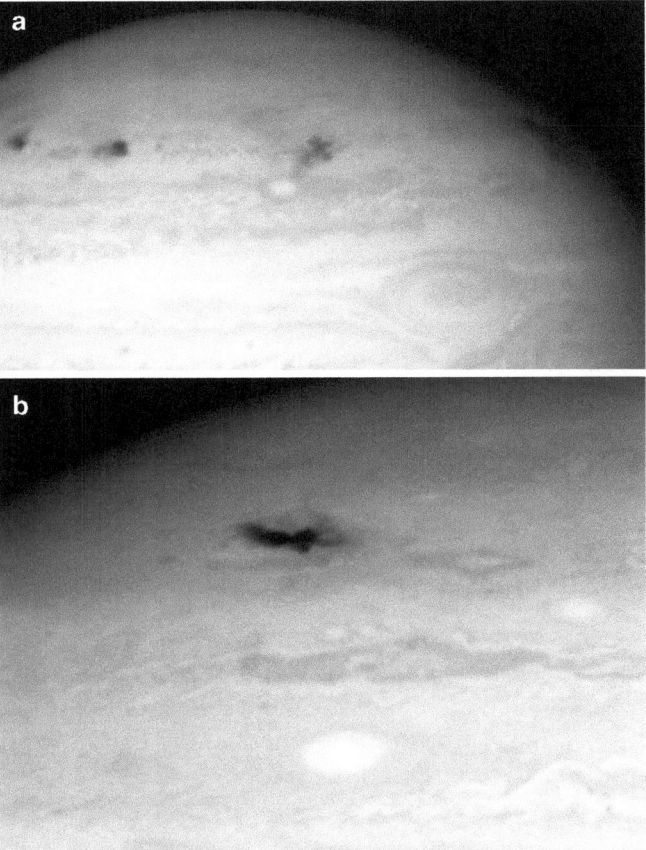

Fig. 3.1. (**a**) Bruises on Jupiter caused by the fragments of Shoe-maker–Levy 9 hitting the planet in July 1994, imaged by the Hubble Space Telescope. Image: NASA. (**b**) Another object, probably a small comet, collided with Jupiter in July 2009 and was first spotted by Australian amateur astronomer Anthony Wesley. The Hubble space telescope secured this high resolution image on 23rd July. Image: NASA, ESA, Hammel (Space Science Institute), and the Jupiter Impact Team.

When I was first becoming interested in astronomy as a child in the late 1960s these numbers were one hundred times smaller! In many ways comet discoveries are simply a professional by-product of the professional NEO patrols: for every new comet discovery there are almost a thousand minor planet discoveries! So how are most professional minor planet and comet discoveries made? Well, the dead give-away for either of these objects is that they move against the background stars and so some sort of moving object detection software is required. Objects in the asteroid belt typically move at a rate of 30–35 arc-sec per hour at opposition (opposite the Sun in the sky) so if you image the same field every half hour or so an object may have moved about 15–18 arc-sec in the intervening period. So if you have an image scale at least as fine as a few arc-seconds per pixel it should be a formality to detect the motion. In practice therefore the most sensible way to patrol for objects moving at this sort of rate is to efficiently image other fields while you are waiting for half an hour to elapse. For example, if you use a 60 s exposure you could take 29 other 60 s exposures in the time between imaging the original fields again. With a 2° field of view you could therefore image a strip 60° long by 2° wide before returning to the first field and then checking all the other fields in sequence. At this rate of patrolling you could, during the course of a 10 h night, patrol ten such strips twice, covering a sizeable 60 by 20° swath of the sky. With apertures of half a meter and larger objects down to magnitude 20 can be detected even with such short exposures. Of course, in practice, slewing to each field and back to the starting point of each strip of 30 fields would take up an additional fraction of the observing time but I am sure you can see what I am describing here and whether the exposures are 40, 50, or 60 s and the CCDs cover 2–3° fields is just fine detail.

Excluding the 1,700 or so SOHO comet discoveries, which are of tiny cometary fragments discovered in very close proximity to the Sun, the most successful automated or semi-automated comet discovery systems at the start of 2010 are those of the LINEAR (191), Catalina/Siding Spring/Mt Lemmon Sky Surveys (150), NEAT (53), LONEOS (42) and Spacewatch teams (41). The numbers in parentheses show the numbers of comets found. Within some of these teams there are often individual human image checkers whose names often appear as the discoverer rather than the name of the survey. For example, the indefatigable Robert McNaught has discovered 54 comets which bear his name but he is also a major part of the Siding Spring patrol.

Let us now have a look at some of these professional patrol systems in more detail. In every case these systems are mainly discovering asteroids and NEOs and that is the reason they are funded. Whenever you scan the sky for really faint comets using CCDs you will inevitably trawl in far, far more asteroid and NEO discoveries. However, most readers of this book would be more than happy with an asteroid or a NEO discovery themselves on the way to possibly finding a comet.

LINEAR

At the start of 2010 the Lincoln laboratories near-Earth asteroid research project (LINEAR) had discovered 191 comets amongst the 110,000 new asteroids it had also swept up. The LINEAR comet discoveries break up into 115 long period comets and 76 short period comets. Of the 76 short period comets 23 had been transferred to numbered status when this chapter was written. During the late 1990s and early years of the twenty-first century LINEAR was the undisputed king of comet and asteroid discoveries, despite being primarily a non-astronomical US Air Force facility set up to develop electro-optical space surveillance technology for searching and detecting military satellites and space junk in Earth orbit. This may seem odd but remember the size of the US Defense budget compared to the US astronomy budget and it makes more sense.

The LINEAR facility under the dark and clear skies of Socorro, New Mexico uses two identical 1 m aperture f/2.15 Ground-based Electro-Optical Deep Space Surveillance (GEODSS) telescopes (see Fig. 3.2)

Fig. 3.2. One of the two 1.0 m LINEAR telescopes at the White Sands Missile Range in Socorro, New Mexico funded by the United States Air Force and NASA. Reprinted with permission of MIT Lincoln Laboratory, Lexington, MA.

for searching which are improved versions of the previously favored Baker–Nunn design cameras. These cameras had, in turn, evolved from the ultra-fast Baker–Schmidt and Schmidt camera designs. In essence the Baker–Nunn design used an ultra-fast (f/0.75!) primary mirror plus a secondary mirror and triplet corrector lens in front of a photographic plate to track artificial satellites in the pre CCD era. In the GEODSS system a $2,560 \times 1,960$ pixel fast download CCD chip, with 24 μm pixels, is placed at the focus of the f/2.15 light cone. Because of the huge size of the secondary mirror obstruction in the GEODSS telescope design the effective aperture of these 1.0 m instruments is nearer to 0.7 m, making the effective f-ratio nearer to f/3.0. The CCDs in the system cover a physical area of 61×47 mm at 2.15 m focal length covering a sky area of $1.6 \times 1.3°$ at a resolution of 2.3 arc-sec per pixel. Despite the 5 megapixel chip size the images can be downloaded via four frame store buffers in only 0.2 s. Such a large aperture used with a sensitive CCD delivers impressive limiting stellar magnitudes in short exposures. The LINEAR GEODSS patrol telescopes can reach magnitude 19 in 2 s, magnitude 20 in less than 10 s, magnitude 21 in 30 s and magnitude 22 in a typical 100 s exposure. The serious LINEAR NEO searches using the big five megapixel CCD started as a trial system in October 1997 and for the next 8 years or so LINEAR was the king of the skies, discovering NEOs and night-time comets (as opposed to SOHO cometary fragments near the Sun) at a greater rate than any other system. There is a different third telescope at the New Mexico site but that is only used for follow up observations, not searching.

The way that the LINEAR systems search the night sky is as follows. Firstly, three to five images are taken of the same region of the sky with a 30 min interval between exposures of the same field. Software then aligns the second through to last frames with the first frame and normalizes the background sky brightness to the same level in all the frames. Checking software then looks for any potential motion between adjacent pixels in the first and last images of the same fields and if potential motion is spotted the intermediate frames are analyzed to see if the suspects were moving within the 30 min time spans too. If not, the suspects are rejected. It goes without saying that for a machine to do this efficiently a sophisticated star matching algorithm and access to a digital star catalog is vital. Moving objects detected are then accurately measured in right ascension and declination. The final results of the night's searches are then transmitted to the Lincoln Laboratory in Lexington Massachusetts for additional processing. To reduce the burden on the modestly staffed and modestly funded Minor Planet Center in Boston any detections of slow moving candidates in the night sky are only submitted to them after another night or two of observations, so that there has been time to link the observations together, ready for generating a preliminary orbit. The fastest moving and the medium speed object measurements are dispatched to the MPC with a higher priority so that potential NEOs and PHAs can be observed by other astrometric groups. Of course, the vast majority of objects detected by LINEAR are asteroids, but as the system has also detected 191 comets and will have detected more than 200 by the time you read this book, LINEAR is of great significance to the comet hunting community. The LINEAR team subdivides moving objects in the night sky into fast, medium and slow moving categories. Fast objects are defined as those moving at 0.4° per day or more, whereas medium speed objects are between 0.3 and 0.4° per day. Slow moving objects are those moving at less than 0.3° per day. In arc-seconds per half-hour (the standard interval between re-imaging with the system) these numbers equate to 30"/half-hour and faster, 22.5 –30"/half-hour and less than 22.5" /half-hour. You may prefer these figures in pixels on LINEAR's detector? OK, the figures then translate to 13 pixels per half-hour, 10–13 pixels per half-hour and less than 10 pixels per half-hour. Most automated patrol twenty-first century comet discoveries bag comets when they are traveling in the slowest LINEAR category as they are often well beyond the inner solar system when discovered.

LINEAR searches are conducted during the half of the month when the Moon is within a week of being new, from the last quarter to first quarter phase, so that background noise is reduced and the search can go deeper without a high sky background complicating the software analysis. As many as 12,000 square degrees of sky can be searched every month (more than a quarter of the 41,000 square degree celestial sphere) which equates to roughly 6,000 CCD fields, and a simple calculation shows that if the average exposure is 60 s then this implies 6,000 min of exposure time multiplied by the number of frames per field taken. So, for three fields of every $1.6 \times 1.3°$ field, 18,000 min of exposure time would be required or 20 h per night for the whole "Moon-dark" half-month period. As there are two identical LINEAR 1.0 m f/2.15 search telescopes this works out at 10 h per "Moon-dark" night per telescope. LINEAR is certainly a workhorse of a discovery machine under the clear New Mexico skies! A preliminary analysis suggests that when LINEAR was at its peak it could discover one comet (typically of about 16th or 17th magnitude) for every 3,000 square degrees of sky imaged down to a stellar limiting magnitude of 21 or so. However, that was in its peak days before the astronomers caught up with the technology used by the US Air Force!

NEAT

Between December 1995 and April 2007 the near-Earth asteroid tracking program (NEAT) discovered 53 comets amongst the tens of thousands of asteroids it had also found. NEAT also detected 86 supernovae between 1999 and 2003 using software developed by Michael Wood-Vasey of Berkeley Lab's Physics Division. The original driving force behind NEAT was the principal investigator Eleanor F. Helin, along with co-investigators Steven H. Pravdo and David L. Rabinowitz.

NEAT had very similar roots to the LINEAR program and a very similar search strategy. One of the main differences was that the NEAT GEODSS telescopes were based at Haleakala, Maui, Hawaii and at Palomar Observatory in California and the project was run by the Jet Propulsion Laboratory (JPL). The NEAT camera was designed and fabricated at JPL in 1995, and consisted of a 4,096 × 4,096 CCD with 15 μm pixels. The camera was attached to a standard 1.0-m f/2.15 GEODSS telescope on Haleakala and the pixel size gave an image scale of 1.4 arc-sec per pixel, resulting in the 4,096 × 4,096 CCD covering 1.6 × 1.6° (compared to the 1.6 × 1.3° LINEAR field). The NEAT detector was a commercial CCD manufactured by Lockheed–Martin Fairchild Systems and the images were downloaded simultaneously in four 2,048 × 2,048 quadrants in a rather lengthy 20 s download time. In theory two of these quad detectors could be squeezed into the focus but not without significant image degradation at the edges.

The first NEAT system was installed via the USAF subcontractor PRC Inc., in December 1995 and imaged on twelve nights per month, around new Moon for the first year. From January 1997 this schedule was reduced to only six nights per month and the camera system was removed from the Hawaii 1.0 m telescope after some 3 years and 2 months of operation in February 1999. The NEAT camera system was then moved to the larger 1.2 m Maui Space Surveillance Site telescope and from February 2000 the program called for eighteen nights per month of searching for near-Earth asteroids and comets. The United States Air Force Space Command provided the money to modify the 1.2 m telescope, giving it a much larger field of view, suitable for NEAT's asteroid survey; the work was carried out by the Air Force Research Laboratory and by Boeing (see Fig. 3.3).

Fig. 3.3. The massive NEAT CCD camera mounted on the secondary mirror support ring of the 1.2 m Maui Space Surveillance Site telescope. Image: Boeing/JPL/NASA.

The JPL Observing System

Like all successful patrols, the JPL NEAT system was fully automated and ran from a script that could be interrupted if cloud intervened. NEAT's basic strategy was to search the night sky close to the ecliptic plane and near to opposition (the area of sky transiting at midnight). The sky was searched along strips, each one being roughly perpendicular to the ecliptic, and separated in longitude from one another by 11.25°. In the 1997–1999 patrols the length of each strip was chosen so it could be searched three times in 45 min, equating to a triplet of 20 fields of 1.6° (so, 32° long) allowing for slewing and image download. Thus, 60 images were obtained every 45 min with an average of 45 s per image. This 45 s value was comprised of a 20 s exposure time plus a 25 s download and slewing overhead. Within a couple of minutes of the third image of a triplet being obtained the 1.6 × 1.6° field was software checked for moving objects by simultaneously comparing each image in the triplet for each of the CCD chip's four 2,048 × 2,048 pixel quadrants.

In 45 min the sky moves by 11.25°, so the next 32° long strip was then on the meridian meaning all the patrols were carried out when the fields were at as high an altitude as possible. In five nights a swath of the ecliptic to roughly ±16° of ecliptic latitude and ±45° of ecliptic longitude could be imaged. The 21° north latitude of the Hawaii facility restricted the southerly declination searches to a declination of −38°, but this was not really a problem bearing in mind the low point of the ecliptic, near the Sagittarius/Ophiuchus border, is at a declination of −23.5°. A strategy was adopted for identifying slow moving objects (less than 30 arc-sec per hour) by checking some fields again on the sixth night of an observing run. The NEAT survey specifically avoided the Milky Way regions (within 15° of the plane of the galaxy) due to the cluttered star fields confusing asteroid detection. It also specifically avoided the Kitt Peak Observatory Spacewatch patrol regions being searched by Scotti, Gehrels, & Rabinow-itz in the 1990s. When the system detected a moving object the team members at JPL used software called PATCHVIEW to check that there was a genuine moving object in the data rather than an image artifact.

As I have already mentioned, the NEAT camera was moved in 2000 to the larger 1.2 m telescope on Hawaii. In addition, in 2001, a NEAT camera system was installed on the Oschin 1.2 m f/2.5 Schmidt Camera at Mount Palomar in California. This NEAT system was comprised of three 4,096 × 4,096 pixel CCD arrays working at an image scale of 1 arc-sec per pixel and so covered a 1.1 × 3.4° field.

All things considered NEAT was a highly productive discovery system and, from the point of view of the topic of this book, its 53 comet dis-coveries were impressive. However, during its peak years NEAT was competing head-to-head with a similar but superior system in the form of the two LINEAR patrol telescopes at Socorro New Mexico. The LINEAR telescopes scanned the sky more efficiently and ruthlessly and were not limited (by funding and politics) to as little as six nights per month. In addition LINEAR's software was just so much better at detecting genuine fast moving objects and the LINEAR US Air Force investigators did not have to relocate their cameras to different telescopes and different sites in the course of the survey. So, the NEAT funding came to an end in 2007, but by then some even more potent astronomical search systems were in operation.

The Catalina, Siding Spring and Mount Lemmon Sky Survey's

As with the LINEAR and NEAT surveys the Catalina Sky Survey (CSS) has a prime mission to discover NEOs that may pose a threat to life on Earth. As I mentioned earlier, the realization that a comet or asteroid had caused the Yucatan impact that accelerated the demise of the dinosaurs

65 million years ago, in addition to the impact of the comet Shoemaker –Levy 9's components with Jupiter in 1994, concentrated even the minds of politicians onto the reality of the situation. No doubt the two blockbuster movies *Deep Impact* and *Armageddon* helped the cause too! Fortunately, in this very same era, large quantum efficient CCDs were just becoming available, as were highly powerful number crunching microprocessors and so trawling in asteroid discoveries in huge numbers was suddenly possible. In 1998 the US Congress directed NASA (and therefore JPL/NEAT) to identify kilometer sized and larger bodies on Earth-crossing trajectories to a 90% confidence level. Such was the success of this directive that in June 2006 Congress raised the bar to carry out the same survey, but to detect objects of 140 m size and larger to the same confidence level. So, LINEAR, NEAT and the Catalina surveys all sprang from this NEO hazard directive and the subsequent funding. As an inevitable spin-off, these projects started trawling out comet discoveries too, which is what the reader of this book is (hopefully) really interested in.

At the time of writing (early 2010) the three survey teams that are part of the Catalina Sky Survey (Catalina, Siding Spring and Mt Lemmon) have discovered or co-discovered almost 150 comets since 2003. The comets bear various names, such as McNaught (37), Catalina (20), Christensen (18), Hill (18), Gibbs (13), Garradd (12), Boattini (9), Siding Spring (7), Mt Lemmon (4), Larson (4), Kowalski (4), Beshore (3) and Grauer (1). Robert McNaught and Gordon Garrrad are professional (and former amateur) astronomers working at Siding Spring Australia, whereas all the other names are of observers at the Catalina Sky Survey in the USA. So despite the fact that these are professional CCD surveys it is good to see human names on the Catalina Sky Survey comets in large numbers. During the late 1990s and early years of the twenty-first century virtually all comets seemed to be called LINEAR or NEAT, a rather depressing and incredibly confusing situation!

The Catalina Sky Survey began its operations on Mount Bigelow in the Catalina Mountains north of Tucson Arizona well before its complete equipment overhaul in 2003. For 5 years prior to that it had been using a 0.4 m aperture Schmidt telescope with a $4,096 \times 4,096$ pixel CCD and it discovered 46 near-Earth objects in that time. From November 2003 the current system has used a 68 cm aperture (76 cm primary mirror) f/1.9 Schmidt telescope for its discovery work (see Fig. 3.4) and can patrol an impressive 800 square degrees of sky in a single night. Each $4,096 \times 4,096$ pixel exposure covers a field of more than two by two degrees with the ability to get well below 20th magnitude in 60 s. The CSS has stated that it favors the ecliptic plane in its searches as well as regions that are elongated some 70° from the Sun. Such regions, while not being at the highest altitude above the horizon in the evening or morning skies are arguably

Fig. 3.4. The Catalina Sky Survey's 0.68-m Schmidt camera. Image by kind permission of Steve Larson, Ed Beshore and Rik Hill of CSS.

more neglected by other patrols and always hold the promise of an asteroid or comet emerging when closer to the Sun and therefore brighter.

Siding Spring

The Uppsala Schmidt telescope, used by Rob McNaught (see Fig. 3.5) and Gordon Garradd to discover asteroids and comets as part of the Catalina Sky Survey team, was originally based at Mt. Stromlo Observatory from 1957 to 1982. Tragically, that observatory was destroyed by fire in 2003, but fortunately the Schmidt telescope had been moved to Siding Spring 21 years earlier. The Schmidt telescope has a relatively modest aperture of 52-cm (corrector plate diameter) and a focal length of 175-cm, making it an f/3.4 system. In 2000 and 2001 this photographic telescope was modernized and the Schmidt focus was diverted outside the tube so that optically (if not mechanically) it is not dissimilar to a 52-cm f/3.4 field-flattened reflector. In other words it is not all that different to some of the largest amateur Newtonians, at least in aperture and focal length. The f/3.4 focus modification enabled a top quality 4,096 × 4,096 pixel CCD detector, cooled to −90°C, to be employed, rather than the old photographic plates. The CCD chip, just like the one at Mt. Bigelow near Tucson, is massive at approximately 60 mm in width, and it features 15 μm pixels. This gives a field of view of slightly more than 2° and a resolution just

Fig. 3.5. Rob McNaught with the modified 52-cm f/3.4 Uppsala Schmidt telescope. Image: Rob McNaught.

under 2 arc-sec per pixel. With short exposures down to 19th or 20th magnitude, and the telescope working throughout the night, considerable swathes of sky can be searched and with little competition from the other competing surveys, which are mainly northern hemisphere based. The southern hemisphere location for this part of the CSS patrol gives it a huge advantage, especially with a tireless discoverer like Robert McNaught involved! McNaught is now the world's leading comet discoverer with more than fifty comet discoveries to his credit (he was discovering objects well before he became part of the CSS team) and, undoubtedly, a lot more comet discoveries by the time this book is published!

Mt. Lemmon

The Mt. Lemmon observatory's role in the CSS team is predominantly as an observational support team for NEO work using the 1.5 m reflector at the facility. The telescope's f/2 mirror was refigured for the CSS work and its control system was upgraded. A three-element field corrector provides a sharp image over a 1.1 by 1.1° field at a scale of 1 arc-sec per pixel and a stellar magnitude of 22 can be reached in short exposures. However, follow up observations are not this telescope's only trick as the Mt Lemmon 1.5 m telescope has discovered four comets as part of the CSS survey.

The Spacewatch Patrol

Of all the successful near-Earth asteroid and comet patrols the Space-watch system is the one which has been running for the longest period of time. The patrol was originally founded 30 years ago by Professor Tom Gehrels and Dr Robert McMillan using the venerable 0.91 m f/5.3 Newtonian telescope at Kitt Peak, Arizona. The Spacewatch patrol has the distinction of being the first to discover a comet using a CCD patrol. That comet was 1991 R2, now renamed 125P/Spacewatch. The comet was discovered in Aquarius by Tom Gehrels on September 8, 1991. The comet was then situated 1.70 AU from the Earth. Jim Scotti measured the initial position and determined the nuclear magnitude as 21.1 on September 8 making it the faintest comet ever discovered. The Spacewatch system is also credited with being the first to make a fully automatic software-checked discovery of a comet, in the case of C/1992 J1 (Spacewatch).

Unusually the original Spacewatch system used the so-called "drift-scan" approach to scanning the sky. This technique is sometimes called time delay integration (TDI) and involves the lines on a CCD being read out very quickly at the rate at which the sky drifts over the pixels. Of course, normally a telescope tracks the stars at the sidereal rate and any drift is highly undesirable, but by slowing or even stopping the sidereal drive the stars move past steadily at 15 arc-sec per second of time on the celestial equator. At first this seems like the craziest plan imaginable, but if the checking system works best with huge, long, east to west strips of the night sky, rolling past at 15° per hour, you can patrol huge swathes of the sky with this method.

The 0.91 m Newtonian was used with a 2,048×2,048 CCD from the early 1990s when it patrolled to about magnitude 21 for 3 weeks per month covering some 200 square degrees of sky. The interval for detecting moving objects in the sky was 30 min with the third image in a triplet being exposed 1 h after the first. However, CCD technology and competition from LIN-EAR and NEAT had moved on apace in the late 1990s and a larger 1.8 m Spacewatch telescope (see Fig. 3.6) was completed in 2001 and designed for the follow-up of asteroids and NEOs. In 2002, a very large CCD mosaic camera made from four $4,608 \times 2,048$ pixel CCDs was added to the aging

Fig. 3.6. The 1.8-m Spacewatch telescope at Kitt Peak, photographed by Principal Spacewatch Investigator Dr Robert S. McMillan. Image: Spacewatch, The Lunar and Planetary Laboratory, the University of Arizona, and NASA. © 2002 The Arizona Board of Regents.

0.9 m Newtonian and the optical system was compressed and flattened to give a 2.9 square degree field. Operations were also switched from the drift scan mode to the conventional sidereal tracking mode. Since 2005 the observatory's Spacewatch team gradually moved its focus to follow-up astrometric work and to searching the outer solar system for Centaur and Trans-Neptunian minor planets. In total the Spacewatch software and team members have discovered more than 10,000 NEOs with more than 20% being in the potentially hazardous category. They have also bagged more than forty comets and continue to detect one or two per year during their astrometric and outer solar system patrols. In 2009 the comet P/2009 SK280 (Spacewatch-Hill) was jointly discovered by the Spacewatch and Catalina patrols as an asteroid, but shortly afterwards its fuzzy nature betrayed it and it was reclassified as a comet.

LONEOS

Mentioning the Lowell Observatory at Flagstaff Arizona usually conjures up a vision of Percival Lowell, canals on Mars, and the discovery of Pluto by Clyde Tombaugh. However, in recent decades the observatory has had a proud tradition of asteroid discovery going right back to the photographic era. LONEOS stands for Lowell observatory near-Earth object search and since 2000 the efficiency of this search using the 0.6-m f/1.8 Schmidt telescope has been enhanced significantly due to software and CCD camera upgrades. Despite the relatively modest aperture of the Schmidt telescope by professional standards the sensitive 4,096 × 4,096 pixel detector and its 2.9 × 2.9° field can reach below 19th magnitude and cover the entire visible sky four times per month. So what the LONEOS system lacks in sheer aperture it makes up for in patrol speed. Admittedly the Siding Spring Survey uses an even smaller aperture but there is far less competition in the southern hemisphere. At the time of writing the LONEOS team had bagged 42 comets and no doubt many more by the time this book appears in print. One of the author's images of a bright LONEOS discovered comet is shown in Fig. 3.7.

Fig. 3.7. Comet 2007 F1 (LONEOS) imaged with a 0.35-m Celestron 14 and SBIG ST9XE at f/7.7 on October 20, 2007. 50 × 20 s co-added exposures. The field is 12 arc-min high. The comet was at 11° altitude and the Sun was 13° below the horizon. Image: Martin Mobberley.

La Sagra Sky Survey

The La Sagra Sky Survey (LSSS) is a relative newcomer to the comet and NEO discovery game but, in 2010, looks like being the only serious European threat to the other main patrols. It is difficult to know whether to class LSSS as a full blown professional survey as it is a unique and modestly funded system. The Observatorio Astronómico de La Sagra (OLS) sits in the Andalusian mountains of Southern Spain not far from Puebla de Don Fadrique. The facility is operated by the Observatorio Astronomico de

Mallorca (OAM) which is financed by the Balearic government's culture, education and sports department. Following the first asteroid discovery in August 2006 continuous equipment upgrades turned the observatory into a major asteroid and NEO facility. Three-quarters of a million astrometric measurements had been submitted to the Minor Planet Center from La Sagra by early 2010 and OLS is now by far the most productive European NEO facility with only the previously mentioned LINEAR, NEAT, Catalina, Spacewatch and LONEOS facilities submitting more data. LSSS employs three 0.45-m aperture f/2.8 patrol telescopes using commercial 36×24 mm SBIG STL-11000 CCDs giving a $1.1 \times 1.6°$ field and an image scale of 9 μm per pixel. The whole analysis procedure is, uniquely, carried out remotely via the Internet, mostly by analysts in Spain, but also by observers in Croatia and even Hong Kong! By the start of 2010 LSSS had discovered 4,000 asteroids, 20 NEOs and two short period comets, P/2009 QG31 (La Sagra), originally labeled as an asteroid and P/2009 T2 (La Sagra). The three 0.45-m f/2.8 field corrected "Centurion" Newtonians used at La Sagra were made by the Astro Works Corporation of the USA (10% of their total production of around 30 units to this design!) and use a prime focus field corrector lens to give acceptable images across a 36×24 mm DSLR sized field. Two of the three telescopes were formerly owned by the amateur discover Bill Yeung of Benson, Arizona.

CHAPTER FOUR

Amateur Twenty-First Century Comet Hunters

Before the era of automated professional CCD patrols the names of amateur comet discoverers who had bagged comets simply by sweeping the skies visually would regularly appear in IAU circulars. Many amateurs, including myself, feel rather sad that hardly any visual comet discoverers have survived the onslaught of the remorseless machines. In my younger days seeing the names of legendary figures like Bill Bradfield, David Levy and Don Machholz appear in astronomical circulars made me feel that there was still a link to the good old days of the amateur beating the professional through a sheer love of the night skies, rather than simply throwing funds at a technical problem. I have met a number of comet discoverers in my life, most of them at the August 1999 International Workshop on Cometary Astronomy (IWCA) in Cambridge, England (see Fig. 4.1). I knew the UK discoverer George Alcock (see Fig. 4.2) very well and filmed a video documentary of his life in 1991 and I have also met Roy Panther, briefly, on a couple of occasions. George succeeded because, to him, studying nature was a compulsive ritual, not a chore. He observed birds, flowers and trees for his whole life and sketched everything he saw. Observing the night sky was a routine for him. If it was clear he would sweep the skies with binoculars. It was not tedious to him. He simply enjoyed it and could not ignore a clear night. Ultimately his fascination with the star patterns enabled him to memorize some 30,000 Milky Way

M. Mobberley, *Hunting and Imaging Comets*, Patrick Moore's Practical Astronomy Series, DOI 10.1007/978-1-4419-6905-7_4, © Springer Science+Business Media, LLC 2011

Fig. 4.1. A huge gathering of comet discoverers at the IWCA conference in Cambridge, England in August 1999. From *left* to *right*: Seargent, Machholz, Stonehouse, Tritton, Hale, Biesecker, Alcock, Liller, Jager, Takamizawa and Cernis. Image: Martin Mobberley.

Fig. 4.2. The legendary George Alcock with his huge tripod mounted 25 × 150 World War II binoculars and handheld 15 × 80 binoculars. Alcock discovered five comets and five novae. The nova discoveries were learnt by memorising the Milky Way to eighth magnitude (and fainter in places).

stars as seen through his 15×80 binoculars! As well as being fascinated by George I was always impressed by the dedication of the Japanese comet hunters and recall that in the late 1960s and 1970s there almost always seemed to be a triple barreled Japanese comet name in the sky!

Well, sadly, those days are now gone and the patrol machines are just too efficient to make a visual amateur comet discovery anything but a rarity. But, as always, the leading amateurs have moved with the times and employed CCDs themselves to hunt down comets. I would like to go back to 1998 in my summary of amateur twenty-first century comet hunters, simply because it was around this time that the leading amateur hunters started to seriously turn to CCDs and the professional LINEAR and NEAT systems started to discover worrying amounts of comets.

1998

The first amateur comet discovery of 1998 was made by sheer luck! Australian amateur astronomer Peter Williams discovered C/1998 P1 (Williams) when he pointed his 30-cm f/6 Newtonian at the field of the variable star EK TrA and studied the region at 72× magnification, only to be amazed to find a new fuzzy object in the familiar field of the variable star. It might be thought that this kind of discovery would be a one-off in the history of astronomy. After all, what is the chance of a bright comet being missed by professional patrols and then spotted inside a 0.7° eyepiece field? Well, variable star observers can be very enthusiastic with many making tens of thousands of magnitude estimates in their lives and a few making hundreds of thousands of estimates. Multiply these numbers by the hundreds of such observers worldwide and the chance of a lucky discovery becomes far more likely. In fact, discovering a comet by good fortune, in the field of a variable star, has occurred many times. In 1936 Kaho-san co-discovered comet Kaho–Kozik–Lis in the eyepiece field of R LMi and, 10 years later, Albert Jones of New Zealand would discover Comet Jones while observing the variable star U Pav in 1946. We have not heard the last of Albert Jones either as this would happen to him twice! We have to wait 51 years for the next lucky variable star observer who was Justin Tilbrook and his comet, C/1997 O1, conveniently placed itself in the field of TV CrV. Then, in 2000, Albert Jones would do it again, aged 80! While studying the variable star T Aps with an 80 mm refractor he would co-discover C/2000 W1 (Utsunomiya–Jones). At 80 this made him not only the oldest ever comet discover, but the longest time-span comet discoverer: 54 years between his two finds! However, I have digressed a bit here and the second amateur comet discovery of 1998, a co-discovery, was made Roy Tucker of Tucson Arizona. Roy co-discovered P/1998 QP$_{54}$ (LONEOS-Tucker) using his pioneering triple 35 cm f/5 Newtonian array which employed the drift-scan imaging method (see Fig. 4.3). The third and final amateur comet discovery of 1998 was made by one of the world's leading comet photographers and imagers, Michael Jaeger of Austria. Michael spotted the new object, P/1998 U3 (Jager), on a photograph he had taken (using hypersensitized Kodak TP 6415 film) with a 0.25-m Schmidt camera. So, two of the three amateur comet discoveries in 1998 were lucky finds, whereas Roy Tucker's discovery was made as part of a deliberate amateur NEO patrol.

Fig. 4.3. One of Roy Tucker's unusual trio of 35-cm Newtonians which scan the sky using the drift scan technique: the telescopes do not track the stars as the image is downloaded at the rate the sky drifts over the CCD sensor.

1999

The first amateur discovered comet bagged in 1999 was the second found by Australian amateur Justin Tilbrook, but his visual discovery of C/1999 A1 (Tilbrook) was a result of a proper search this time, using a 20-cm f/6 Newtonian at 70×. Korado Korlevic and Mario Juric of Croatia captured the second amateur discovery of that year. Originally classed as asteroid 1999 DN$_3$, the object captured with the Croatian 0.41-m f/4.3 reflector on a CCD image was renamed as P/1999 DN$_3$ (Korlevic–Juric) once it revealed a coma. Australian Steven Lee would make the next amateur comet discovery, C/1999 H1 (Lee); it was spotted while using a 0.40-m Dobsonian for deep sky observing at the Mudgee Star Party.

Remarkably, a third Australian, Daniel Lynn, would discover C/1999 N2 (Lynn) at magnitude 7.5 while sweeping the sky near the horizon with a pair of 10×50 binoculars. Korado Korlevic then captured his second comet of 1999 with the 0.41-m f/4.3 reflector and CCD in the form of a moving dot, originally named as asteroid 1999 WJ$_7$, but then re-classified as a short period comet P/1999 WJ$_7$ (Korlevic). Finally, the sixth amateur comet discovery of 1999, in a great year for amateur comet discoveries, was made by U.S. amateur astronomers Gary Hug and Graham Bell. One month after taking delivery of an SBIG ST-9E cameras they spotted the magnitude 18.8 comet, later named P/1999 X1 Hug–Bell, while using a 30 cm f/6.3 Schmidt–Cassegrain. This was the second faintest comet ever discovered by amateur astronomers. The discovery was made while blinking 6 min exposures taken in search of an asteroid. Subsequent 10 and 20 min exposures revealed the comet's tail. The comet has a period of 7 years so P/1999 X1 Hug–Bell will regularly return to the solar system.

2000

In contrast to 1999 the year 2000 was almost a completely barren one for amateur comet discoveries but then, as the year drew to a close, comet C/2000 W1 (Utsunomiya–Jones) was discovered in late November. I have already mentioned this comet in connection with it being discovered in the field of the variable star T Aps by the 80 year old New Zealand variable star observing legend Albert Jones, an amazing 54 years after his accidental discovery of his first comet. The co-discoverer was the Japanese amateur astronomer Syogo Utsunomiya who had been sweeping the skies with a pair of 25×150 Fujinon binoculars, so popular amongst

Japanese amateurs, when he first spotted the eighth magnitude comet a week earlier, but no other observatories had been able to confirm his suspect which had been moving very rapidly. It later transpired that on November 25, the day of Jones' independent discovery, the comet had been at its closest to the Earth, at a distance of just 0.28 AU.

2001

As in 2000 there was just a single amateur comet discovery in 2001 and, as with Steven Lee's 1999 discovery, it came while using a large reflector to study deep sky objects at a star party. Vance Petriew of Regina, Saskatchewan in Canada was observing with a 0.51-m reflector at a star party at Cyprus Hills, Saskatchewan, when he discovered the comet on August 18. It was near the star Beta Tauri and Petriew said the comet was discovered accidentally. He was looking for the Crab Nebula and had actually started from the wrong star in Taurus, much to his enormous benefit! Petriew described the comet as eleventh magnitude and 3 arcmin in diameter. Two other observers at Cyprus Hills, Huziak and Campbell confirmed the find. The comet was soon determined as periodic and now has a periodic number and is named 185P/Petriew. It was originally designated P/2001 Q2 (Petriew).

2002

After the two very barren amateur comet discovery years preceding it 2002 would prove to be a highly successful one with eleven amateurs involved in seven comet discoveries. The first comet was originally reported as a twentieth magnitude asteroid discovery by William Kwong Yu Yeung of Benson, Arizona who was taking CCD images from near Apache Peak on January 21 using a 0.45-m f/2.8 reflector that runs on wheels! The object was linked to other observations by Spacewatch and LINEAR and it was realized that it had initially been given a designation of 2001 CB40 in the previous year when first seen. However, the re-discovery enabled the object to be determined as having a cometary orbit. In response to this Timothy Spahr took images with the 1.2-m telescope on Mt Hopkins which showed it to be a fuzzy comet with a total integrated magnitude nearer to 17. As well as the asteroidal designation it was also named P/2002 BV (Yeung) and finally 172P/Yeung, a short period comet with a period of 6.6 years which reached perihelion in mid March at 2.24 AU from the Sun.

Then, on February 1, came the most exciting comet discovery of 2002, the discovery of C/2002 C1 (Ikeya–Zhang) which would prove to be the return of a comet last seen 341 years earlier and so would ultimately be labeled as the periodic comet with the longest proven period. Kaoru Ikeya of Mori, Shuchi, Shizuoka, Japan was already one of the legendary figures of amateur astronomy because, as a young man he discovered or co-discovered five very bright comets in the 1960s. At that time he was competing with another Japanese legend Tsutumo Seki and both men would ultimately discover six comets with Seki going on to be a prolific asteroid discoverer and periodic comet recoverer using a 60 cm reflector at Geisei observatory. Ikeya and Seki co-discovered the brilliant Kreutz sungrazer C/1965 S1 (Ikeya–Seki) and, as of 2002, Ikeya had not discovered a comet for 35 years. However, he had still been searching during that period and, after such an incredible wait he was rewarded with a gem of a discovery on February 1 while sweeping the sky with a 25-cm reflector at 39×. His co-discoverer, Daqing Zhang was, rather appropriately, a Chinese comet hunter, because on its first recorded visit to the inner solar system in 1661, the comet was spotted by the Chinese. Zhang spotted the magnitude 8.5 comet just 90 min later using a 20-cm f/4.4 reflector at 28× from Kaifeng in Henan province. The comet was provisionally designated C/2002 C1 (Ikeya–Zhang) but was later assigned periodic number 153P making it 153P/Ikeya–Zhang, the comet that re-defined the boundary between short and long period comets (see Fig. 4.4a, b). It arrived at perihelion on March 18 at 0.5 AU from the Sun and reached third magnitude, making it the brightest comet discovered since the amateurs Alan Hale and Thomas Bopp discovered the amazing comet Hale–Bopp in 1995. It was good to see such an important comet discovered not by a machine, but by two amateurs observing visually. But the amateur comet discoveries of 2002 would not end there. Sadly, the Japanese amateur comet discoverer Yuji Hyakutake died of a heart aneurysm, aged 51, shortly after the comet Ikeya–Zhang, co-discovered by his fellow countryman, reached perihelion. The next amateur comet find of 2002 was C/2002 E2 (Snyder–Murakami) which was a joint discovery shared between Doug Snyder of the USA and Shigeki Murakami of Japan. Both observers had been using large Newtonian reflectors visually, on March 11, when they spotted the new comet. Snyder had used a 0.50-m f/5 Newtonian and Murakami-san had used a 0.46-m f/4.5 reflector at 68×. The comet was magnitude 10.5 at discovery and near the border of the densely star-packed constellations of Scutum and Aquila, close to M11.

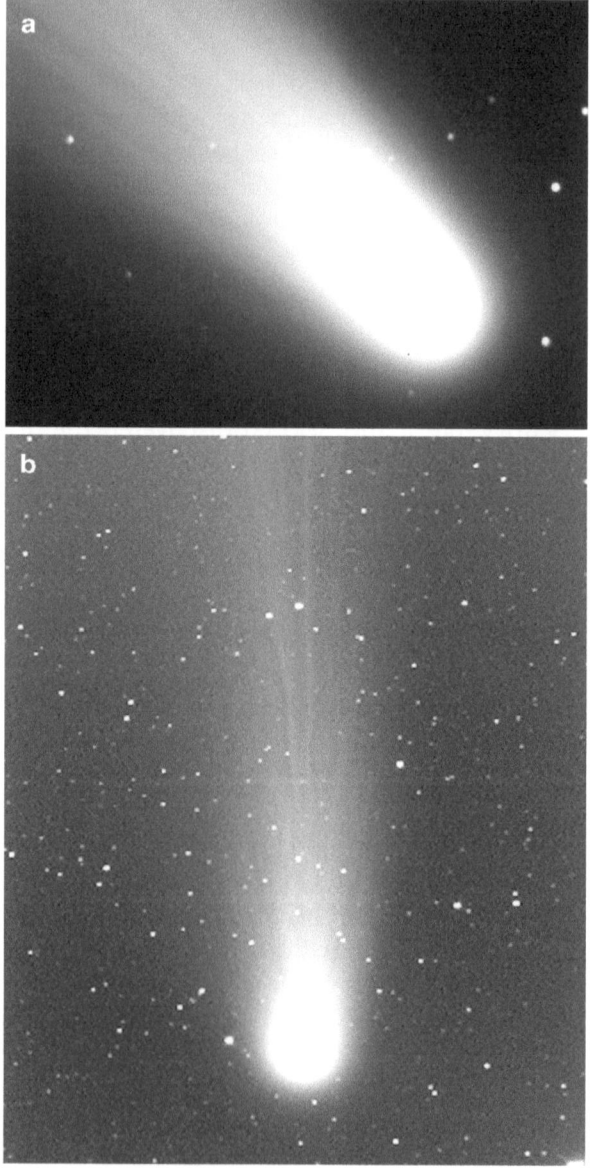

Fig. 4.4. (**a**) Comet 153P/Ikeya–Zhang imaged in a 60 s exposure with a 0.49-m f/4.5 Newtonian on March 21, 2002; 14 × 11′ field. Image: Martin Mobberley. (**b**) Comet 153P/Ikeya–Zhang imaged with a 530-mm focal length Takahashi E-160 astrograph (160-mm aperture f/3.3) and Starlight Xpress MX916 CCD on April 6, 2002; a 1° high field. Three 30 s exposures were co-added. Image: Martin Mobberley.

The comet showed a 3 arc-min coma with a faint condensation and had already passed perihelion on February 22.

The fourth comet discovery of 2002 and the third visual discovery of that year was another comet found by the Japanese comet hunter Syogo Utsunomiya (Japan). Once again Utsunomiya-san swept the comet up with his trusty pair of 25 × 150 Fujinon binoculars, so popular amongst the Japanese patrollers (see Fig. 4.5).

On July 22, 2002 the amateur astronomer Sebastian Hoenig became the first German, from an observing site actually within Germany, to discover a comet since 1946, although he had been one of many amateurs to spot cometary fragments in the SOHO satellites images prior to his "proper" comet find. Sebastian had apparently been unable to sleep and so drove with his 25-cm Schmidt–Cassegrain to a dark site in the Odenwald woods near Heidelberg. While deep sky observing he came across a faint twelfth magnitude 2 arc-min fuzz just above the Square of Pegasus

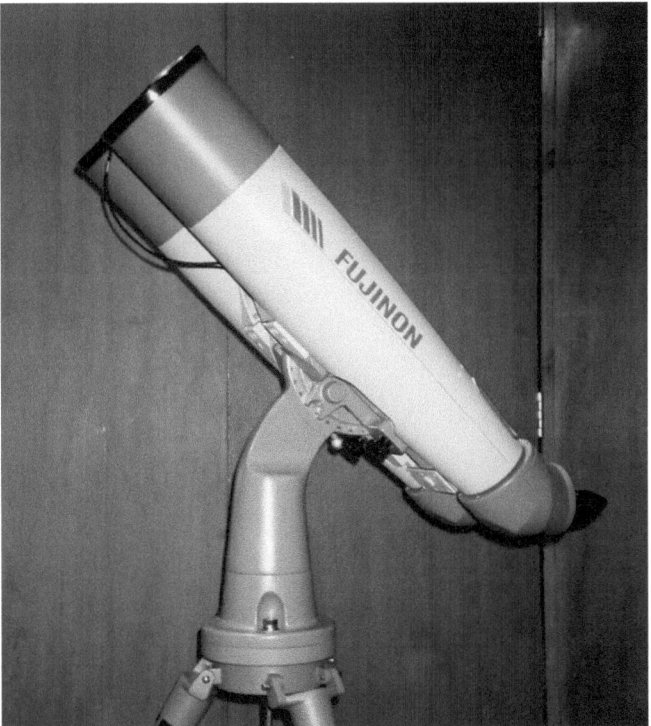

Fig. 4.5. Fujinon's classic 25 × 150 binoculars have always been especially popular amongst Japanese comet hunters. Image: Martin Mobberley.

and knew there were few bright galaxies in that region. However, Hoenig was traveling light and, not surprisingly, had not expected to stumble across a comet! So he had no star atlas and no notebook with him. The telescope was only roughly polar aligned but he used the GO TO hand controller to give him the approximate position and, having no paper available, he used the label of a water bottle on which to sketch the star field! Observing the fuzzy object carefully he detected a motion of some 3 arc-min per hour due north. After returning home the next day he e-mailed the Central Bureau for Astronomical Telegrams (CBAT) with his description but also his concern that the position could be a degree or two in error. Hoenig and others failed to find the comet in the following nights but then the Japanese came to the rescue. After an appeal from Nakamura-san, K. Kadota of Ageo, Saitama took many CCD images of the region with an 18-cm reflector on July 27 and found the twelfth magnitude comet, complete with a 1.8 arc-min tail. The comet C/2002 O4 (Hoenig) had, surprisingly, been missed by the professional patrols and yet another amateur astronomer had bagged a discovery. The comet brightened to ninth magnitude and moved into Cassiopeia as a ninth magnitude object in mid-August (see Fig. 4.6).

The sixth amateur comet discovery of 2002 was C/2002 X5 (Kudo–Fujikawa) a joint Japanese amateur visual discovery. Kudo-san of Nishi Goshi-machi, Kikuchi-gun, Kumamoto-ken used 20 × 120 mm binoculars

Fig. 4.6. Comet C/2002 O4 (Hoenig) imaged on September 1, 2002. Fourteen 30 s exposures were taken with an SBIG ST7 CCD and a 0.3-m Schmidt–Cassegrain working at f/3.3; 12 × 8′ field. Image: Martin Mobberley.

Fig. 4.7. Comet Kudo–Fujikawa on Christmas Day 2002. This is a single 60 s exposure with a 30-cm f/10 Schmidt–Cassegrain and an SBIG ST9XE CCD; 12 × 12′ field. Image: Martin Mobberley.

to sweep the skies and collect the comet on December 13 while the ninth magnitude object was in the constellation of Boötes (see Fig. 4.7). Independently the veteran comet discoverer Shigehisa Fujikawa (Oonohara, Kagawa, Japan) also spotted the comet on the following night. This would be Fujikawa-san's sixth comet discovery in an impressive discovery period spanning 1969–2002! Shigehisha Fujikawa used a 16-cm reflector for his discovery.

Finally, in a truly inspiring year for amateur comet hunters the American Charles Juels (1944–2009) and the Brazilian Paulo Holvorcem collected the seventh amateur find of 2002 when they spotted a fifteenth magnitude comet on a first light image taken with a 12-cm refractor and a CCD camera in a collaborative Internet patrol on the night of December 28. It might be thought that a discovery on the first night of use of a piece of equipment must be unique, but, mysteriously, it has happened a few times before and at a similar time of year! One example was the discovery of comet Candy 1960 Y1 on December 26, 1960. On that occasion the British amateur astronomer Mike Candy (1928–1994) was testing out a brand new 5-in. aperture comet sweeping refractor (made by Horace Dall) from an upstairs window of his house in Hailsham. He pointed the new instrument towards the constellation of Cepheus and there it was, an eighth magnitude fuzz! Forty-two years and 2 days later the "first light" refractor plus CCD system of Juels and Holvorcem captured C/2002 Y1 (Juels–Holvorcem). The comet passed perihelion 4 months later and was not expected to get much brighter than tenth magnitude. However, it surprised everyone by reaching sixth magnitude and sporting a 1° tail (see Fig. 4.8).

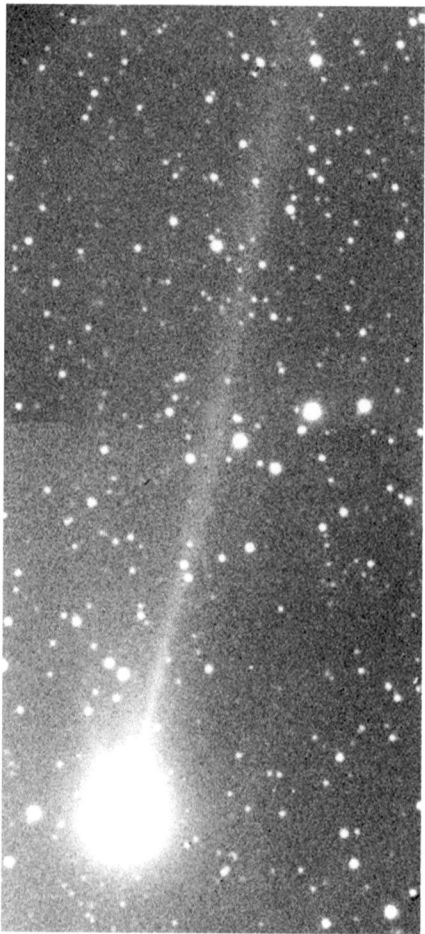

Fig. 4.8. Comet Juels–Holvorcem imaged on March 22, 2003 with a 30-cm Schmidt–Cassegrain and an SBIG ST9XE CCD working at f/6.3. Two 100 s images were used to create this half-degree high field. Image: Martin Mobberley.

So ended the magnificent amateur comet discovery year of 2002, a year in which amateurs successfully stole seven comets, including the longest period "short period" comet known, from under the digital eyes of the machines. Excellent!

2003

Well, sadly, but rather predictably, the 2002 amateur comet blitz would not continue. Only one comet discovery by an unpaid non-machine entity was made in 2003. This was the discovery of C/2003 T3 (Tabur) by Australian amateur Vello Tabur who was searching for comets with a 140-mm f/2.8 Nikon lens attached to an SBIG ST8 CCD, giving a field of view of $5.6 \times 3.8°$. He detected the magnitude 11.5 fuzz near the constellation border separating Pavo and Telescopium on October 14. It peaked at magnitude 9.6 in April 2004.

2004

This was a far more successful year for amateur comet hunters and it was particularly gratifying to see two of the greatest visual comet discoverers of recent decades, Bill Bradfield and Don Machholz, making a comeback with visual discoveries. What can anyone say about Bill Bradfield, apart from him being one of the greatest visual comet discoverers of all time and the greatest visual discoverer of the twentieth century? The man known formerly as "The Wizard of Dernancourt" after the town where he lived for most of his comet discovering life had bagged his eighteenth comet from Yankalilla and his first discovery of the twenty-first century, despite being 76 years of age. As we have already seen Albert Jones of New Zealand discovered his second comet aged 80 and the UK's George Alcock discovered IRAS–Araki–Alcock in 1983 when 70 years old. Although most of Bill Bradfield's discoveries, since 1972, had been made with a 15-cm aperture f/5.5 refractor he tended to use a 25-cm Newtonian in later life. This telescope, which I mention again in the next chapter, was heavily counterbalanced at the eyepiece end so that the altitude rotation point was centered on the eyepiece, to minimize the eyepiece movement while sweeping the sky in strips. Comet C/2004 F4 (Bradfield) like all of his discoveries was his find alone and not shared with any other observer. Although the comet was discovered at eighth magnitude on March 23 an astrometric position was not acquired until Terry Lovejoy imaged it on April 12, when only 5 days from perihelion, which occurred only 0.17 AU from the Sun. As with so many of Bradfield's comets C/2004 F4 was a splendid performer and a treat for photographers, imagers and visual observers as it reached third magnitude and developed a fine 2° tail.

I have already mentioned Roy Tucker of Tucson Arizona in connection with his first comet discovery, P/1998 QP$_{54}$ (LONEOS-Tucker). Roy bagged his second comet C/2004 Q1 (Tucker) with the unusual drift scan imaging 35-cm f/5 Newtonian triplet system at his Goodricke–Pigott observatory on August 23, 2004. This magnitude 14 comet was snatched during a brief and welcome lull in the awesome LINEAR system's patrols caused by torrential rain at the normally dry New Mexico site. By mid-November the comet had reached magnitude 10.5 (see Fig. 4.9).

The final amateur comet discovery of 2004 was made by the veteran Californian comet discover Don Machholz (see Fig. 4.10a) with a 15-cm f/8 Newtonian at 35× magnification; the telescope being first

Fig. 4.9. Comet C/2004 Q1 (Tucker) imaged on December 16, 2004 with a 0.35-m aperture Celestron 14 at f/7.7 mounted on a Paramount ME and an SBIG ST9XE CCD. One hundred and sixty second exposure and a 13 × 13′ field. Image: Martin Mobberley.

Fig. 4.10. (a) The prolific comet discoverer Don Machholz photographed on a visit to Cambridge England for the second International Workshop on Cometary Astronomy. Image: Martin Mobberley. **(b)** Comet C/2004 Q2 (Machholz). A 1° wide mosaic imaged on December 31, 2004 with a 0.35-m Celestron 14 at f/7.7 and SBIG ST9XE CCD. Three 60 s images were used for the head and five for the tail. Image: Martin Mobberley.

acquired in 1968. C/2004 Q2 (Machholz) would be his tenth comet and was discovered almost 10 years after his ninth. Don is on record as saying that he had searched for 1,458 h since his previous comet and had spent 7,047 h comet sweeping since 1975! The discovery was made at 4:12 AM local time on August 27 while sweeping in the southerly constellations of Fornax and Eridanus, where it was found, at −22° declination. At discovery the comet was magnitude 11. Comet Machholz (see Fig. 4.10b) reached perihelion at 1.2 AU from the Sun on January 24, 2005 and attained a very northerly declination of +85° in the first week of March. It would peak at magnitude 3.7 around January 8. This comet will always be a special one for this author as it is the only comet I have ever observed using the author Patrick Moore's 15-in. Newtonian. I had been invited to the great man's home and traveled down with the supernova discoverer Tom Boles who lives in the same UK county (Suffolk) as me. A TV crew was there and I appeared a few weeks later, on TV, observing Don Machholz' comet!

2005

The year 2005 was looking rather barren for amateur discoveries until the American Charles Juels (1944–2009) and the Brazilian Paulo Hol-vorcem collected their second comet on July 2. The transcontinental imaging team discovered the magnitude 14.5 comet C/2005 N1 (Juels–Holvorcem) in the northern constellation of Perseus and it was already showing a 70 arc-sec coma and a 120 arc-sec tail. The discovery was made with a modest 70-mm aperture refractor plus CCD. The comet peaked at magnitude 11.5 in the morning sky in the following weeks.

Amateur astronomer John Broughton of Reedy Creek Observatory, Queensland, Australia (see Fig. 4.11), ensnared the second amateur comet discovery of 2005. Broughton was already a highly successful aster-oid discoverer when his computer controlled 0.51-m f/2.7 Newtonian captured the eighteenth magnitude short period comet that would be named P/2005 T5 (Broughton). At the time of writing, since 1997, he has discovered some 800 asteroids making him the sixteenth most prolific asteroid discoverer of all time and no doubt in the coming years this tally will increase. With more than 400 asteroid discoveries to his credit by 2005 it was almost inevitable that a comet bearing the name Broughton would appear at some point. He also discovered a potentially hazardous NEO, 2004 GA1, on April 11, 2004 which was one of the first amateur discoveries of an asteroid that posed a threat to the Earth. Broughton was one of five astronomers who secured a "Gene Shoemaker Near-Earth

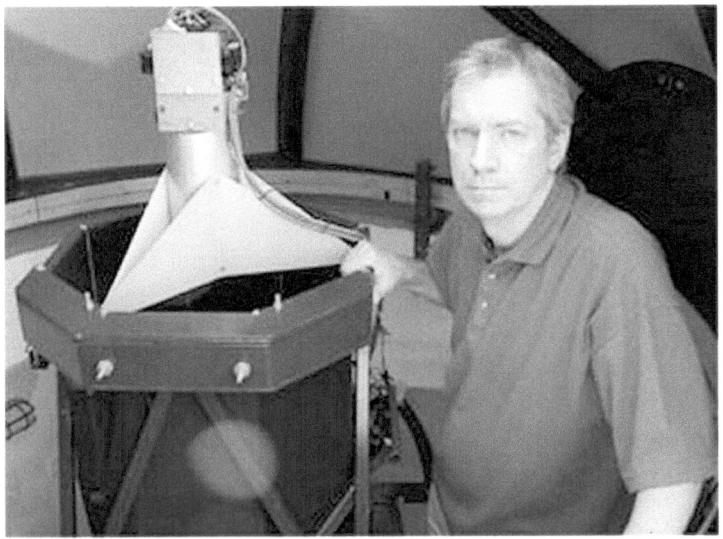

Fig. 4.11. The amazing John Broughton of Reedy Creek, Queensland, Australia, is the discoverer of two comets, 800 asteroids and several Near Earth Objects (NEOs). He is pictured here at the CCD camera prime focus position of his 0.51-m f/2.7 reflector. For much more on John Broughton's amazing home-made telescope see Chap. 12. Image: John Broughton.

Object Grant" from the Planetary Society in 2002 and he used the grant to purchase a sensitive CCD camera. His unique home-made 0.51-m f/2.7 Newtonian is described in detail in Chap. 12.

2006

John Broughton did not waste much time between his first and second comet discoveries as his second asteroid that turned out to be a fuzzy comet, C/2006 OF2 (Broughton), was captured from Reedy Creek Observatory on June 23, with the 0.51-m f/2.7 reflector + CCD. The comet was discovered when at a very distant 8 AU from the Sun. After images on July 17 by C.W. Hergenrother with the 1.54-m Catalina reflector showed the object to have a coma the cometary elements were published showing that C/2006 OF2 (Broughton) would not reach perihelion until September 15 at a distance of 2.4 AU from the Sun. However, with a healthy absolute magnitude between 6.0 and 7.0 the comet put on a decent show reaching tenth magnitude at perihelion and being observable for many months in amateur telescopes.

Excluding SOHO discoveries the second and final comet discovery of 2006 was secured by the famous veteran Arizona comet discoverer David Levy and was made visually with his 0.41-m Newtonian. David swept up the short period P/2006 T1 (Levy) on October 2 just before the start of morning twilight when the tenth magnitude comet was just 40 arc-min north of Saturn! Remarkably this would be David's 22nd comet discovery. In addition to his previous eight visual discoveries starting with comet C/1984 V1 Levy–Rudenko of November 1984, he had co-discovered 13 "Shoemaker–Levy" comets too. These comets were found while collaborating with Gene and Carolyn Shoemaker in their photographic patrols using the 0.45 m Palomar Schmidt camera between 1990 and 1994. Of course, by far the most dramatic of these was D/1993 F2 (Shoemaker–Levy) whose fragments collided with Jupiter in July 1994. After a gap of 12 years, since his visual co-discovery of Comet C/1994 G1 (Takamizawa–Levy) on April 15 of that year, David Levy's drought had ended. Sadly tenth magnitude was as good as comet Levy achieved. It reached perihelion a few days after discovery and it faded very quickly. The comet had undergone an impressive outburst just before discovery and Levy had swept the region at just the right time. P/2006 T1 (Levy) follows a 5.3 year orbit around the Sun and, in theory at least, it could approach the Earth to within 0.006 AU if perihelion occurred on December 31. However, this is not going to happen in the foreseeable future.

2007

Only two amateur comet discoveries were made in 2007 but both were by the same observer and were his first discoveries. The man in question was the well known Australian imager Terry Lovejoy (see Fig. 4.12) whose name was familiar to many astronomers already as a pioneer of early digital SLR deep sky and comet imaging. His two discoveries in 2007 were C/2007 E2 (Lovejoy) and C/2007 K5 (Lovejoy). Terry's comet search began in mid-2004, shortly after the highly affordable Canon 300D DSLR became available. He also acquired a Canon 350D. Terry used 200 mm f/2.8 lenses with his digital patrols and his technique was to expose 10–16 90 s images of a single field and then stack them to make two blink-able images that would reveal moving objects. The patrol was automated by using a software script that controlled imaging and the telescope drives. Terry has stated it takes

Fig. 4.12. Terry Lovejoy of Brisbane, Queensland, Australia discovered two comets in 2007 and is shown here with his Digital SLRs and twin 200-mm lens patrol system. Image: Terry Lovejoy.

him about 5 min to blink-compare each $6.5 \times 4.5°$ field. His discovery of the ninth magnitude C/2007 E2 on March 15 took approximately 1,400 h of patrolling spread over almost 3 years whereas finding the thirteenth magnitude C/2007 K5 on May 26 only took an additional 20 h!

2008

I cannot help thinking that the discovery of comet C/2008 C1 (Chen–Gao) on February 1, 2008 was greatly influenced by the success of Terry Lovejoy in the previous year as C/2008 C1 was also discovered using a Canon 350D digital SLR equipped with a 200 mm f/2.8 lens. However, I could be wrong here because Tao Chen (who took the images) and Xing Gao (who checked them) were quoted as apparently not carrying out a joint comet patrol but a joint nova patrol from their homes at Suzhou City in the Jiangsu province of China and Urumqi in Xinjiang province. Apart from the fact that virtually all amateur astrophotographers latched onto the power of the digital SLR it is a fairly obvious fact that the 200 mm f/2.8 lens is the best compromise between light grasp and cost for anyone planning wide field imaging of the night sky. Move up to 300 mm f/2.8 and 100 mm slabs of exotic glass combinations start to get very pricey! C/2008 C1 (Chen–Gao) appeared to be magnitude 13 when it was discovered amongst the dense star fields of the constellation of Cepheus, close to the border with Cassiopeia. However, CCD images of the comet shortly after discovery indicated that it had a total magnitude nearer to magnitude 11.

The other three amateur comet finds of 2008 were made using equipment at completely the other end of the scale from a 200-mm focal length f/2.8 lens. When the first of these comets, C/2008 N1 (Holmes), was discovered on July 1 it seemed to cause momentary confusion. Less than 9 months earlier the historic comet 17P/Holmes had outburst from 17th to second magnitude in <2 days and even the mention of comet Holmes would cause instant discussion when in the company of other astronomers. Suddenly, from no-one having heard of a comet Holmes there were two around! Was this new object in some way identical to 17P? Had the original Edwin Holmes, who died in 1919, been resurrected from cryogenic suspension? Well, no! By sheer coincidence there was a new comet discoverer on the block: Robert Holmes from Charleston, Illinois. As with John Broughton, Robert Holmes was a keen asteroid hunter and his name was already fairly well known amongst the world's most advanced amateur astronomers. Comet C/2008 N1 was discovered by Robert at his "Astronomical Research Institute" based at Charleston, Illionis, while he was making routine astrometric measures of the new NEO 2008 JZ30 using his 0.61 m reflector. The new object in the same field appeared to be another asteroid at magnitude 20.2. The object was immediately placed on the MPC's NEO confirmation web page and 7 days later, after images had been secured from eleven other observatories, a coma was detected, confirming its cometary nature. Perihelion occurred on September 25 at a distance of 2.8 AU from the Sun and the comet peaked at magnitude 17 with a tiny tail. Robert Holmes had already discovered numerous asteroids and six supernovae (between 2000 and 2006) prior to discovering C/2008 N1. At the time of writing his Astronomical Research Institute owns, in addition to the 0.61 m f/5.2 Newtonian, a 0.81 m f/4.6 Newtonian; a 1.2 m Newtonian is under construction too. As well as that impressive equipment list a 0.76 m f/6.65 telescope, which is a twin of the Lick Observatory KAIT supernova telescope, is being resurrected by ARI with four other collaborators. On January 31, 2009, Robert Holmes discovered a Potentially Hazardous Asteroid (PHA) named 2009 BD81, roughly 300 m across, which may, in 2042, pass within 5.5 Earth radii of the Earth. He describes that find as his greatest discovery.

The third comet discovery of 2008 was made from Slovenia by Stanislav Maticic, using the 0.60-m f/3.3 Cichocki Sky Survey Telescope at Crni Vrh Observatory. This fast telescope uses an f/3 primary mirror with a 1.1× triplet lens Wynne corrector to create the so called "Deltagraph" wide field design, made popular by the Astro Optik Company's Phillip Keller. A 1,024 × 1,024 pixel CCD with 24 μm pixels gives a field of 42 arc-min with this system and a scale of 2.5 arc-sec per pixel. IAU Circular 8966, issued on August 20, 2008 announced the discovery by Stanislav Maticic of this new comet, named C/2008 Q1 (Maticic). It had been discovered in the course of the Comet and Asteroid Search Program at Crni Vrh Observatory and was magnitude 18 at discovery. After the discovery of a moving object was posted on the Minor Planet Center's "NEOCP" webpage a number of imagers noted its non-stellar appearance defining it as a comet. The comet reached a perihelion distance of just under three AU from the Sun on December 30 and crept up to about magnitude 17 in April 2009.

The final amateur comet discovery of 2008 was another European find and was made by Michael Ory of Switzerland with another huge telescope by non-professional standards, the 0.61 m f/3.9 "Bernard Comte" telescope at the Observatoire Astronomique Jurassien (Jura Observatory) near Vicques in the Canton of Jura. The observatory was constructed in the 1990s and is operated by the Société Jurassienne d'Astronomie. P/2008 Q2 (Ory) was discovered on August 27 at magnitude 17.6 and reached perihelion on October 19 at 1.38 AU from the Sun. It soon became apparent that this was a short period comet with a period of 5.84 years and it peaked at fourteenth magnitude in November. Michael Ory has 68 asteroid discoveries to his credit, including the NEO 2009 KL2 as well as the supernovae 2006ev and 2003lb.

2009

The first amateur comet discovery of 2009 was made by the prolific Japanese supernova discoverer Koichi Itagaki (see Fig. 4.13). Most of Itagaki-san's patrolling has been made with a huge 0.6-m f/5.7 reflector but during additional relentless patrolling with a 21 cm f/3 astrograph, having a wide 2.2° field, Itagaki-san has discovered other objects too. He bagged two novae in Aquila and Ophiuchus, recovered (with Hiroshi Kaneda) the returning comet P/1896 R2 (Giacobini) which had been lost for 111 years and then discovered the long period comet C/2009 E1 (Itagaki). In 1968 Itagaki-san had narrowly missed out on being a co-discoverer of the triple-barreled Japanese comet Tago–Honda–Yamamoto but, unfortunately, four names

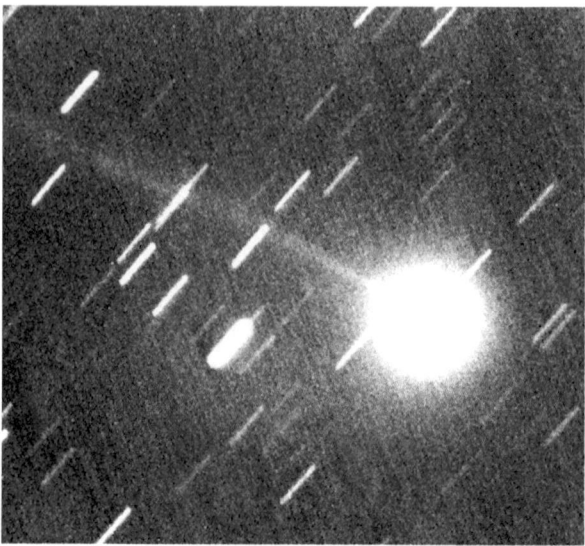

Fig. 4.13. Comet C/2009 E1 (Itagaki) imaged on March 2009 27.830 UT with a 0.35-m Celestron 14 Schmidt–Cassegrain working at f/7.7 and an SBIG ST9XE CCD. This is a stack of 30 exposures, each of 60 s duration, stacked with respect to the comet's central coma. Twelve arc-min field. Image: Martin Mobberley.

are not allowed and so he was never credited with that find. Now in his sixties Koichi Itagaki shows no sign of reducing his prolific discovery output. He is by far the most successful Japanese supernova hunter with more than fifty to his credit and many of his supernova discoveries have been as bright as fifteenth or sixteenth magnitude. His supernova 2004dj, discovered in NGC 2403 reached a spectacular eleventh magnitude! However, before I get side-tracked, let us get back to his comet discovery, C/2009 E1 (Itagaki). This was captured in a CCD image with the 21-cm f/3 reflector at Takanezawa, Tochigi, Japan on March 14, 2009. The discovery magnitude was a bright 12.8. Itagaki-san was using software developed by H. Hiroshi Kaneda of Sapporo, Japan to automatically detect moving objects when he found C/2009 E1. The new comet was only about 20° above the western evening horizon in Cetus when found, just 2° below the constellation boundary with Aries and moving at 210 arc-sec per hour NNW. Itagaki's comet was just 6° below the ecliptic plane when swept up, in an area full of moving asteroids, but was not a short period comet and its orbital inclination to the ecliptic was 127°. At the time of its discovery it was only 0.80 AU from the Sun and 0.93 AU from the Earth. It was also only 24 days from reaching perihelion (on April 7) at just 0.6 AU from the Sun and so it was brightening steadily to ninth magnitude, even though it was already moving away from the Earth. Its proximity to the solar glare prior to discovery almost certainly enabled such a relatively bright comet to evade detection by the major patrols.

Only 12 days after Itagaki's comet discovery another amateur astronomer from the Far East captured a comet. This time the discovery was made with the simplest optical equipment imaginable. Dae-am Yi of Yeongwol-kun, Gangwon-do, South Korea spotted this comet on two 60-s exposures taken on March 26, 2009, with his Canon 5D DSLR and a modest 90-mm f/2.8 lens. The two images were taken 80 s apart and the comet was in Lacerta. Yi estimated the magnitude as 12.5 and described the suspect as a blue–green object about 1 arc-min across. This was the very first time a comet had been named after a Korean. Yi sent his observation to H. Yamaoka (Kyushu University, Fukuaka, Japan). During the delay in reporting the object the Central Bureau in Boston USA was notified by Rob Matson (USA) that he had found images of a comet on ultraviolet SWAN detector images obtained by the SOHO spacecraft. So the comet became known as C/2009 F6 (Yi-SWAN). The new comet was closest to the Earth on April 7, 2009 at 1.76 AU and reached perihelion on May 5 at 1.29 AU from the Sun. During April Yi-SWAN reached ninth magnitude. Once again a comet had evaded detection by the machines. However, once the orbit of the comet was known it was clear to see that C/2009 F6 had spent the last few months crawling through the densest Milky Way fields of Lacerta, Cygnus, Vulpecula, Sagitta and Aquila; regions of the sky so cluttered with stars that even the cleverest software finds it difficult to detect moving objects in such cluttered fields.

Remarkably, the third amateur comet discovery of 2009 was another made by an Oriental team. I have already mentioned the Chinese comet C/2008 C1 (Chen–Gao). This was discovered in the previous year, during the course of a nova patrol carried out using a Canon 350D digital SLR equipped with a 200 mm f/2.8 lens. Well, P/2009 L2 (Yang–Gao) was co-discovered by the same Xing Gao of Urumqi in Xinjiang province, but this time working with Rui Yang, Hangzhou, Zhejiang, China. The team was now referred to as the Xingming Comet Survey and, as before, a Canon 350D digital camera was used, but with a "10.7-cm f/2.8 lens" from Mt. Nanshan. The comet was found on June 15 and appeared to be magnitude 14 at discovery when it was already a month past its perihelion encounter of 1.3 AU from the Sun. It turned out to be a periodic comet with a 6.3 year orbital period. The comet was close to the Earth when it was discovered (the minimum distance was 0.303 AU) and already moving away from us, but it briefly achieved twelfth magnitude in July (see Fig. 4.14).

Fig. 4.14. Comet P/2009 L2 (Yang–Gao) imaged on June 2009 23.948 UT with a 0.35-m Celestron 14 Schmidt–Cassegrain working at f/7.7 and an SBIG ST9XE CCD. Sixty second exposure; 13 arc-min field. Image: Martin Mobberley.

The Edgar Wilson Award

Amateur comet discoverers are a rare breed, because the task is long and grueling and demands an infinite amount of patience as well as, for the visual hunter, a love of sweeping the night sky in all weathers and climbing out of a warm bed at 3 AM to do the pre-dawn patrol. Even if your search is carried out indoors from scrutinizing CCD images opening up the observatory still requires you to venture outside in the early hours on a regular basis. Any comet discovery is going to be a momentous day in an amateur astronomers' life and for a few individuals, like Alan Hale or Thomas Bopp, it will guarantee them immortality. So, the discovery itself is the biggest prize imaginable and any award will always be just an extra "icing on the cake." Nevertheless, awarding prizes for comet discovery is not a new phenomenon.

Joseph Jerome Le Français de Lalande (1732–1807) put up 600 Francs for the first comet discovery of the nineteenth century and, surprisingly, it was clinched by a newcomer in 1801, Jean Louis Pons (1761–1831), who went on to be the most prolific comet discoverer on record until, almost 200 years later, the Shoemakers and Rob McNaught came along. Pons stole the 600 Francs from under the noses of his countrymen Charles Messier and Pierre Méchain. The Imperial Academy of Sciences at Vienna announced, in 1870, that it would award up to eight gold medals per year for at least 3 years to anyone discovering a comet. Other scientific awards were sometimes given to comet discoverers too. For discovering two comets in 1881 the discoverer Lewis Swift received the French Academy of Sciences' Lalande Prize, consisting of a silver medal and 540 Francs. In that same year, 1881, following the termination of the Vienna scheme, the "patent medicine king" of the USA, Hulbert Harrington Warner, who was building an observatory for Lewis Swift, decided to start his own comet reward system. The strangely named Warner Safe Remedy Prizes consisted of a gold medal and $200 in cash for the first American discoverer of each comet. If less than five comets were discovered the $1,000 outlay would be divided amongst the discoverers equally. This was a huge amount of money in the 1880s and William R. Brooks, who discovered 21 comets in his lifetime, half in the Warner award era, earned thousands of dollars in reward money from the scheme.

In June 1998 the IAU announced the creation of the Edgar Wilson award which would be allocated annually among the amateur astronomers who, using amateur equipment, had discovered one or more new comets. Only comets officially named for their discoverers would be included in the annual count. The annual pot of money would be divided equally amongst the discoverers or co-discoverers of all the amateur comets discovered, so each individual would earn more per comet, but less if there were plenty of other discoverers and co-discoverers. The award would be administered by the Smithsonian Astrophysical Observatory (SAO), as the beneficiary under the last Will and Testament of Edgar Wilson. This administration would be carried out through the International Astronomical Union (IAU) Central Bureau for Astronomical Telegrams (CBAT), which, with the advice of the Small Bodies Names Committee (SBNC) of IAU Division III, had the responsibility for naming comets. The first year was defined as running from 1998 June 11.0 to 1999 July 11.0 and the funds for that first year were to be $20,000. Edgar Wilson was a Lexington Kentucky businessman, keen on promoting amateur astronomy, who died in 1976; 22 years before the award that bears his name was founded. The trustees of his funds inaugurated the award, in collaboration with the IAU/SAO/CBAT.

The winners of the Edgar Wilson awards since June 1999, along with the associated comets, are as follows:

1999: Peter Williams, Heathcote, NSW, Australia (C/1998 P1); Roy A. Tucker, Tucson, AZ, USA (P/1998 QP$_{54}$); Michael Jaeger, Weissenkirchen, Wachau, Austria (P/1998 U3); Justin Tilbrook, Clare, S. Australia (C/1999 A1); Korado Korlevic and Mario Juric, Visnjan, Croatia (P/1999 DN$_3$); Steven Lee, Coonabarabran, NSW, Australia (C/1999 H1).

2000: Daniel W. Lynn, Kinglake West, Victoria, Australia (C/1999 N2); Korado Korlevic, Visnjan, Croatia (P/1999 WJ$_7$); Gary Hug and Graham E. Bell, Eskridge, KS, USA (P/1999 X1).

2001: Albert F. A. L. Jones, Stoke, Nelson, New Zealand (C/2000 W1); Syogo Utsunomiya, Minami-Oguni, Aso, Kumamoto, Japan (C/2000 W1).

2002: Vance Avery Petriew, Regina, SK, Canada (P/2001 Q2); William Kwong Yeung, Benson, AZ, USA (P/2002 BV); Kaoru Ikeya, Mori, Shuchi, Shizuoka, Japan (C/2002 C1); Daqing Zhang, Kaifeng, Henan province, China (C/2002 C1); Douglas Snyder, Palominas, AZ, USA (C/2002 E2); Shigeki Murakami, Matsunoyama, Niigata, Japan (C/2002 E2); Syogo Utsunomiya, Minami-Oguni, Aso, Kumamoto, Japan (C/2002 F1).

2003: Sebastian Florian Hoenig, Dossenheim or Heidelberg, Germany (C/2002 O4); Tetuo Kudo, Nishi Goshi, Kikuchi, Kumamoto, Japan (C/2002 X5) ; Shigehisa Fujikawa, Oonohara, Kagawa, Japan (C/2002 X5); Charles Wilson Juels, Fountain Hills, AZ, U.S.A.; and Paulo R. Holvorcem, Campinas, Brazil (C/2002 Y1).

2004: Vello Tabur, Wanniassa, A.C.T., Australia (C/2003 T3); William A. Bradfield, Yankalilla, S. Australia (C/2004 F4).

2005: Roy A. Tucker, Tucson, AZ, USA (C/2004 Q1); Donald Edward Machholz, Jr., Colfax, CA, USA (C/2004 Q2).

2006: Charles Wilson Juels, Fountain Hills, AZ, USA and Paulo R. C. Holvorcem, Campinas, Brazil (C/2005 N1); John Broughton, Reedy Creek, Qld., Australia (P/2005 T5).

2007: John Broughton, Reedy Creek, Qld., Australia (C/2006 OF2); David H. Levy, Tucson, AZ, USA (P/2006 T1); Terry Lovejoy, Thornlands, Qld., Australia (C/2007 E2 and C/2007 K5).

2008: Tao Chen, Suzhou City, Jiangsu province, China and Xing Gao, Urumqi, Xinjiang province, China (C/2008 C1).

2009: Robert E. Holmes, Jr. (Charleston, IL, USA) for C/2008 N1; Stanislav Maticic (Crni Vrh Observatory, Slovenia) for C/2008 Q1; Michel Ory (Delemont, Switzerland) for P/2008 Q2; Koichi Itagaki (Yamagata, Japan) for C/2009 E1; Dae-am Yi (Yeongwol-kun, Gangwon-do, Korea) for C/2009 F6.

CHAPTER FIVE

Finding the Next Hale–Bopp with your Gear

So, what lessons can be learned from the amateur discoveries of recent years covered in the previous chapter? Well, let us not be under any illusions here: discovering comets is not easy for the twenty-first century amateur astronomer! I have ambitiously titled this chapter "Finding the next Hale–Bopp with your gear." Obviously I live in a fantasy world you might think? Well, maybe, but the title probably grabbed your attention for starters. Let us be honest here; for most amateur astronomers any comet discovery is a major achievement. There are tens of thousands of amateur astronomers worldwide, but only a fraction of a percent of them will ever discover a comet. Nevertheless, in the CCD era it is obvious that faint comets can be discovered while carrying out routine astrometric or photometric work on comets and asteroids in the ecliptic plane. The advanced amateur can reach a limiting stellar magnitude of 20 with ease and there are still objects brighter than that magnitude waiting to be discovered. Many of the amateur comet discoveries of recent years have been detected with amateur CCD systems during the course of routine astrometric work on NEOs discovered days earlier by the professional patrols. At the other end of the scale we can see that discoveries with digital SLRs using 100–200 mm lenses have been made too. The remorseless machines *can* be beaten because even the best systems can be clouded out and even the best software can be confused by dense Milky Way backgrounds and

M. Mobberley, *Hunting and Imaging Comets*, Patrick Moore's Practical Astronomy Series, DOI 10.1007/978-1-4419-6905-7_5, © Springer Science+Business Media, LLC 2011

by moonlight. So while you may not discover the next Hale–Bopp you may discover something.

As I mentioned earlier in this book I have been fortunate enough to know quite a few amateur discoverers in my years in the hobby. Standing head and shoulders above the rest was the late George Alcock whose five bright comet and five bright nova discoveries, all achieved using binoculars, were a remarkable achievement. To have memorized 30,000 stars, in patterns is quite astounding. I have also known the supernova discoverers Ron Arbour, Mark Armstrong and Tom Boles too. All these characters have had one thing in common, namely total dedication to their searching. For the greatest patrollers hunting the night sky simply becomes an instinctive ritual, replacing the typical twenty-first century ritual of sitting in front of the television. Huge amounts of time and effort need to be directed towards a single goal and this, for most people, is just not possible. Many adults have demanding day jobs as well as a family to support and entertain. Trying to combine hunting the night sky with a career and a family is never going to work. For this reason the vast majority of successful patrollers are either early-retired, wealthy, single, on the brink of a divorce that they did not see coming, or have very forgiving partners! The bottom line here though is that to become a successful discoverer you must have far more free time than the average person or be so superhuman that you can survive on a few hours sleep every night. The alternative is to so efficiently automate your CCD patrols that your telescopes can search the night sky on their own and just flag up suspects for you to check. There is no easy solution but I do not think craving a discovery badly will help. You have to enjoy the hunt, even if you are not successful; otherwise you will at some point just burn out from the effort. According to David Levy there is a story that the great Japanese comet and nova discoverer Minoru Honda advised the young (in the 1960s) Kaoru Ikeya along the same lines. Namely, that if Ikeya was simply desperate to discover a comet he should pack up immediately, because there were no guarantees, but if he enjoyed sweeping the skies, regardless of any discovery, he should continue.

I have been in this hobby for more than 40 years now and I have learnt that the really best observers who achieve the most are those who love whatever aspect of their hobby they undertake. Those observers who crave success or fame, but do not actually enjoy what they do under the stars, may achieve something, but most will burn out after a few years and never return to observing at the same level. You have to love what you do to sustain it for a lifetime, not simply crave that fickle commodity called "fame"! George Alcock was Britain's most successful comet and nova discoverer and he loved sweeping the skies, but hated media attention.

He once told me that after making a discovery he would often spend a week bird watching in Norfolk, out of contact with the media. His phone would ring and ring at home, with TV and newspaper people literally apoplectic that he could not be contacted and had no interest in talking to them, while George chilled out watching wildlife. TV and media people can often be leeches, relying on the fact that no-one will ever turn down a TV interview and 5 min of fame and no-one will ask for a fee either. The fact that George did not want publicity was stressing out these "very strange TV and radio people who always got their facts wrong" and annoying them simply enhanced his pleasure at his latest discovery. After the BBC mis-reported him in 1959 as having discovered two new *planets* in a week, his view of TV people was pretty much fixed!

Searching for the Next Kreutz Sungrazer

We have already seen that some of the most spectacular comets in the nineteenth and twentieth century's have been caused by massive fragments of a sungrazing comet which may have originally broken up at perihelion in the years 214, 426 or 467 AD. We have also seen that a huge chunk of that progenitor formed the Great Comet of February 1106. In addition we know that the superb comets designated C/1882 R1, C/1945 X1, C/1963 R1, C/1965 S1 and C/1970 K1 can be traced back to that 1106 AD comet breaking up and that C/1843 D1, C/1880 C1 and C/1887 B1 are closely related to that 1106 AD comet, or its 1100 AD twin fragment. These are not the only comets involved either as a huge number of small SOHO Kreutz fragments have been found and orbital analysis indicates that further fragmentation may have occurred in 1102, 1700, 1749 and 1847 AD. Bearing in mind that spectacular Kreutz sungrazers have returned across a time spread of 127 years, from 1843 to 1970, it is obvious that there are lots of big chunks all around the 750–900 year orbit of this comet family and another big chunk may be heading our way soon. So, maybe one search strategy should be to look out for returning "Kreutz chunks" along the typical orbit track of these past comets? It might be thought that as these comets have broken up so many times they would all have completely different orbits, but a simple check of the orbital elements of the brightest Kreutz sungrazers in recent years shows this is simply not the case. There are small differences, yes, but the cometary fragments are simply spread along a very similar orbital path. So, where should we search to stand the best chance of bagging an incoming Kreutz monster? Well, the aphelion of the Kreutz comets is somewhere between 160 and 190 AU away in space, or roughly 25 billion km, or, if you prefer light travel times, about one light day. This is approximately six times the distance of Neptune from the Sun. When at aphelion the invisible Kreutz chunks live not far from the brightest star in the sky, Sirius. To be more precise they live just above a line from Sirius to Saiph (the lower left bright star in Orion) at roughly one-third of the way from Sirius (see Fig. 5.1). In coordinates this is somewhere near 6 h 33 m in right ascension and −13° in declination. Of course, the Earth's orbit has a radius of 1 AU or 149.6 million km and so as the Earth swings around in orbit from September to March this makes objects near to the inner solar system appear to move significantly due to parallax. At 25 billion km distant (the Kreutz aphelion point) this parallax swing will be small at roughly plus or minus 20 arc-min from a mean position, but obviously as any potential comet enters the inner solar system the parallax swings can become enormous. So, the trick to spotting any incoming

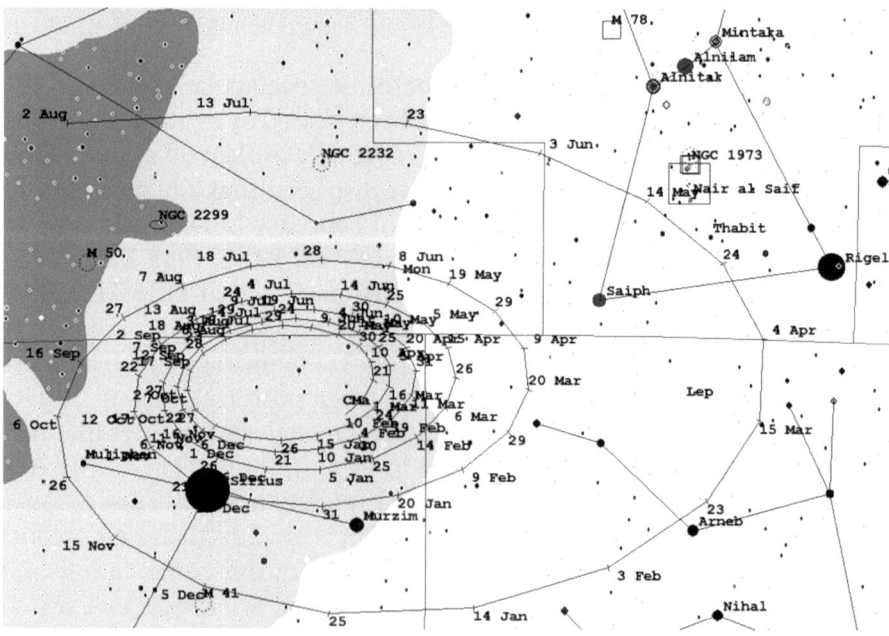

Fig. 5.1. The so-called vanishing point, or more correctly, emerging point, for the Kreutz sungrazing comet C/1965 S1 (Ikeya–Seki) lay just to the northwest of the brilliant star Sirius. The spiral pattern represents the position of the comet in the years before perihelion with each circle covering 1 year of travel. These tracks, centered northwest of Sirius, are typical for all incoming Kreutz sungrazers.

Kreutz blockbuster will be trying to pick it up when it is a long way away, very faint, but not moving very much due to parallax, while still being bright enough for an amateur CCD image to have a chance of registering the object. If you look at the so called "vanishing point" for previous Kreutz sun grazers like C/1965 S1 (Ikeya–Seki), by plotting the comet's path backwards in time, you can see the incoming paths form a spiral pattern with each spiral (due to the parallax) getting smaller every year prior to discovery. Kaoru Ikeya and Tsutomu Seki discovered C/1965 S1 on September 18, 1965, when it was 10° west of the star Alphard (alpha Hydrae). The comet was around magnitude eight at the time, just 5 weeks prior to perihelion on October 21. By the start of October it had reached fifth magnitude and it was second magnitude by mid-month. On the day of perihelion it was photographed near the Sun with a coronagraph at the Tokyo Observatory Norikura Corona Station and the magnitude was then estimated at a staggering −10 or −11. Comet C/1965 S1 had executed a crafty hiding game in the months prior to discovery, keeping well out of discovery range in the solar glare. In mid-June 1965 the comet, then probably about magnitude 13, had been close to the Orion/Monoceros border only about 27° due south of the Sun making Antarctica perhaps the only place from which it might have been discovered! In addition, the comet was already not far from that Kreutz "vanishing point" region of the sky. Now go back to a full year before perihelion, on October 21, 1964, and Ikeya–Seki would have been in Canis Major, not far from Sirius at, perhaps, 16th magnitude, 5.6 AU from the Sun, and already only 9° or so east of its aphelion "vanishing point." The motion would have been close to 17 arc-sec per hour. Go back another year, to October 1963, and a 19th magnitude Ikeya–Seki would still have been in Canis Major, near Sirius, at 8.8 AU from the Sun and crawling along at 10 arc-sec per hour, just 6° east of its vanishing point. Finally, let us just go back one further year. In October 1962, 3 years prior to perihelion, a magnitude 21 Ikeya–Seki would have been 11.5 AU from the Sun and travelling at a leisurely 8 arc-sec per hour against the stellar backdrop of Canis Major. The comet would have been within 5° of the orbital "vanishing point."

So what does this tell us about a strategy for detecting an incoming Kreutz comet of a similar size to the nucleus of C/1965 S1 (Ikeya–Seki)? Well, my estimated magnitudes are, of course, no more than an educated guess, but if we assume they are not way off then searching for a Kreutz sungrazer more than 3 years prior to perihelion is probably pointless. Few amateurs can hope to image a moving dot fainter than 21st magnitude. So, the radius of our search needs to be at least 5° from the "vanishing point." Of course, the Kreutz fragment may only become detectable a year or so before perihelion in which case the search radius needs to be at least 9°. One point I have not yet made clear is that the vanishing point spirals are wider than they are high due to the position of the vanishing point with respect to the ecliptic. At 1 year from perihelion a safe strategy would be to repeatedly patrol a rectangular region roughly 20° wide in right ascension and 10° high in declination from 5 h 50 m at the western edge to 7 h 10 m at the eastern edge and −8° at the north edge to −18° at the south edge. This may seem like an area that can be patrolled easily but it is a sizeable 200 square degrees which, on a 30 or 40 cm telescope will take hundreds of images to cover, unless you own a very fast system with a very large CCD chip. Of course, one could argue that the dead centre of the image does not need covering as the Kreutz aphelion point is hardly important. No-one is going to detect a 35th magnitude comet 25 billion km away and even if they did it would be some 400 years from perihelion! One problem with this Kreutz strategy is that the region of sky that needs patrolling is near Sirius and is quite low down for far northern latitudes. Yes, even UK observers can see the region easily enough, but if you are trying to image to 21st magnitude it is a good idea to be looking at objects near the zenith, not at 20° altitude!

Now you may well think that searching for such Kreutz sungrazers is pretty pointless because surely Catalina, LINEAR or another patrol will pick these objects up first? Well, maybe, but as we shall see in the next chapter the tireless machines have missed quite a few comets in recent years. Cloud, a very slow motion against the stars, a field less than 90° from the Sun or just a cometary outburst when an amateur is imaging the field can all bring morsels of hope the amateur astronomer's way. In addition, the vanishing point spirals for the Kreutz orbits curve through some very dense starfields, especially on that eastern patrol edge, around 7 h of right ascension, and in Canis Major, a few degrees below the border with Monoceros. Two years prior to perihelion passage a Kreutz comet can be placed very close to Sirius in the sky too, the light from which can thwart any search attempts, professional or amateur! Moving object detection software that looks for three or more equispaced dots in a triplet of images taken, say, 30 min apart, will always struggle to cope in densely packed starfields. The entire Kreutz vanishing point spirals are in densely packed star regions but the star density is especially high on that eastern edge. This is an area where inspection by a human eye and brain often works far better than automated mindless software. We may see sentient "AI" (Artificial Intelligence) machines in the twenty-second century, but they are not here yet! The chance of discovering an incoming Kreutz sungrazer may be slim, but it is definitely not zero and we know where to look. In addition, discovering another C/1965 S1 (Ikeya–Seki) clone would guarantee astronomical immortality for its discoverer.

Comet Discovery as a Secondary Aim

Of course, the Kreutz comets are a very special case, not only because of their near suicidal encounters with the Sun but also because we know where to look along the orbit. I will say more about finding sungrazing chunks in the SOHO satellite data in Chap. 8. However, most amateurs with comet discovery aspirations will want to have more than one option and discovering an incoming Kreutz monster is a bit of a long shot. Undoubtedly endless patience is the most important virtue to have when sweeping the skies but it also helps to be able to combine a productive astronomical pursuit where a comet discovery is a spin-off and not the sole intended aim of the project. Many of the amateur discovered comets in recent years have been found as a by-product of NEO follow up work. The usual sequence of events here is that a professional patrol, like the Catalina Sky Survey or LINEAR, flags up a fast moving object which corresponds to no known asteroid or comet. It may be a chunk of rock a few hundred meters across within a few million kilometers of the Earth: in other words a potentially hazardous asteroid or PHA. It is vital to measure the positions of these objects as soon as they are found because, by their very nature, they are small, have a very low absolute magnitude and will fade rapidly as they move away from the Earth's vicinity. Such objects can quickly drop below 22nd magnitude within days of discovery, meaning that they are lost even to 1 m aperture instruments. Every lost NEO is a potential future threat to the Earth unless you can pin the orbit down with lots of precise measurements. So, a number of dedicated backyard astronomers around the world have accepted the demanding challenge of tracking down these painfully faint fast moving targets every clear night and, now and again, during this work, new asteroids, NEOs, or even comets can be found. Carrying out astrometry on these faint objects is demanding work, requiring a remarkably high precision and attention to detail.

To detect fast moving objects of 20th magnitude and fainter not only demands apertures of at least 30 or 40 cm, but precise focusing, tracking and crystal clear nights too. In addition, by their very nature, their orbits are unknown and where they will be on the night after discovery is often uncertain by many arc-minutes. It is often necessary to take many images in the predicted area of the suspect to bag it at all. Also, recording the precise time of the exposure to the nearest second is essential when NEOs are close to the Earth and travelling, in extreme cases, at thousands of arc-seconds per hour. By pinpointing a NEO's position in orbit when it is very close to the Earth and time-stamping the observation to 1 s accuracy, you have a very good idea of whether it poses a near-future threat. Remember, the Earth moves its own diameter in orbit every 7 min and so knowing the position of a NEO a few hundred years ahead, to an accuracy of a few minutes of time, is very useful to avoid future catastrophes. I am digressing a bit from comet hunting here but if you are carrying out useful science every time you go outside your ultimate goal of making a discovery will be less demoralizing. At least you will have something to show for your efforts!

Anyone who is planning to discover a comet, either as their main aim or as a spin-off from other research needs to be fully clued up on what is currently in the sky, have plenty of knowledgeable friends in the comet world and needs to know that there are various Internet places to check out for info. The top priority Internet sites to visit are those of the Minor Planet Center (MPC), a body which is closely affiliated with the International Astronomical Union (IAU) and the Central Bureau for Astronomical Telegrams (CBAT). Specifically, the following pages are crucial:

Ephemerides for all comets brighter than magnitude 23 (or so!)
http://www.minorplanetcenter.org/iau/Ephemerides/Comets/
Comet and NEO/Distant Minor Planet Designations Assigned in the Past Year
http://www.minorplanetcenter.org/iau/lists/LastYear.html
NEO Confirmation Page
http://www.minorplanetcenter.org/iau/NEO/ToConfirm.html
NEO and Comet checker
http://scully.cfa.harvard.edu/~cgi/CheckNEOCMT
Minor Planet & Comet Ephemeris Service
http://www.minorplanetcenter.org/iau/MPEph/MPEph.html
Orbital elements for popular planetarium software packages
http://www.minorplanetcenter.org/iau/Ephemerides/Comets/Software Comets.html

In addition to these pages there are three *Yahoo!* based User Groups for comet observers which will certainly be of interest. These are the Comets Mailing List, the Comet Images Group and the Comet Observers Group which can be found at the following Internet addresses:

http://tech.groups.yahoo.com/group/comets-ml/
http://tech.groups.yahoo.com/group/Comet-Images/
http://tech.dir.groups.yahoo.com/group/CometObs/

A fourth *Yahoo!* group also exists for sharing comet information, observations and techniques and to discuss comets and comet observing in general. This group can be found at:

http://tech.dir.groups.yahoo.com/group/CometChasing/

There are other websites in cyberspace that will be of great interest too. Seiichi Yoshida has a web page about currently visible comets which can be found at:

http://www.aerith.net/comet/weekly/current.html

He also has a comet rendezvous page and a comet recovery page. The former lists upcoming dates when comets will be within 3° of each other or within 3°of deep sky objects and the latter lists comets that are returning to perihelion and crying out for someone to be the first to image their return. I will have more to say on this subject in Chap. 7. Yoshida-san's rendezvous and recovery pages can be found at the following websites:

http://www.aerith.net/comet/rendezvous/current.html
http://www.aerith.net/comet/recovery.html

The comet pages of the British Astronomical Association can also be found at:

http://www.ast.cam.ac.uk/~jds/

The UK magazine called "The Astronomer" has a comet images page at:

http://www.theastronomer.org/comets.html

I may have drifted well away from the subject of comet discovery, but what I am trying to emphasize here is that a deep knowledge of comets and the contactable people in the comet community is essential if you want to make a discovery. If you do not have a comprehensive knowledge of what is happening in the sky, and good contacts, you may make many false discovery claims.

Astrometry

One skill that needs to be acquired before you can claim a discovery is that of astrometry: the ability to precisely measure the position of an object with reference to the background stars. I am covering comet imaging in depth later in the book but to digitally hunt for comets the reader of this chapter will already need to have acquired a CCD camera and the proprietary software that goes with it. However, that software rarely includes any astrometric features and, even if it does, there is one piece of software that stands head and shoulders above the rest for measuring positions on images and that is Herbert Raab's *Astrometrica* (see Fig. 5.2). The software is free for a trial period (shareware) and very affordable (25 Euros) if you purchase it. More importantly it is intuitive to use and very accurate. Ideally your CCD image needs to be in the astronomical

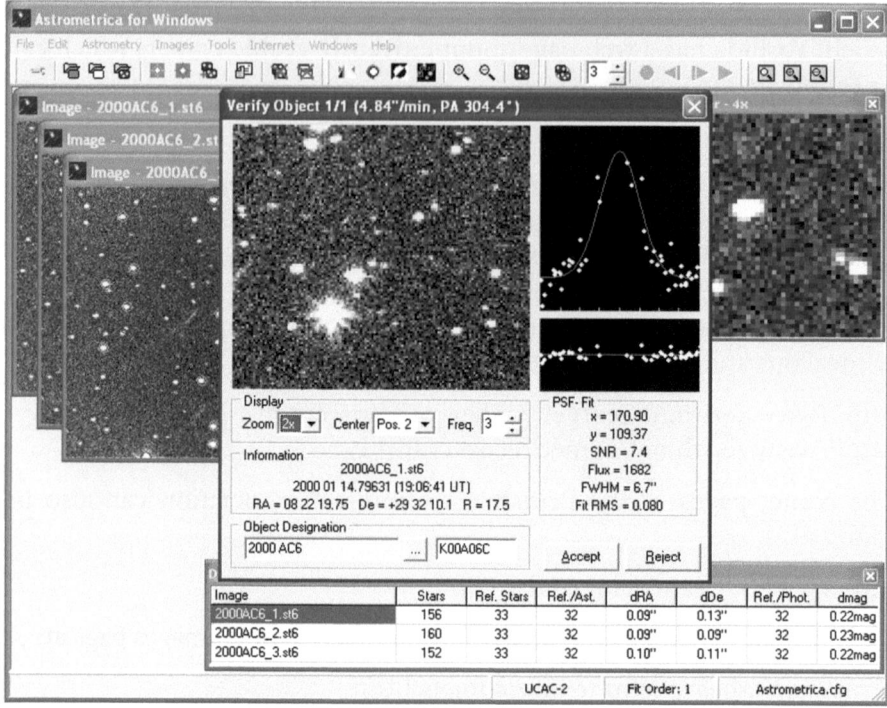

Fig. 5.2. Herbert Raab's excellent *Astrometrica* software enables the positions of asteroids and comets on FITS format images to be measured with ease and includes many useful features.

"FITS" (Flexible Image Transport System) format which almost all astronomical CCD cameras support, but, some other image formats used by SBIG cameras are supported too. FITS is the standard astronomical data format endorsed by both NASA and the IAU. It is much more than an image format (such as JPG, BMP, TIF or GIF) and is primarily designed to store scientific data sets consisting of multi-dimensional arrays (1-D spectra, 2-D images or 3-D data cubes) and 2-dimensional tables containing rows and columns of data. FITS images never compress the data and so are big, but no information is lost. In addition, the FITS file has a header which can contain vital data from the CCD camera relating to the time, camera, telescope, image scale, camera temperature and where the telescope thought it was pointing.

With any astrometric software you need access to a detailed digitized star catalogue as it is only by reference to such a catalogue that an object's precise position can be measured. The UCAC2 catalogue, produced by the US Naval Observatory, has proved the most useful in this regard, with the entire 2 GB database, in three chunks, easily fitting onto a hard disk. A specific request to the US Naval Observatory is needed to acquire it as the original 1,000 CD production run soon dried up! However, the UCAC3 double sided DVD which supersedes it was released in late 2009 to UCAC2 users who had already contacted the USNO. Software Bisque's *The Sky 6* and *The Sky X* contain the UCAC2 catalogue and its 48 million stars and, used with *CCDSoft* can carry out slick astrometry. But neither of these packages is cheap! The UCAC2 and the massive USNO B1.0 data are available for the downloading of specific sky areas at:

http://vizier.u-strasbg.fr/viz-bin/VizieR/

For up to date information on UCAC3 visit the US Naval Observatory Home page at

http://www.usno.navy.mil/. Further data may be found at:

http://cdsarc.u-strasbg.fr/viz-bin/ftp-index?/ftp/cats/aliases/U/UCAC3 and e-mail requests for the UCAC3 DVD can be sent to:ucac3@usno.navy.mil

If you have chosen to use *Astrometrica*, you first click on File/Settings/ Environment/Star Catalogue and point the software to your preferred star catalogue. You then load your FITS/SBIG format image into *Astrometrica* and, to get north at the top, you can flip the image vertically and horizontally if it has south at the top or is mirror-imaged. You then select Astrometry/ Data Reduction and Click "OK." *Astrometrica* will auto-identify the field (using the FITS header information regarding approximate image position and telescope scale) and WHAM!, it identifies every catalogue star in the field, instantly. Mouse-clicking on the object of interest then reveals the precise RA and Dec (and the magnitude) of the object you are measuring.

The next step is e-mailing your results, as plain text, to the Minor Planet Center professionals. *Astrometrica* can do this for you but it is important to understand the precise (and complex) format required. Once you have provided good measurements and your precise latitude, longitude and observatory height above sea level the MPC will allot your humble observatory a professional observatory code. All this may sound complex, and there is a significant learning curve, but the MPC web pages explain all and *Astrometrica* makes the measuring process tolerable. An example of an astrometric measurement of comet P/2009 T2 (La Sagra) this author submitted to the MPC in October 2009 is shown below. The message format is precise and consists of lines which describe the observer's observatory code (COD), as issued by the IAU Minor Planet Center, the observer's name (OBS), the star catalogue used (NET), the telescope details (TEL), an acknowledgment type code (ACK) and acknowledgement e-mail address for the observer (AC2). Following these lines there is a space and then the IAU Minor Planet Center optical astrometric format which includes the so-called packed provisional format which codes up a cometary identity. The astrometric line includes the object's identity in this packed format system along with the type of observation, precise time of observation, right ascension, declination, and magnitude and, again, the observatory code. The format is very strict, right down to the precise column spacing as the observations are handled by an automatic system due to the vast amount of data received. More information on the format can be found at http://www.minorplanetcenter.org/iau/info/OpticalObs.html

COD 480
OBS M. P. Mobberley
NET UCAC2
TEL 0.36-m f/7.7 Schmidt-Cassegrain + CCD
ACK MPCReport
AC2 martin.mobberley@btinternet.com

PK09T020 C2009 10 19.87867 01 42 41.57 +19 01 30.9 17.2 R 480

Following the dispatch of the above e-mail to mpc@cfa.harvard.edu you should, usually within a matter of minutes, receive an acknowledgement that reads:

The receipt of a message (probably containing observations) is hereby acknowledged.

Your message's ACK identification string is: MPCReport. The formatting code returned the following statistics:
Number of header lines read = 6
Number of observation lines read = 1

Before we leave the topic of orbits and astrometry it is worth mentioning that if you really get hooked the Project Pluto software "Find_Orb" enables you to compute orbits from your own measurements rather than just trusting the professionals to do it. More information can be found at: http://www.projectpluto.com/find_orb.htm,

Blink Comparators

Unlike the automated military grade software used by LINEAR and the software used by professional astronomical NEO patrols there is no all-powerful checking software of the same standard available to amateur astronomers to detect moving objects. However, there are a number of options for blinking images of the same star field to detect if anything like an asteroid or a comet has moved. The human eye and brain is very good at detecting changes in patterns and so the power of a good, slick, intuitive image blinking system cannot be over-emphasized where a comet discovery is the aim.

If your image is in FITS format then the aforementioned *Astrometrica* package by Herbert Raab has a great blink comparator tool under "Tools" on the main Menu bar (see Fig. 5.3). It is slick because it has a powerful alignment feature that instantly aligns two frames without fuss even if there is a bit of x and y difference in the two frames, as there always will be at long focal lengths.

Fig. 5.3. Astrometrica's blink comparator tool can show objects that move between frames.

However, the large color images captured by DSLRs cannot always be loaded into astronomical applications. If a large image you have taken fails to load into an astro-package there are a number of steps you can take. Firstly, you can try resaving the file in a different format. BMP, TIF and JPEG images should all be compatible with most image processing applications. If that fails, you can try simply reducing the image size; some older programs simply do not like multi mega-pixel images! Resizing the master and patrol images to 50% or smaller is easy in Photoshop or Paint-shop Pro. If you still get no joy, converting the image to a greyscale (i.e., Black & White) sometimes works. An associated problem that sometimes complicates the blinking process is if two images are a slightly different size. This can happen if the image has had to be rotated by a degree or two because your camera was tilted on its mount. When a rotation is applied and the image cropped, it may end up with a slightly different number of pixels. Careful cropping of the image, or increasing the background canvas size to a standard value, can sometimes solve this problem. I have successfully used the "align and blink" software in Richard Berry's *AIP4Win* software and this has a useful two star alignment routine which successfully eliminates rotational differences (see Fig. 5.4). Software Bisque's *CCDSoft* is fairly slick at aligning and blinking images that have no rotational misalignment. Of course, downloading freeware to do the job can save you an expensive disappointment, as you might find the pricey package you purchased will not align and blink your images. Fortunately most commercial packages now allow you to download a trial version which expires after a month so you can test the software prior to purchase. The *IRIS* software package is, perhaps, the most popular astronomy freeware and can be downloaded from:

http://www.astrosurf.com/buil/us/iris/iris.htm

IRIS is a partly command line driven system, that some amateurs, used to a more graphics oriented menu system may not like, but, it is powerful and designed by amateurs, for amateurs. For example, to quickly register two images simply by a horizontal and vertical alignment using one star in the field you use the command line:

QR [NAME1] [NAME2], where NAME1 and NAME2 are the image file names.

I have even come across amateurs who use Microsoft *Powerpoint* to blink images, treating them as if they were consecutive slides in a slide show. However, this is a tedious method too and, of course, *Powerpoint* only understands JPEG's BMP's and TIF's, not astronomical FITS format images. Ajai Sehgal, who is a member of Tim Pucket's Supernova Search

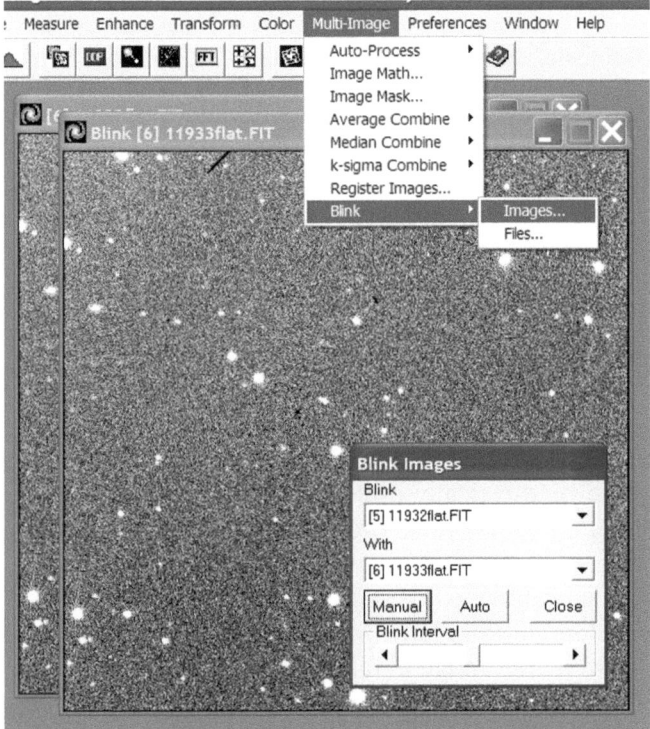

Fig. 5.4. AIP4Win's blink comparator can be found under the multi-image menu option.

Team has written a useful tool that is now incorporated into Diffraction Limited's *Maxim DL* image processing software. Sehgal's software requires you to have "before and after" images with the same file names in separate "before and after" directories. Selecting auto-blink will align and blink images of the same image scale. Pressing next (or back) will move to the next pair, in the two directories, that need checking. Alternatively, and free (!) you could download Dominic Ford's blink software *Grepnova* which can be found at his Cambridge University website at:

www-jcsu.jesus.cam.ac.uk/~dcf21/software.shtml

Grepnova was designed for supernova patrolling but will blink FITS files of starfields whether or not a galaxy is present.

The DC3 Dreams Visual Pinpoint software is also worth considering and more information can be found at:

http://pinpoint.dc3.com/

Reporting a Discovery

To get a comet discovery accepted as genuine you just have to convince one organization, namely the Central Bureau for Astronomical Telegrams, often abbreviated to CBAT. No-one uses telegrams or telex machines these days but the name has stuck. Apparently CBAT receives about five false comet discovery claims for every genuine discovery; this is a statistic that does not surprise me that much because, in connection with the UK magazine called "The Astronomer," edited by Guy Hurst, I used to check out loads of false alarms in the 1980s and early 1990s. The vast majority of these false discoveries were caused by the same problem: ghost reflections in optical systems caused by imaging near to bright stars. This issue is very common in lens based systems and, more often than not, the ghost image is found diametrically opposite a bright star, with respect to the field centre. Of course, comets move against the stellar background and the vast majority of newly discovered comets move faster than 10 arc-sec per hour, so at a typical CCD image scale of 2 arc-sec per pixel it does not take long to detect motion. However, ghost images can move with respect to the stars too if the image is taken with the bright star at a fractionally different position in the image. So, utmost caution is required when imaging anywhere near bright stars.

As any regular astro-imager will know certain parts of the sky are packed full with fuzzy galaxies that can look exactly like faint comets. Fortunately the Space Telescope Science Institute's Digitized Sky Survey (STScI DSS) offers an online archive of Palomar Sky Survey images to about magnitude 20 which can easily reveal whether a fuzzy patch should be in any part of the sky (see Fig. 5.5). This can be found at:

http://archive.stsci.edu/cgi-bin/dss_form

In addition, at any one time there may well be 20 or more distinctly non-stellar comets within typical amateur CCD imaging range, especially if apertures of 30 cm and larger are being used with a quantum efficient CCD camera. The presence of a comet in any region of the sky can easily be verified using the CBAT NEO and Comet checking page at:

http://scully.cfa.harvard.edu/~cgi/CheckNEOCMT

If you are convinced that the object in your image is not a ghost of a bright star, a galaxy, or a deep sky object, or even a known comet and that the object is definitely moving against the background sky then your next step should really be to contact some knowledgeable comet experts, especially those who are active CCD observers. It often transpires that a

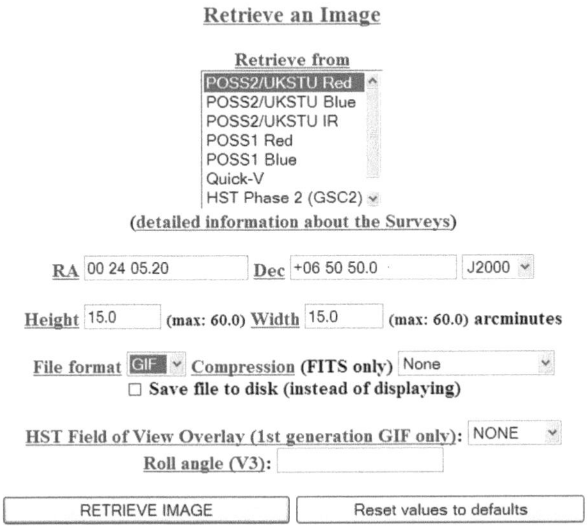

Retrieve an Image

Retrieve from

```
POSS2/UKSTU Red
POSS2/UKSTU Blue
POSS2/UKSTU IR
POSS1 Red
POSS1 Blue
Quick-V
HST Phase 2 (GSC2)
```

(detailed information about the Surveys)

RA `00 24 05.20`   Dec `+06 50 50.0`   `J2000`

Height `15.0` (max: 60.0) Width `15.0` (max: 60.0) arcminutes

File format `GIF` Compression (FITS only) `None`
☐ **Save file to disk (instead of displaying)**

HST Field of View Overlay (1st generation GIF only): `NONE`
Roll angle (V3):

`RETRIEVE IMAGE`   `Reset values to defaults`

Fig. 5.5. The Space Telescope Science Institute's Digitized Sky Survey is an invaluable online resource allowing images of any part of the sky to magnitude 20 to be downloaded.

second night's image is needed to confirm the reality of an object and in the days before the high speed Internet a series of cloudy nights after a discovery could cause endless frustration. However, now that we are living in an era of robotic telescopes that can be controlled via the Internet this is not a problem as remote imaging facilities located in very dark and cloud-free sites across the world can be accessed very easily. I will have much more to say about this in Chap. 13. If you can get some of the aforementioned knowledgeable comet experts to have a look at your discovery images, and they think you definitely have something unknown in your pictures, it will then be time to compose an e-mail to CBAT. For this purpose you will need to supply as much accurate information about the suspect as possible. For a start you will need to supply a very precise position of the object. If you are familiar with carrying out astrometric measurements that's great, but if not it should be possible to judge the suspect's position to an arc-minute or so of declination and a tenth of a minute of time in right ascension within the 2000.00 equinox system, using planetarium software such as Guide 8.0. Some sort of estimate of the object's appearance is also important, such as the diameter and appearance of the fuzzy coma, the approximate magnitude and the length and position angle of any tail. Also, you must supply the precise Universal Time (UT) when you made the observation. In Microsoft Windows™ the "Created" time (under mouse right click/properties/) of your original image will tell you when that image was created and, if your image was saved in the astronomical FITS standard the header information should also tell you the time relating to the start of the exposure and its precise duration. Remember, these times are only as accurate as your PC's clock so it is worth checking that it is correctly set to UT. The time should be quoted to decimals of a day to the nearest 0.001 of a day (each 0.001 of a day is equivalent to 1.44 min). You will also need to supply your name, your observing location and your telescope aperture and whether a CCD was used, along with the exposure duration. The checks that you have carried out to determine that the suspect is not a known object are worth listing too as well as the name of any reputable world-class observer who has been able to verify the object. In addition, if a second image was taken proving the suspect was moving then the motion should be given in units such as arc-seconds per hour, along with the direction. Note, in astronomy position angles are defined as: north = 0°; east = 90°; south = 180°; west = 270°. Please be aware that, in the heat of the moment it is easy to confuse north and south in astronomy and to forget if a star diagonal has been used thus mirror-flipping the view. When you are happy that all possible checks have been carried out you should contact CBAT. This can be achieved by e-mailing to cbat@cfa.harvard.edu, although at the time of writing the CBAT address seems likely to change to cbatiau@eps.harvard.edu.

Although most amateur discoveries have been made with CCDs and digital SLRs in recent years the occasional visual comet discovery is also made. With a visual discovery there is no "proof" until you can contact another comet observer of high renown or until you can convince a friendly CCD imager to help image the part of the sky you were looking at. Of course, if you have a proven track record of visual comet discovery that is an entirely different matter but if you do not have a surname like Bradfield, Levy or Machholz then that does not apply to you!

Of course, if you are using a portable telescope, like a Dobsonian, and there are no R.A. and Dec digital readouts then working out precisely where you are looking when that fuzzy patch makes your heart skip a beat is very tricky. These were the circumstances that presented themselves to Sebastian Hoenig when he discovered comet C/2002 O4, as described earlier in Chap. 4. However, with a bit of care and help from the Japanese that comet was confirmed. Nevertheless if you are seriously searching for comets you need to have a foolproof system for knowing precisely where your telescope is pointing and even Dobsonians can be equipped with digital setting circles.

Prior to the era of the remorseless discovery machines like LINEAR an amateur astronomer could pretty much discover a comet in any part of the sky by sweeping visually. However, now that the sky is being scoured remorselessly every clear night the rules of the game are rather different. The visual observer stands a far better chance of discovering a comet if he or she sweeps the sky at low altitude close to the twilight zone. The major patrols tend to avoid regions within 60 or even 90° of the Sun. The night sky only becomes astronomically dark when the Sun has set to at least 18° below the horizon. In addition, objects at less than 20° altitude will suffer greatly from atmospheric extinction and light pollution too, so even for objects directly above the Sun (above in this context meaning with respect to your horizon) you cannot seriously sweep closer than about 35–40° to the Sun. However, this is far closer than the machines dare to go simply because their software needs sharp high contrast images to compare with each other and any fields where the background sky becomes brighter or the star images have become distorted will be impossible for the machines to check reliably. Remember, the software in these automated patrols looks for moving star-like objects on a black background. If, by sheer luck, a comet's trajectory into the inner solar system, mixed with the position of the Earth in its orbit, means that a comet has appeared to "sneak up," hidden in the twilight glare then this is great news for the visual comet sweeper. Despite the highly technical era we live in a sentient human can totally outwit a machine when patrolling for a fuzzy object in a twilight sky.

Lessons from Hale–Bopp

I actually entitled this chapter "Finding the next Hale–Bopp with your gear" and I would like now to return to that particular comet as a case in point. Hale–Bopp was discovered visually by Alan Hale and Thomas Bopp on the night of July 22/July 23, 1995 in the USA at a magnitude of about 10.5. At the time of its discovery it was 7.1 AU from the Sun, 6.2 AU from the Earth and at a declination of −32°.

Alan Hale (see Fig. 5.6a) was, and still is, a highly experienced visual comet observer who had spent some time sweeping for comets prior to co-discovering Hale–Bopp. As I type these words he has observed over 450 comets visually! On the night of the discovery he was apparently not comet sweeping, but just making routine observations with a 0.41-m Newtonian from his observing site at Cloudcroft, New Mexico. He spotted the comet near to the globular cluster M70 in Sagittarius (see Fig. 5.6b). After making a few essential checks he contacted CBAT

Fig. 5.6. (**a**) Alan Hale, on the left, is one of the world's leading visual comet observers and the co-discoverer of comet Hale–Bopp. He is seen here talking to the prolific photographic era comet, nova and variable star discoverer Kesao Takamizawa, at the second International Workshop on Cometary Astronomy in Cambridge England. Image: Martin Mobberley. (**b**) When discovered, the monster comet C/1995 O1 (Hale–Bopp) was near the globular cluster M70 at the base of the famous Sagittarius tea pot shape.

alerting them that there was a suspected new comet near to the Messier object. The co-discoverer, Thomas Bopp, was not as experienced as Alan Hale but he had been observing deep sky objects from Stanfield Arizona with a 0.44-m f/4.5 Dobsonian when he too spotted the new comet near to M70. After checking a few star maps he contacted CBAT as well. Very quickly others confirmed the object and the comet Hale–Bopp became big news. Undoubtedly the great comets C/1995 O1 (Hale–Bopp) and C/1996 B2 (Hyakutake) were both discovered in a lull between the photographic patrol era (typified by the work of the Shoemakers, with David Levy working from Palomar) and the CCD era. LINEAR and NEAT were in the very near future but had not yet arrived on the discovery scene. Nevertheless, there are certain lessons to be learnt from the discovery of the most massive comet to come around since 1811.

When C/1995 O1 (Hale–Bopp) was discovered it was only moving at 23 arc-sec per hour in a westerly direction. It was close to opposition so 23 arc-sec per hour would have been its fastest speed that year as the relative motion due to the Earth's orbital velocity would have been at its most significant. This is pretty slow for a comet of magnitude 10.5 which, under normal circumstances, might have been expected to be moving at more than 100 arc-sec per hour as most comets of that brightness are well within the inner solar system and not midway between the orbits of Jupiter and Saturn. It was also immersed in the most star-packed constellation in the sky, namely Sagittarius.

Before long Rob McNaught of the Anglo-Australian observatory had located an image of Hale–Bopp on a 50-min UK Schmidt survey plate obtained by C.P. Cass on April 27, 1993. The comet had then been 13.1 AU from the Sun, around magnitude 18 and was moving at less than 5 arc-sec per hour. At its peak velocity in 1993 Hale Bopp would have been traveling at 12 arc-sec per hour.

It is interesting to consider whether Hale–Bopp might have been discovered by amateur CCD imagers in the twenty-first century, especially when it was around 13 AU from the Sun. Most automated patrols would have missed a motion of a few arc-seconds in the half hour span of their imaging sessions. In addition the Milky Way is so cluttered that the software may not even have detected the object, and some of the patrols positively avoid the Milky Way fields for this reason. Another factor at the time would have been that in 1995 there were no successful dedicated professional patrols searching that far south of the celestial equator anyway. In the twenty-first century amateur CCD imagers are discovering magnitude 19 and 20 NEOs and even a few NEOs that faint which have later revealed themselves as comets. So Hale–Bopp would have been detectable in the current era by any amateur with a large telescope and a sensitive CCD patrolling for very slow moving objects where many of the pro's daren't go, namely deep into the Milky Way. So discovering your own Hale–Bopp may not be as far-fetched as it first seems. There may well be a strategy here to beat the machines. Carry out deep nightly patrols within the Milky Way and then image the same regions a few hours later and blink compare them to weed out any really slow moving comets at a great distance. There could be another Hale–Bopp out there, or even a Transneptunian Pluto-like world hiding in the Milky Way. It is just a wild thought, but who knows where it might lead?

Visual Sweeping Techniques

Visual comet hunters are, not surprisingly, a dying breed. With professional CCD patrols routinely picking up 17th magnitude comets with ease (and they are by no means the faintest professional survey discoveries) how can a visual observer with a practical comet discovery limit of magnitude 12, possibly compete? Well, as we have already seen, visual discoveries are still made, although in many cases luck played an enormous part. When a tireless comet sweeper like Don Machholz takes six years to bag his eleventh comet you know things are grim. Even so, visual comet sweepers still have a huge advantage when sweeping close to, or within, the twilight zone and the fact that Bradfield, Levy, Machholz and Ikeya have succeeded in the LINEAR era shows that visual comet discoveries are still possible.

So, given that visual sweeping can bear fruit and is a low cost option it is well worth considering the methods used by the keenest visual comet discoveries that were so successful up to the late 1990s. The fact that professional machines remorselessly patrol the night skies in 2010 does not mean they will do so forever. Remember the NASA directive of June 2006: to discover 90% of Earth orbit crossing asteroids larger than 140 m in diameter. At some point this target will be seen to be achieved (whether it is or not) and, at the slightest excuse, politicians will slash budgets and pull the funding plug. So, the era of the visual comet discoverer could return. In addition, sweeping the skies with a decent aperture telescope can be a very relaxing way to enjoy the hobby, provided the observer does not become obsessed with the discovery and become totally de-moralized after every unsuccessful night. I spent several years comet sweeping in the early 1980s and I was hungry for discovery at first. However, after a while I learned how to love the act of sweeping in itself. You pick up lots of deep sky objects doing it and really learn your way around the sky, so much so that you realize after a few years that you just instinctively know where things are in the sky without even using the finder telescope. If you can enjoy sweeping the skies for pleasure, regardless of any success, then you can survive as a comet sweeper. If a discovery is the be-all and end-all you will become demoralized and stop sweeping after a year has elapsed.

So, bearing these facts in mind, what can we learn from the comet sweepers of the past? The first thing to consider is what was the most popular telescope type and size to use for discovering comets in the visual era? Well, undoubtedly, the 25×150 Fujinon binoculars, so popular amongst the prolific Japanese hunters were the gold standards. Apart from the fact that anyone with good stereo vision will feel more comfortable using both eyes there is also the fact that these sort of binoculars have eyepieces angled at 45° such that, unlike with normal binoculars, you do not have to crick your neck to painful angles to sweep the sky. When sweeping for comets at altitudes of 45° above the horizon you will be looking horizontally into the eyepieces and at lower altitudes you will be looking slightly down. This is a far more sustainable activity than staring upward. Other comet sweepers have used mountings where a telescope is used but where the eyepiece barely moves at all. The ultimate version of this is called the Coudé mounting where, typically, a long focus refractor has its light diverted by mirrors through the declination and polar axes such that the observer sits comfortably in a chair and looks through a right angled star-diagonal sticking out of the bottom of the polar axis. This is all very nice, but expensive to make and fiendishly complex to make yourself. In practice the simplest way to make a telescope in which the eyepiece barely moves is to use a Newtonian on an alt-azimuth mounting where the altitude axis is attached to the tube directly opposite to the eyepiece position. I mentioned this concept in Chap. 4 when discussing Bill Bradfield's 2004 comet discovery and his specially counter-balanced 25-cm Newtonian (see Fig. 5.7a, b). With an alt-azimuth design, once the optimum eyepiece height is determined a mount can be designed where a pipe or shaft is attached to the tube, opposite the eyepiece such that as the elevation is varied the eyepiece simply rotates. For azimuth rotation the observer will still have to walk around the telescope but for the vast majority of the slow vertical-sweeping time the observer does not have to move and can even remain seated as long as the mirror end of the telescope does not hit his or her legs. At the top or bottom of each slow sweep, the azimuth position can be moved by a field width (say, a degree or two) and the next sweep can commence. The New Zealand amateur Rodney Austin's second comet sweeping telescope utilized this fixed eyepiece method. It was a relatively

Fig. 5.7. (**a**) Bill Bradfield and his 250-mm alt-az mounted reflector which is balanced around the eyepiece point by a concrete counterweight for maximum ease of use. His last four discoveries were made with this instrument. Formerly known as "The Wizard of Dernancourt" he has now moved from Dernancourt to Yankalilla in South Australia, a nearby town. Image by kind permission of Reinder Bouma. (**b**) Another view of Bill Bradfield and his 250-mm alt-az mounted reflector. Image by kind permission of Reinder Bouma.

lightweight 20-cm f/4 Schmidt–Newtonian which used a 2× Barlow lens to achieve focus at f/8 (the commercial telescope was primarily designed for photography) and a 40 mm eyepiece (50 mm barrel diameter) at 41× magnification. The usable field was approximately 1.4°. Of course, Newtonian telescopes are heaviest at the mirror end and so if the azimuth axis is as high up the tube as the eyepiece a counterweight of even greater mass than the telescope mirror needs to be hanging ahead of the eyepiece point, depending on exactly where that counterweight is positioned.

There are various ingenious ways to balance such a telescope but the lighter the optics the less you need to acquire in weights. Most sports shops or sporting internet sites will sell bodybuilders weights at a reasonable cost. Many of the modern dumbbells are made from a dense plastic material which is less painful if you drop it on your foot than a lead weight and the weights are relatively affordable too. They are ideally suited for lashing up into an alt-azimuth counterweight system. Precisely how fast a visual comet hunter sweeps the sky is down to them and their equipment, but it goes without saying that the faster you sweep the quicker you can cover those vital regions bordering on the twilight, but the higher your chance of missing a comet too. Before the machines started their remorseless scans of the sky in the late 1990s there was a growing trend for amateurs to use large 250–400 mm aperture Newtonian or Dobsonian reflectors, rather than the relatively small 100–150 mm apertures of the 1950s, 1960s and 1970s. Undoubtedly the aperture fever of the 1980s was partly responsible but, as well as that factor, the old idea that big reflectors covered too small an area of the sky had been proved wrong by discoverers like David Levy, Rolf Meier and Kazimieras Cernis who all had successes with large apertures. The simple fact was that there were so many more comets available to find if you could sweep down to 12th magnitude.

The famous American comet discoverer Leslie-Peltier (1900–1980), who discovered 12 comets and two novae, built himself a very cozy "Merry-go-round" observatory which must have been the ultimate comfortable comet sweeping (and variable star observing) arrangement. A short focus 6-in. f/8 refractor was attached to the observatory structure and it pivoted in altitude about a point where the observer's ears would be, to minimize head movement. In addition the entire 6-foot by 6-foot by 5-foot high observatory moved in azimuth by means of a steering wheel in front of the observer: sheer observing bliss!!

The choice of magnification when sweeping the skies is important too. The pupil of the human eye can only dilate to around 8 mm diameter as a maximum value when dark adapted and for many observers, especially those in middle age, this value will be much less. In practice this means that you cannot ever use a lower magnification than 1× for every 8 mm of telescope aperture or the bundle of rays trying to enter your pupil will be too big. A much safer value to adopt, especially if you are not fully dark adapted, would be 5 mm in which case you should avoid lower magnifications than 1× for every 5 mm of telescope aperture or, roughly, 5× per inch. So, for a 15-cm comet sweeping telescope a minimum magnification of 30× is sensible and for a 20-cm telescope, 40×. Of course, anyone sweeping the skies will be tempted to use the minimum magnification available, because lower powers mean a wider field of view for the same eyepiece. Another factor here is whether to acquire an eyepiece with a 2-in. (50 mm) barrel or to stick with the standard 1.25-in. (31.7-mm barrel) size. In practice, if your telescope is faster than f/4.5 or so you can save money by not going to the larger eyepiece size. A few simple calculations will show that at f/4.5, with a magnification of 5× per inch and a 60° apparent field eyepiece, the 27-mm field lens diameter of a 31.7-mm diameter eyepiece barrel is just wide enough to deliver that 60° apparent field. So, with this information, is there a commercial eyepiece that would deliver a 60° apparent field and 5× per inch magnification at f/4.5? Well, the nearest quality eyepiece that fits the bill is the TeleVue Panoptic 24-mm eyepiece. With an 8-in. (20.3 cm) f/4.5 Newtonian it would deliver a magnification of 38×, and with its 27 mm field lens placed at the focus it would capture a 1.7° field, equivalent to a 64° apparent field. OK, this is a fraction under the quoted apparent field of 68° but let's not quibble! Most comet hunters would be more than happy with a 1.7° field on an 8-in. telescope, especially with a 5.3-mm diameter bundle of light rays entering their pupil. At f/4.5 this specific eyepiece would work efficiently with any telescope in delivering a 64° apparent field. Double the aperture to 16 in. and the real field halves to 0.85° but there will always be an inverse ratio between aperture and real field of view. It's a choice: you can spot fainter comets but scour less sky, or scour more sky but lose those fainter comets. The decision is yours. In recent years the bigger apertures have tended to bag more comets, especially when lucky discoveries with giant Dobsonians, used for deep sky work, have been made at star parties.

The only other thing to say about visual observing is try to keep warm! Here in the UK at least, visual observing in January and February can be nothing less than a battle to keep frostbite at bay, global warming or not! Some quality thermal clothing of the type mountaineers wear can be a sound investment and keep you going even when the telescope tube is glistening with frost.

Sometimes you Just Get Lucky!

Anyone who knows me well would know that I would be the last person to depend on good luck for a comet discovery, or for anything at all. I assume that bad luck and sod's law rule the universe and any day where they don't strike me is a good one! The advantage of being a pessimist in life is that you can never be disappointed, as you expect things to go wrong! However, as cynical as I am I have to admit that good luck has played a part in quite a few comet discoveries. We saw earlier in the book that comets have been found while casually gazing at deep sky objects. Hale–Bopp was the ultimate example I guess, but there is also Vance Petriew's comet, discovered by offsetting from the wrong star to try and find the Crab Nebula, and bagging comet Petriew. Then there is Steve Lee's 1999 Mudgee star party discovery of C/1999 H1. Then there are the two lucky variable star observer finds of Albert Jones (twice, while observing U Pav and T Aps), Kaho (who bagged Kaho–Kozik–Lis in the eyepiece field of R LMi in 1936) and Justin Tilbrook and his comet, C/1997 O1, which conveniently placed itself in the field of TV CrV. There are also those occasions where a brand new instrument, on its first night of use, has magically conjured up a comet at Christmas time, specifically Juels and Holvorcem's December 28, 2002 find and Mike Candy's December 26, 1960 discovery. Maybe new telescopes have special powers when first turned onto the sky? As far as lucky comet discoveries are concerned I suppose we might even include planetary imager Anthony Wesley's observation of July 19, 2009 where he bagged the bruise on Jupiter after a mystery comet hit the giant planet for the second time in 15 years. OK, he actually discovered a comet just after it had committed suicide, but maybe it is simply a special case?

A truly amazing set of lucky breaks surrounded the discovery of comet C/1985 T1 (Thiele) which was originally designated as 1985m when discovered by Ulrich Thiele of the Max-Planck-Institute of Astronomy in the early hours of October 9, 1985. He was taking photographs of comet 1P/Halley with the Hamburg Schmidt telescope based at the joint Spanish-German Calar Alto observatory in the Sierra de Los Filabres region of Andalucía in Southern Spain. Of course, loads of amateurs and professionals were also imaging comet Halley, at that time. It was brightening rapidly and on that date it was precisely four months before perihelion. Comet Halley was a tenth magnitude object in the extreme northeastern corner of Orion on October 9 and it was not far from the border with Gemini. On the night in question it was passing just below the open cluster NGC 2175 and its associated nebulosity.

With considerable cloud and fog at Calar Alto that night Ulrich decided to shut the observatory down just after 0400 UT. However, as is so often the case in these circumstances, as soon as he stepped outside the skies had cleared and so he went back into the dome and powered the 0.8-m aperture f/3 Schmidt back up. In his rush to take some pictures, before the cloud and fog re-formed and before dawn twilight commenced, Ulrich forgot to allow for precession from the 1950 coordinates of the comet to the 1985 values. Ulrich exposed two photographic plates centered on 0435 and 0513 UT. The field of view of the Schmidt was a wide five and a half degrees but the precession error would move the field center from Halley's precise position to 30 arc-min west of the field center. Of course Ulrich was totally unaware that another, as yet undiscovered comet, was in the region of Halley and precisely half the width of the Schmidt plate field west of that famous comet. Here's the scary bit: if Ulrich had remembered to allow for precession then the new undiscovered comet would have fallen exactly on that western edge of the plate or maybe beyond it! As things turned out though the 30 arc-min error meant the new comet lay on the plate, some 30 arc-min east from that western edge and Halley lay 30 arc-min east from the field centre. In addition, the very fact that Halley was within the nebulosity associated with the star cluster NGC 2175 caused Ulrich to search for Halley on the plate and so spot the new comet. In normal circumstances he would have spotted the tenth magnitude Halley instantly and not searched the plate at all! The other bit of luck was that immediately after Ulrich exposed the photographic plates the sky was cloudy and foggy once more. He had just bagged the new object in a lucky break in the weather and if he had not gone back into the dome just after 0400 UT, and not forgot to precess the 1950 coordinates there would have been no comet named Thiele 1985m alias C/1985 T1 (Thiele). Sometimes sod's law goes away for the night and something opposite to it comes into force.

    The normal sod's law rule applied to me on December 25 (yes, Christmas day!), 1980. I had taken delivery of a 14-in. (356 mm) telescope only 18 days earlier and that was my first clear moonless night of use with the telescope and the third night of use in total. Having an f/5 Newtonian focus I decided I would use the telescope partly for comet sweeping at around 44× magnification. I decided to start my first ever comet sweep by centering the telescope field on the famous double–double star epsilon Lyrae. However, the eyepiece position in that part of the sky was just too tricky and so after a few attempts to get comfortable I decided to sweep in a more comfortable part of the sky. I found out a few days later that Roy Panther, of Walgrave, Northampton (see Fig. 5.8), had bagged his

Fig. 5.8. A rare conjunction of two British discoverers in December 2000. Roy Panther, on the left, discovered his comet on Christmas Day 1980, 20 years (minus 1 day) after Mike Candy's 1960 comet discovery. Roy swept up Comet Panther 1981 II (C/1980 Y2) very close to the double star Epsilon Lyrae; he was using his 0.3 m altazimuth Newtonian. After 33 years and over 600 h sweeping he was due for a discovery. The comet reached perihelion ($q = 1.66$ AU) on January 27, 1981 and passed extremely close to the pole star. On the right is the supernova discovery phenomenon Tom Boles who, at the time of writing, is the world's most successful individual supernova patroller with an amazing 126 discoveries to his name.

first comet discovery at the same time I was trying to point my telescope at epsilon Lyrae. The comet had been just a degree southeast of that famous multiple star and a sitting duck if I had persevered and moved the telescope a degree to the southeast on my first ever comet sweep! Is it just me or is there something strange about brand new telescopes, comet discoveries and the Christmas period! I might add that Roy Panther was not connected to the telephone network when he made his discovery on Christmas day 1980. He wrote his discovery down on a piece of paper and managed to get a local taxi driver to deliver the discovery details to the editor of "The Astronomer" magazine, Guy Hurst, who lived in the same town, Northampton!

I cannot end this chapter without relating another lucky comet discovery that took place right near that famous double-double star Epsilon Lyrae, some 34 years earlier. The eagle-eyed comet observer Anton Weber made an independent discovery of the sixth magnitude comet C/1946 K1 (Pajdusáková-Rotbart-Weber) while he was actually sitting on the toilet in his Berlin-Steglitz home on 1946 May 31, and gazing through a hole in the lavatory wall at a point just a few degrees northeast of the bright star Vega! The hole had been created by military activity during World War II that had blown out the lavatory window which, fortunately, had never been replaced. So, it is now obvious what is needed to become a successful comet discoverer. You need to blow your lavatory window out with high explosive!

CHAPTER SIX

Comets that Have Been Missed by the Pros

By definition, any comet that has been discovered by an amateur astronomer must have been missed by the professional patrols, but there have been quite a few cases in recent years where the automated discovery machines really should have bagged the comet first. Examining a few of these intriguing misses by the likes of LINEAR, NEAT and the Catalina Sky Survey should help in understanding how amateur astronomers can exploit the weaknesses of the professional sky patrols. As I mentioned in Chap. 4 some of the years since 1998 have been especially successful for amateur comet discoverers in a period where both LINEAR and NEAT were proving highly efficient. So let us now examine some of those discoveries again from the viewpoint of how they were missed by the pros. There will be some duplication with Chap. 4, but now that we have examined amateur search techniques in detail it is time to revisit a few of the recent comets that have sneaked past the professional nets. This time we will try to work out just how they sneaked in the back door.

2000

I would like to start in the year 2000 when LINEAR and NEAT were at their most productive and, to begin with, look once more at the discovery of C/2000 W1 Utsunomiya–Jones. This was a comet with a very faint absolute magnitude ($H=11$) that moved into visual detection range in twilight, due to exponential brightening caused by its very small perihelion distance of 0.32 AU (reached on December 26, 2000). In April 2000, this feeble comet would have been about 20th magnitude as it sank within 90° elongation of the Sun, placing it out of detection range of LINEAR and NEAT, because professional telescopes search in dark skies and well away from the horizon. With the comet south of the Sun, in the northern hemisphere summer, LINEAR and NEAT would have had little chance of bagging it unless its elongation improved or it moved north or, at least, north of the Sun. The fact that the Siding Spring arm of the Catalina Sky Survey had not yet been established made the southern hemisphere a fairly safe place to hide in 2000. C/2000 W1 had a sneaky trajectory and as its solar elongation improved in autumn 2000, and it started to brighten rapidly and move out of twilight, it plunged southward. When Utsunomiya-san spotted it at magnitude 8.5 it was already beyond LINEAR's limit (LINEAR cannot search below Dec −30); a week later, when Jones spotted it, it had plunged to Dec −77. So C/2000 W1 avoided detection by being feeble, keeping close to the Sun, and plunging far into the deep southern sky, as it rapidly brightened towards its small perihelion point (see Fig. 6.1). A few years later it may well have ended up with the name McNaught, Garradd or Siding Spring, but it would be Utsunomiya and the 80 year old Albert Jones who saw it first.

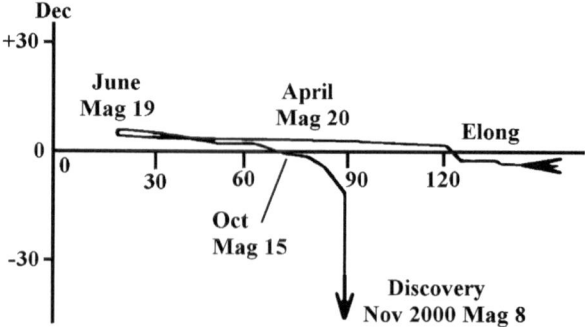

Fig. 6.1. The track of comet Utsunomiya–Jones' position with respect to its elongation from the Sun (x-axis) and declination (y-axis) in the months leading up to its discovery.

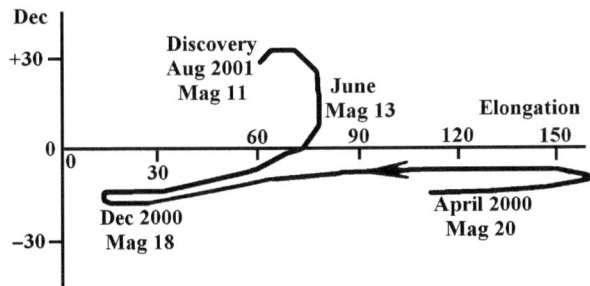

Fig. 6.2. The track of comet Petriew's position with respect to its elongation from the Sun (x-axis) and declination (y-axis) in the months leading up to its discovery.

2001

C/2001 Q2 Petriew would have been about magnitude 19 when it dropped below 90° solar elongation in the autumn of 2000, keeping just south of the Sun. As the comet brightened to magnitude 15 or 16 in the spring of 2001, it was still too close to the Sun for detection by LINEAR or NEAT, although it is perhaps a bit surprising that by June, at nearly 90° elongation and at a similar declination to the Sun, it was still not picked up (see Fig. 6.2). Of course, comets sometimes brighten more rapidly than expected just before discovery, and some have an abnormal outburst, which could be a factor. However, there does seem to be a general rule that both LINEAR and NEAT did not usually nab comets within 90° of the Sun, even though, in theory, they could patrol down to quite faint magnitudes in regions elongated by only 50°. Anyway, as it transpired Vance Petriew bagged his comet at Dec +28 and 11th magnitude at a solar elongation of about 60°. This was another intrinsically feeble comet and with a q of only 0.95 AU, so it is just as well he was using a large aperture at the time!

2002

Now we come to the most spectacular and famous comet to avoid detection in the LINEAR and NEAT era: the superb C/2002 C1 Ikeya–Zhang. With an absolute magnitude of around 6.5, this was certainly not a feeble comet, so it was well within LINEAR's detection magnitude range

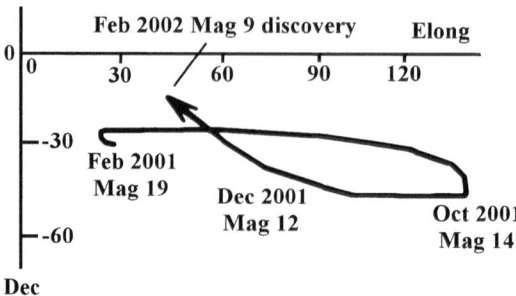

Fig. 6.3. The track of comet Ikeya–Zhang's position with respect to its elongation from the Sun (x-axis) and declination (y-axis) in the months leading up to its discovery.

for about a year prior to discovery. It was also well elongated from the Sun (over 130° in the fall of 2001, at about magnitude 14). However, it was just on, or below LINEAR's southern limit at this time and by the time it moved back above Dec −30, it was only 40° from the Sun (see Fig. 6.3). This is yet another comet that surely would have been found by McNaught or Garradd in later years.

As we saw in Chap. 4 C/2002 E2 Snyder–Murakami was swept up visually. The telescopes used were big ones though, with a 50-cm f/5 Obsession Dobsonian and a 46-cm f/4.5 Newtonian being used when the comet was magnitude 12. Snyder was, apparently, in his 70th hour of deliberately sweeping for comets. This comet was determined to avoid detection! For one full year before its discovery, the comet C/2002 E2 either kept *well* below Dec −30, or well within 90° solar elongation, and often both. It was just crossing into the northern hemisphere when it was discovered, at just more than 60° solar elongation (see Fig. 6.4). A very sneaky comet indeed!

C/2002 F1 Utsunomiya was just as crafty too. It metaphorically teased LINEAR and NEAT as it spent most of its pre-discovery year crawling just under their southern limit. In January 2002, C/2002 F1 was only elongated a few degrees from the Sun, even if it had crawled above Dec −30. Like Snyder–Murakami, it was discovered as it moved into the northern hemisphere, but at a much smaller solar elongation of 30°.

Now we come to that German comet discovery, C/2002 O4 (Hoenig). Well, even now, quite how the automated patrols missed Sebastian Hoenig's comet is a bit of a mystery. The vanishing point of the comet's orbit, and therefore the origin of the incoming spirals marking its inward track, lives in the far southern hemisphere, around 21 h 50 m and Dec −65

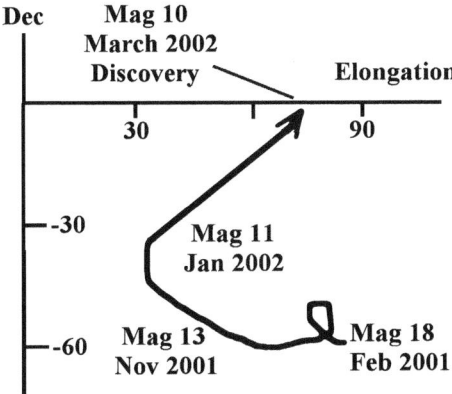

Fig. 6.4. The track of comet Snyder–Murakami's position with respect to its elongation from the Sun (x-axis) and declination (y-axis) in the months leading up to its discovery.

in the constellation Indus. From October 2001 the comet moved steadily north. From March to June 2002 it would have tracked north through Aquarius passing within a degree of the Helix nebula on April 2, but at a solar elongation of only 42°. In June comet Hoenig would have moved north through Pisces and from mid-June its solar elongation would have been more than 90° from the Sun. With the new Moon occurring on June 10 and the comet then moving at 65 arc-sec per hour it was surely an easy catch for LINEAR and NEAT? It was also sailing due north through especially barren starfields at the time, right through the Square of Pegasus (see Fig. 6.5), and yet it was found by an amateur as late as July 22 while at a solar elongation of 109°. Possibly this comet discovery, more than any other, suggests that the automated patrols do occasionally miss sitting ducks. After all, even New Mexico is not cloud free every night and heavy rains can fall in the region in July and August. The Catalina Sky Survey, along with many other observatories in the American Southwest, use the July and August "monsoon season" to carry out maintenance work on the patrol telescopes so they can be maintenance and upgrade free in the September to June clear sky periods. This is an important lesson to be learnt by any amateur comet hunter. Even places like New Mexico do have cloud and rain so the July to August period should be exploited to the full by amateur hunters. Keeping a daily check on the weather in these regions is very worthwhile as in any competition you have to exploit your opponents weaknesses to the maximum! In addition comets can, near to perihelion, suddenly jump in brightness by a few magnitudes in

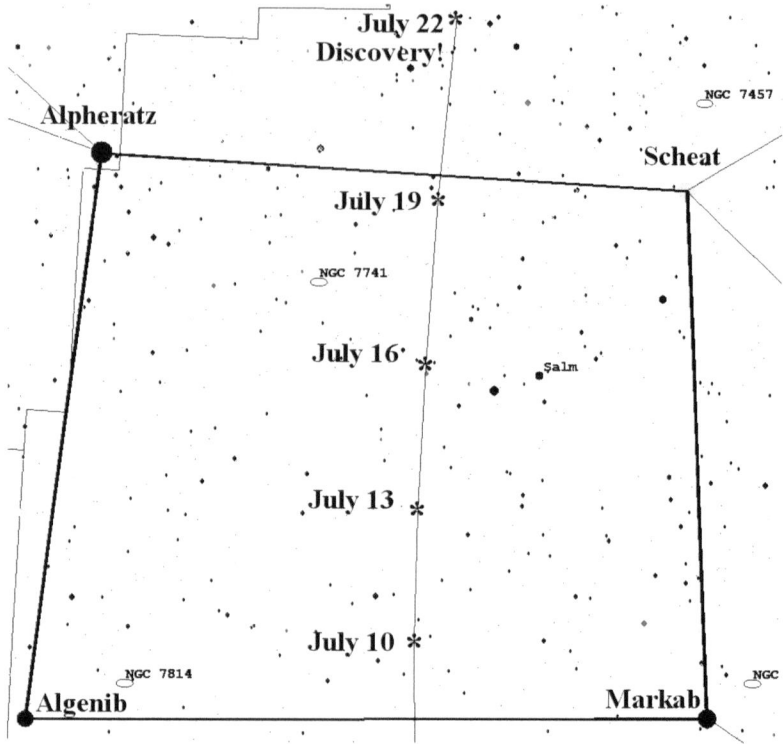

Fig. 6.5. In the weeks prior to its discovery, in July 2002, comet Hoenig tracked right through the Square of Pegasus and should have been a sitting duck for the automated patrols.

a week. Also, sometimes equipment maintenance downtimes, poor skies and a comet entangling itself with other deep sky objects, can simultaneously play into amateurs' hands. Nevertheless, it is by no means clear how C/2002 O4 beat the machines.

Then, 5 months later, we have the strange case of C/2002 X5 (Kudo–Fujikawa) which was a bright ninth magnitude object near the Bootes/Hercules border when discovered in a dark sky on December 13. In CCD images taken immediately after discovery it already had a 5.5 arc-min coma and an 18 arc-min tail and it was moving at a very rapid 230 arc-sec per hour! Terry Lovejoy of Queensland Australia, soon to become a comet discoverer himself, searched back over the SWAN images taken by the SOHO satellite detector 1 month earlier and found faint images of 2002 X5 on November 6 to November 13 when it appeared to be approximately magnitude 10 or 11 near the Canes Venatica/Ursa Major border.

Its motion was roughly 100 arc-sec per hour due east at that time. It should be stressed that the comet was not near the Sun this time which is where the main SOHO detector concentrates; SWAN is a different instrument. So for the previous 5 weeks or more 2002 X5 had been an obvious target. For that whole pre-discovery period the comet had been between +45° and Dec +48 although its elongation from the Sun had arguably been less favorable at around 75°. Analyzing 2002 X5's inward journey in detail prior to discovery it is fair to say that it sneaked in over the top of the Sun from our perspective (with an orbital inclination of 94°) and from April to December it stayed closer to the Sun than 90°. Back in March 2002 it would have been well placed and 110° from the Sun but was probably below magnitude 18 and so right on the limit of LINEAR's detection threshold for a diffuse non-stellar object. But it is interesting to note that this comet did not avoid detection by sneaking in from the southern hemisphere skies.

We saw in Chap. 4 how C/2002 Y1 (Juels–Holvorcem) had been discovered on December 28, the first night of use of a 12-cm refractor with a CCD covering a 2.3° field. At 15th magnitude this was a much fainter comet than Kudo–Fujikawa: 250 times fainter in fact! However, C/2002 Y1 did have a similar solar elongation at discovery, namely 84° in the morning sky. The motion at discovery was 88 arc-sec per hour northeast. Go back 2 months prior to discovery and the solar elongation was less than 40° and the magnitude would probably have been below 18, so Juels and Holvorcem bagged it pretty shortly after it had brightened enough and escaped the twilight sufficiently to be discoverable.

2003

Ten months later, at a declination of −57°, Vello Tabur snapped up comet C/2003 T3 from Australia on October 14. Tabur's find was in the year prior to the regular patrols being carried out by Rob McNaught and Gordon Garradd as part of the Catalina Sky Survey's Siding Spring operation. However, it should not be thought that there was zero competition in the southern hemisphere at that time; after all, Bill Bradfield was still active. Nevertheless an efficient professional CCD machine would not be fully patrolling the skies from Australia until the Siding Spring Survey began in April 2004. Vello Tabur's comet was moving due north at 48 arc-sec per hour when discovered and was elongated by 89° from the Sun in the evening sky. For virtually all its incoming life comet Tabur had been way south and below a declination of −60. Its vanishing point, the radius of its incoming spiral track across the sky, is centered near to Beta Centauri, close to Dec −60. So, one can safely say that C/2003 T3 beat the machines because LINEAR and NEAT could not reach below roughly Dec −30. and the Siding Spring Survey would not be up and running for another year.

2004

Whether the legendary Bill Bradfield deliberately decided to intensify his efforts to bag an eighteenth comet in the year leading up to the Catalina Sky Survey's Siding Spring outpost starting its patrols, I know not, but he finally ensnared a twenty-first century comet in the weeks before that patrol started. However, C/2004 F4 was not discovered in the far southern hemisphere, as on March 23 of that year it was actually found, at eighth magnitude, just 1° below the celestial equator in northern Cetus. It was moving at 46 arc-sec per hour, but only 25° away from the Sun. This is most definitely territory where the pro's cannot search and, even for the "Wizard of Dernancourt," it was a remarkable find. Remarkably, this comet did not sneak under the automated surveys' "radar" by keeping well south of their patrol regions; it had spent most of the years prior to its discovery in the northern hemisphere constellation of Aries with its vanishing point close to the star Metallah near the Triangulum/Aries/Pisces border. From late August to December of 2003 Bradfield's comet should have been a sitting duck for all the automated patrols of that era (see Fig. 6.6a) as it would have traveled between 180 and 100° from the Sun and should have been above magnitude 17! Easy fodder for LINEAR, surely? Maybe one can find an excuse around the new Moon period in late August and early September, in as much as comet Bradfield momentarily reached a stationary point in its apparent motion northeastward, prior to swooping southeast during September to January? Any patrol images taken around that time would have imaged the comet crawling at just a few arc-sec per hour, a speed that would have been imperceptible to the software? These stationary points are caused when the effects of the comets incoming path are briefly negated by parallax due to the Earth's own motion in orbit. However, that can only explain away the comet being missed for one patrol month. Then again, as David Levy once said "Comet's are like cats: they have tails and do what they want." Maybe comet Bradfield had an outburst prior to discovery and was very faint even as late as December 2003? C/2004 F4 certainly beat the machines, but, like Sebastian Hoenig's comet it is hard to know why! An image of C/2004 F4 (Bradfield) in the field of the SOHO LASCO detector is shown in Fig. 6.6b and I will have more to say about SOHO in Chap. 8.

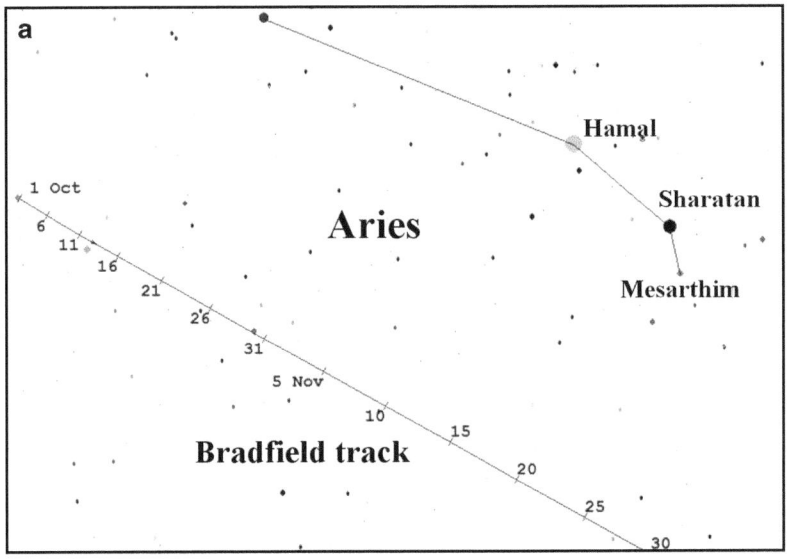

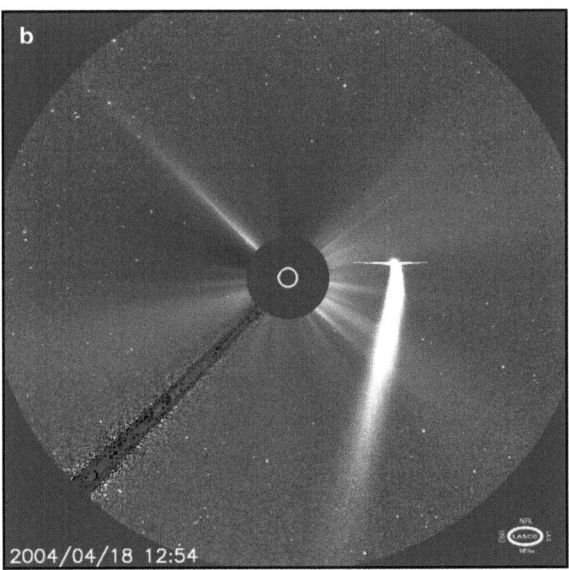

Fig. 6.6. (**a**) During October and November 2003 comet C/2004 F4 (Bradfield) was well-placed in Aries and should have been bright enough for the automated patrols to detect. (**b**) Comet C/2004 F4 (Bradfield) on April 18, 2004 at 11:54 UT, as imaged by the SOHO LASCO detector (see Chap. 8) just 4 weeks after discovery. The comet had reached perihelion the day before at 0.17 AU (25 million kilometer) from the Sun. Credit: SOHO Consortium, LASCO, ESA, NASA.

I have already dealt with Roy Tucker's discovery of C/2004 Q1 (Tucker) in Chap. 4 and it is very likely the machines were beaten on this occasion by torrential rain in New Mexico. This brings us once more to Don Machholz's tenth comet C/2004 Q2 which was at a solar elongation of 96° when discovered and traveling at 48 arc-sec per hour. In the 6 months prior to discovery of the comet, it had journeyed through Sculptor and along the Fornax/Cetus border region before finally arriving in Eridanus. Its elongation had been quite small since February, when it was closer to the Sun than 40°, only clearing 80° elongation from July onwards. This comet, like many missed by professional patrols, had hugged the twilight zone for many months prior to discovery and with the Siding Spring survey in its infancy the comet evaded their early searches, only to finally be captured by a Californian amateur.

2005

C/2005 N1 (Juels–Holvorcem) was another twilight hugging comet, a veritable master at hiding itself in the solar glare. For the early months of 2005 it never strayed more than 50° from the Sun and dropped below 20° for much of the February to April period. It was finally bagged on July 2 when it had brightened to magnitude 14.5 even though still only 46° from the Sun, while in the southern part of Perseus. Yet again it is obvious that searching the twilight zones must be one of the most powerful techniques available to amateur comet hunters.

On the other hand the next amateur comet discovery, P/2005 T5 (Broughton), was discovered in a dark sky, 144° from the Sun, on October 9 of that year and was a faint 18th magnitude at discovery. It was crawling along at 22 arc-sec per hour at the time, while a full 2.4 AU from the Earth and 3.3 AU from the Sun. The reason an amateur bagged that find was that Broughton's home-made 0.51 m f/2.7 imaging system (see Chap. 12) has enough light grasp to compete with the professional surveys.

2006

The same light grasp advantage applied to Broughton's next find, C/2006 OF2 (Broughton), another 18th magnitude comet close to opposition and close to the zenith from his Australian site.

A really superb comet that sneaked past the professionals was C/2006 M4 (SWAN) and I will have more to say about the discovery and performance of this comet in Chap. 8. In the year prior to its discovery in early July 2006, comet C/2006 M4 (SWAN) spent most of its time in the deep south, between Dec −50 and −60 where patrols are less intensive; more importantly though, from mid-July 2005 to the start of 2006 the comet stayed within 80° of the Sun. It briefly swung out to a maximum 111° from the Sun in March 2006, but then swiftly closed back down to less than 40° elongation at discovery. Another factor here was that the comet appears to have had a less than exciting absolute magnitude of around seven prior to discovery, even though in late October 2006 it outburst to an actual visual magnitude of 4.3, with a superb ion tail. This was a rare discovery in images taken by the SWAN detector on the SOHO spacecraft and the main patrols seem to have missed it because it traveled from the less well patrolled southern hemisphere and, like so many comets missed by LINEAR et al. kept close to the twilight glare.

David Levy's visual discovery of a periodic comet, P/2006 T1 (Levy), was, as mentioned previously, almost certainly due to the comet experiencing a significant outburst just prior to David's comet sweep across the region. Its elongation of only 48° at discovery, while in Leo in the morning sky, would also have precluded a professional survey picking it up.

2007

So what about Terry Lovejoy's two discoveries of 2007? What factors enabled them to be captured by an amateur using nothing more than a digital SLR and a 200 mm lens? Well, his first discovery, C/2007 E2 (Lovejoy), had been within 90° of the Sun for 4 months prior to discovery, despite passing not far from the southern celestial pole. When discovered on March 15 its elongation had increased beyond the 60° point. Prior to its passage into that safe zone, within 90° of the Sun, it was probably magnitude 16 or fainter (in late 2006) and may have been close to the limit of the Siding Spring survey. Terry's second discovery, C/2007 K5 (Lovejoy) stayed less than 40° from the Sun from February to mid-May 2005, and when discovered at 13th magnitude, on May 26, had just started to emerge from the twilight at an elongation of 44°. So both his discoveries were bagged as they started to clear the twilight and attain sufficient altitude above his horizon in a dark sky. You will not find any professional Earth-based patrols searching at 44° elongation from the Sun.

At this point I would like to mention the discovery of comet C/2007 N3 (Lulin) despite the fact that it was discovered by professional astronomers and so, if working strictly by the title of this chapter, was not missed by them.

Comet Lulin turned out to be an unusual and bright comet and the best comet of 2009. It reached a magnitude of 4.5, with a tail several degrees long and a head some 25 arc-min across (see Fig. 6.7), but even rarer than that was its orbital inclination of 178.4° which meant it exhibited a permanent anti-tail type dust spike for the months of January to March 2009. It orbited virtually in the ecliptic plane (going around the Sun backwards) despite being a long period comet. On February 24, 2009 it passed only 0.41 AU (61 million kilometer) from the Earth. I would like to include it here simply because it was a bright and unusual comet that avoided the major patrols and therefore could just have easily been discovered by an amateur. The name of the comet comes from the Lulin Observatory at Nantou, Taiwan.

Comet Lulin was discovered a full 18 months prior to perihelion, on July 11, 2007, and was initially thought to be an asteroid by Quanzhi Ye of Sun Yat-sen University, Guangzhou, China, on three CCD images obtained by Chi-Sheng Lin of the Institute of Astronomy, National Central University, Jung-Li, Taiwan. Chi-Sheng Lin had been using the

Fig. 6.7. Comet Lulin was the best comet of 2009 but was not discovered by the major sky patrols. The figure shows the comet passing to the west of the bright star Regulus in Leo on February 28. Imaged remotely using a Global Rent-a-Scope Takahashi FSQ 106ED f/5 refractor sited in New Mexico. This was a 180 s exposure with an SBIG STL11000M CCD. The field is 2½° wide. Image: Martin Mobberley.

41-cm Ritchey–Chretien at Lulin Observatory (Nantou, Taiwan). Hence the name Lulin is that of the observatory and is a team name under the IAU naming rules, rather than the name of the discoverer. At discovery the presumed asteroid was described as being magnitude 18.9, but 6 days later J. Young of Table Mountain Observatory, CA, USA, noted a coma 2–3 arc-sec across, with a bright central core.

At the time of discovery comet Lulin was 6.4 AU from the Sun and 5.7 AU from the Earth. The comet was in Aquarius (roughly 12° north of the Helix nebula) at 22 h 34 m − 8° 43′, and at a healthy elongation of 132° from the Sun while transiting at 3 a.m. It was moving at 22 arc-sec per hour in a west-south-westerly direction. Aquarius is a fairly barren constellation with a low star density and comet Lulin would have spent its entire incoming trajectory in that constellation up to the point of discovery. However, by the very nature of its ecliptic based orbit, comet Lulin approached the inner solar system in the disc of the asteroid belt, teeming with hundreds of thousands of faint moving objects, but also a prime hunting ground for the major patrols. I suspect Lulin was not discovered by the professional patrols like LINEAR simply because it was very faint, but maybe also because it was moving relatively slowly in the period prior to discovery. In mid-May, due to the Earth's relative motion canceling its own apparent motion against the stars, it would only have been crawling along at 3 arc-sec per hour and would probably have been fainter than magnitude 19.0. Discovering such slow moving quarry could be carried out by an amateur patrol especially if pictures were blinked within the space of a few hours between imaging the same field; there is potentially a professional loophole to exploit here as the main patrols are used to detecting much faster moving objects.

2008

Moving into 2008 now, another oriental discovery, C/2008 C1 (Chen–Gao), hid in the dense starfields of the Milky Way constellation of Cepheus in the 6 weeks prior to its February 1 discovery (see Fig. 6.8) and, arguably, in December and November, its high northerly declination would have placed it at only 30° altitude from the latitudes of the main professional patrol sites, despite probably being around magnitude 15 at those times. The remaining three non-professional discoveries of 2008, as discussed in Chap. 4, were made using large, computer controlled, 0.6-m class instruments which can compete with the likes of LINEAR when searching for faint NEOs and comets.

Fig. 6.8. Comet C/2008 C1 (Chen–Gao). A 120 s exposure taken on February 8, 2008 788 UT using a Celestron 14 at f/7.7 and SBIG ST9XE. The field is 13′ × 13′. Image: Martin Mobberley.

2009

Finally, in the most recent complete year prior to writing these words, Itagaki-san's discovery of C/2009 E1 (Itagaki) with a fast 21 cm f/3 astrograph and CCD was another comet which was discovered at a relatively small solar elongation (48°) and by an expert patroller using motion detection software. In contrast, for the comet discovered in the same month, C/2009 F6 (Yi-SWAN), its ability to travel along the densest Milky Way starfields for almost all of the incoming years prior to its discovery is quite remarkable. When you look at the path Yi-SWAN took before and after perihelion you realize that the plane of its orbit is aligned very closely to the plane of our own galaxy. Thus, apart from on the final approach, when the Earth's orbit alters our perspective, it can hide perfectly within the Milky Way!

The comet P/2009 L2 (Yang-Gao) seems to have decided to partially emulate C/2009 F6 in this regard as it was picked up by the Chinese amateur patrol on June 15 near the Serpens/Sagittarius border, having spent the previous months in Sagittarius, Scorpius, Ara and Norma, the most densely packed stellar regions in the entire sky.

Beating the Pro's: Conclusions

So, we have now looked at the ways in which a few comets discovered in the first decade of the automated patrol era have managed to evade those same machines and be captured by amateur patrollers. What can we deduce from this? Well, to be honest there are few surprises once you have read this far into the book and re-analyzed all the orbits involved. While there are certainly some unanswered questions it appears that amateur hunters can still bag comets in a number of specific ways.

Firstly, the good old tried and trusted method of sweeping in regions close to the Sun still works, even in the CCD era. This is the technique that amateurs have practiced for more than 200 years, because the simple fact is that comets brighten exponentially as they get near to the Sun. Unfortunately, the NEO patrols like Catalina and LINEAR are now capable of searching so deeply, so quickly, and without tiring, that many of the comets that might have been bagged in twilight are being discovered 6 months or more prior to perihelion when they are in a dark sky. However, it is obvious that some comets, by virtue of their incoming path relative to the Earth, can still, even in the CCD patrol era, stay on the opposite side of the Sun long enough to avoid the patrols, disappearing when they are brighter than 19th magnitude for as long as a year before finally being caught by an amateur on the edge of the twilight glow, close to perihelion. So, sweeping those regions between 90 and 40° from the Sun is still a very good strategy, whether you are a visual comet sweeper of the old school or a CCD/DSLR imager. The simple fact is, professional checking software, tireless though it is, simply cannot cope with varying degrees of background sky brightness, especially twilight. The software gets confused by strong moonlight too, and all the major professional patrols avoid the skies 5 days either side of the full Moon. Amateurs, checking for suspects visually, can patrol right through the full Moon period. Yes, their limiting magnitude will be severely hampered and long exposures within 30° of the Moon will not be possible, but they can still continue patrolling. Countless amateur discovered supernovae have been captured with a bright Moon in the sky, at the time of the month when the professional patrols have a break.

Secondly, amateurs with big apertures and sensitive CCD cameras can get really deep; faint enough, from dark sites, to challenge the professional patrols. Remember the competing apertures we are dealing with here, namely LINEAR, at a highly obstructed 1.0 m (equivalent to about 70 cm), Catalina, at 68 cm, LONEOS, at 60 cm, and Siding Spring, at 50 cm. These apertures are big, yes, but should not be off-putting. Many amateurs own 35-cm instruments and, as we have seen in the previous chapters, some have access to observatory class instruments of 50 or 60-cm aperture which enable them to discover asteroids and NEOs to magnitude 20 and fainter. The exposures used by the professional patrols are short, typically 60 s or so. Yes, they can reach as faint as stellar magnitudes of 21 or 22 in that time, but then so can 35-cm apertures in exposures of a few minutes. With quality mountings like the Paramount ME an amateur system can easily reach magnitude 21, unguided, on regions near the zenith from a really dark site on a clear night. The professional systems are geared to very fast optics and wide sky coverage. Their exposure times are short to enable most of the celestial sphere to be imaged every month. An amateur does not have to patrol such large areas of sky so rapidly. He or she can concentrate on longer exposures over specific areas of the sky. This may only lead to a NEO or comet being discovered every few years, but I think most amateurs would settle for that!

Thirdly, the major patrols do not discover comets within the Milky Way. LINEAR appears to lightly patrol it occasionally but without success, whereas NEAT avoided it and LONEOS appears to avoid it to. The edge-on disc of our galaxy is just too cluttered with stars for the checking software to work. Detecting a line of stars in a neat row of consecutive exposures will not work when there are countless thousands of stars in the frame! In addition, there are myriads of variable stars in the Milky Way to confuse the issue even further.

Precisely where the major patrols are searching each week can easily be checked on the CBAT/MPC site at http://scully.cfa.harvard.edu/~cgi/Sky Coverage.html. This is an incredibly useful tool for the amateur wishing to avoid scouring identical regions to the professionals.

Amateurs looking for something distinctly fuzzy, or with a greenish hue on color DSLR images, can win out over the automated monochrome patrols, especially in densely packed star fields as those patrols are looking for the motion of a stellar object, not a fuzzy or diffuse and misty greenish patch. Remember, the main aim of the professional surveys is to detect NEOs and PHAs, not comets. The comets are a by-product of those patrols. For southern hemisphere observers it is worth bearing in mind that LINEAR cannot patrol much below a declination of −30°.

Admittedly the Siding Spring Survey now covers those regions well, but the absence of LINEAR is always going to be good news! A tiny region of the north celestial sphere, above +80°, is avoided by LINEAR too due to mechanical design issues. The region of sky that close to the pole is only small in area terms but anywhere that is out of the reach of LINEAR is worth patrolling. Many fork mounted telescope systems find patrolling at that high a declination is a problem because in a Cassegrain type design the camera equipment can often fall foul of the fork mount at very high declinations. Taking every advantage of your latitude is important. The main Catalina and LINEAR patrols may well live in the northern hemisphere, but if you live at, say, +52° north, as I do, the regions higher than Dec +50, low in the northwest after sunset and low in the northeast before sunrise are inaccessible to the Catalina and LINEAR patrols in mid-winter. The local Sun sets much later and rises much earlier in Arizona and New Mexico during December and January than it does at UK latitudes.

Finally, as we have seen, sometimes comets are missed by the professional NEO surveys for inexplicable reasons. Even the world's major observatories have cloud and rain and equipment occasionally breaks down. Key members of staff occasionally leave projects for better paid roles and, as occurred with NEAT, sometimes funding is withdrawn from government allocations, meaning an entire facility disappears. Also, distant but large comets, like Hale–Bopp, can crawl painfully slowly against a stellar background and fail to trigger a software alert. In addition, even nearby comets can momentarily appear to stop traveling against the background sky when the parallax effect of the Earth's motion in orbit temporarily negates the comets own motion. When this happens the object will not be flagged as a moving one. There is still hope for amateur comet hunters. The decades old sage advice offered to visual observers regarding concentrating more on the morning sky regions is still valid in the twenty-first century too. You stand a much higher chance of bagging a comet before the pre-dawn twilight than in the post-dusk evening skies. The morning sky is the most productive discovery area even for machines, because we are looking in the direction the Earth is moving and objects will be more likely to be brightening as the Earth approaches them, rather than fading as it moves away. Also, looking near the pre-dawn twilight means the observer is uncovering areas that have been completely lost in twilight to the patrols for many months. There is also the fact that only the hardiest human observers do the pre-dawn patrol stint so it always has been a place where only the most determined amateur comet searchers sweep. This aspect is less important in the robotic patrol era as machines do not get tired. However, humans still do, whether at the eyepiece or at the monitor of a CCD system.

Unfortunately, the automated patrols are getting ever more powerful, so there is no room for complacency.

Future Threats

At the time of writing NASA's Wide-field Infrared Survey Explorer (WISE) spacecraft was just starting its program to map the entire sky in infrared light and could, potentially, discover a few comets in the process. Far more worrying though is the fact that two additional and potentially hugely powerful ground based patrols are set to scour the skies in the coming years and it is worth looking at the threat they pose before completing this chapter.

The two patrols in question are called Pan-STARRS and the Large Synoptic Survey Telescope (LSST). Pan-STARRS is an abbreviation for Panoramic Survey Telescope And Rapid Response System. It should eventually consist of four 1.8-m telescopes, situated on Hawaii (at Mauna Kea or Haleakala) at a cost estimated to be around 100 million dollars. The aim is to regularly survey three-quarters of the night sky to magnitude 24. The first telescope was completed and activated on December 6, 2008 so, at the time of writing, had been in operation for more than a year, although, as yet, it had not made any impression on the comet discovery scene. The estimated deadline for completion of the whole system is 2012. The project is a collaborative one between formidable technical groups with a proven track record and the MIT/Lincoln Laboratory/USAF teams who were responsible for the LINEAR project are part of the system. Numerous universities and electro-optical technical experts worldwide have a stake in the operation of the first (PS1) telescope. The plan is that ultimately all four 1.8-m instruments will point at the same area of sky together. The unique aspect of the Pan-STARRS system is the field of view and sheer size of the CCD detector. With a $38,400 \times 38,400$ pixel grid covering a 3° field there are some 1.4 billion active pixels per image sampling the sky at a very fine 0.3 arc-sec per pixel. The CCD array in each telescope is approximately 40 cm across! With its light grasp and field of view it might, in theory, dominate solar system discoveries across the whole spectrum of asteroid, comet and Kuiper belt objects. Imaging of the entire night sky, as seen from Hawaii, four times per month, to magnitude 24 is the project aim. Maybe the days of the amateur discoverer are indeed running out? However, do not get too worried just yet as the death knell of the amateur comet hunter has been announced prematurely on many other occasions and there is another issue here too. Technical problems affecting the quality of star images taken with the first Pan-Starrs telescope meant that it was de-activated for repairs in late 2009. At the time of writing Pan-Starrs telescope number one had discovered a total of 18 supernovae. The first, SN 2008id, was discovered by Steve Rodney of the University of Hawaii during a test of the system in November 2008. The other 17 were discovered in two runs during June and October 2009. If the Pan-STARRS system does become fully operational then it could put all the other surveys out of operation for regions of the sky above Dec −30. However, it would still not be able to detect comets which brightened in the far south of the sky or comets that hugged the twilight. It is unlikely that it would fare any better with patrolling the Milky Way regions either. At present it appears that Pan-STARRS telescope number one can discover faint supernovae in superclusters very efficiently, much like the Sloan Digital Sky Survey (SDSS), but its comet discovering ability has yet to be proven.

The Large Synoptic Survey Telescope (LSST) is not due to experience first light until 2015. Its specification is even more impressive than that of Pan-STARSS and this monster is due to be installed in the southern hemisphere in northern Chile, at 2700 m altitude, alongside the Gemini South and Southern Astrophysical Research Telescopes. This will be a giant world class telescope with an aperture of 8.4 m and an unheard of (for that aperture) 3.5° field of view! As with Pan-STARRS the CCD detector is extraordinary comprising some 3.2 billion pixels in a 64 cm wide detector! The aim of the system is to take 15 s exposures down to 25th magnitude and map the entire sky visible from northern Chile, at that magnitude, every few days, thereby discovering far more faint solar system objects and supernovae than any other system. Scary! However, it is acknowledged that examining the 1.3 PB (1.3 thousand million million bytes) of data per year that the telescope collects may be the toughest challenge, as will raising the 400 million dollars for the project in these post credit crunch years!

CHAPTER SEVEN

Recovering Returning Periodic Comets

If discovering comets sounds like just too much hard work, or if you simply do not have the time in your life to allocate hundreds of hours a year to a project that may never bear fruit, then you might like to consider searching along a known comet's orbit as it returns to the inner solar system.

1P/Halley

Of course, the most famous cases of comet recoveries in history have been those of comet 1P/Halley which was first recognized as periodic when Edmond Halley realized the comets of 1531, 1607 and 1682 all had similar characteristics. Halley died in 1742 but predicted that the comet named after him would return again in 1758, which it duly did. A German farmer and amateur astronomer, Johann Georg Palitzsch, recovered it on Christmas day of that year. At comet Halley's next return, in 1835, it was recovered on August 5 by Dumouchel, observing from Rome, and by de Vico just a few minutes later; the comet was in Taurus at the time. Prior to its next return photography had been invented and

M. Mobberley, *Hunting and Imaging Comets*, Patrick Moore's Practical Astronomy Series, DOI 10.1007/978-1-4419-6905-7_7, © Springer Science+Business Media, LLC 2011

with this extra power astronomers had started searching for Halley as early as 1908. The two men with the best chance of recovering it photographically were Max Wolf, at Königstuhl Observatory, Heidelberg, Germany and the tireless astro-photographer Edward Emerson Barnard of Yerkes Observatory, near Chicago. Of course, astronomers needed to know where to search and there were surprisingly few serious attempts to determine the 1910 perihelion return date and corresponding ephemeris for Halley in the years building up to its return. In 1863 A.J. Angstrom had devised a prediction method which seemed to favor a January 1913 perihelion date. A year later, P.G. Le Doulcet had arrived at a date of May 24, 1910. More than 40 years after that, and with no other predictions being made, and with Halley's return imminent, in 1907 and 1908 A.C.D. Crommelin and P.H. Cowell of the Royal Greenwich Observatory tackled the problem and arrived at a perihelion date of April 8, 1910. Finally, A.A. Ivanov, as late as 1909, predicted a perihelion date of April 22, 1910. As it turned out Max Wolf at Heidelberg recovered the comet on the night of September 11/12, 1909. According to Morehouse (one of Barnard's colleagues at Yerkes observatory) Barnard appeared to be devastated and "white-faced" when he heard the news. He had been searching for Halley since October 1908 but his old rival Dr. Wolf had snatched it first. After Wolf's recovery of Halley it soon became apparent that Crommelin and Cowell, and Ivanov too, had all got the month right. In fact, the perihelion date would be April 20, 1910, astonishingly close to Ivanov's prediction! Following Wolf's recovery of Halley some very faint images of the comet were found to have been recorded on plates exposed at the Royal Observatory, Greenwich and the Khedivial Observatory at Helwan, Egypt on September 9/10 and August 24/25 respectively. An earlier (August 29) plate by Wolf, also contained a faint recording of the comet as did a September plate by Barnard. With the comet's position in its orbit now pinned down it did not take long for the salient points of this apparition to be calculated.

Even as recently as the 1980s there was still considerable kudos associated with detecting Halley on its way into the inner solar system and this time it was returning at the very dawn of the CCD era. It was first picked up at that most recent return on October 16, 1982 when David Jewitt and Edward Danielson recorded a 24th magnitude smudge in the comet's predicted position using an early CCD detector on the Palomar 200-in. (5 m) reflector. The comet had been recovered when it was eight magnitudes fainter than in 1910, which equates to a factor of 1,600 times!

Tsutomu Seki

In recent history the renowned Japanese amateur Tsutomu Seki, famous for his discoveries of bright comets in the 1960s turned his hand to recovering periodic comets from the early 1980s onward, primarily using a 60 cm f/3.5 Newtonian reflector made available to him at Geisei Observatory in Japan's Kochi prefecture. This 60 cm f/3.5 Newtonian was donated to Kochi observatory and education centre by Seizo Goto, the director of the Goto telescope company. Using this fine instrument Seki-san recovered 28 periodic comets, a Japanese record (see Table 7.1) and discovered 223 asteroids too, between 1981 and 1998. At one point Brian Marsden described the middle-aged Seki of the 1980s and 1990s as the most important all round amateur astronomer in the world. His final recovery was a joint one, of comet 122P/de Vico in 1995, which had not been seen since 1846, despite returning once in the interim, in 1922 (see Fig. 7.1).

Although Seki-san no longer recovers comets there can be no doubt that his influence spurred many future amateurs on to spot returning comets in the CCD era.

Some of the brighter short period comets with low eccentricity orbits (so their aphelion and perihelion distances are not hugely different) can be followed around a complete orbit. In these cases detecting the comet first after it passes the aphelion point may be, technically, a recovery, but in practice it is not of great importance. However, a large number of periodic comets simply disappear below the threshold of 1 or 2 m aperture instruments at some point below magnitude 23 or so, and there can be significant pride in being the first observer to recover them, post-aphelion. The prestige is especially great if a comet is recovered with a modest amateur telescope of 0.4-m aperture or less. Obviously the world's largest telescopes and the Hubble Space Telescope have the muscle to recover incredibly faint comets, but their observing time is booked solid studying galaxies, supernovae and gamma ray bursts in deep space, not nearby solar system objects. It should be stressed that when objects are typically 20th magnitude and fainter a minimum of two separate comet recovery images, on different nights, will be required to satisfy the MPC that a comet has been recovered, along with a precise astrometric measurement of the position of the comet to arcsecond accuracy.

Table 7.1. Periodic comets recovered by the legendary Tsutumo Seki of Japan, all in the era of photography.

1	Finlay	1974
2	Wolf–Harrington	1977
3	Tsuchinshan 2	1978
4	Clark	1978
5	Tuttle–Giacobini–Kresak	1978
6	Honda–Mrkos–Pajdusakova	1980
7	Longmore	1981
8	Tempel 1	1982
9	Pons–Winnecke	1983
10	Arend–Rigaux	1984
11	Tsuchinshan 1	1984
12	Giclas	1985
13	West–Kohoutek–Ikemura	1987
14	du-Toit–Neujimin–Delporte	1989
15	Lovas 1	1989
16	Swift–Gehrels	1991
17	Wirtanen	1991
18	Arend	1991
19	Tsuchinshan 1	1991
20	Faye	1991
21	Giclas	1992
22	Wolf	1992
23	Schuster	1992
24	Daniel	1992
25	Schaumasse	1992
26	Holmes	1993
27	Brooks 2	1994
28	de Vico (joint recovery)	1995

Tsutumo Seki's first comet recovery, Finlay was made with a modest 21-cm f/5 Newtonian. His second to sixth comet recoveries were made with a 40 cm-f/5 Newtonian. His seventh to 27th recoveries were made with the 60-cm f/3.5 Newtonian at Geisei observatory. His final recovery, of comet de Vico in 1995, was made visually while sweeping for comets with a 20-cm refractor, just minutes after three other Japanese amateurs (Nakamura Tanaka and Utsunomiya) also discovered it

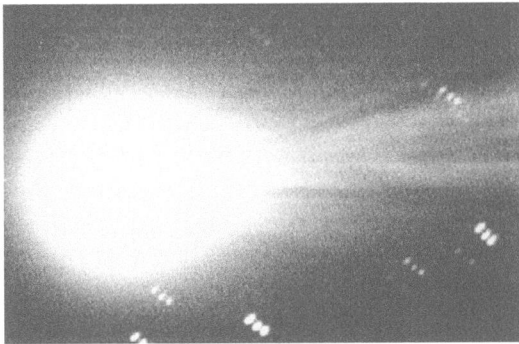

Fig. 7.1. Comet 122P/de Vico returned in 1995 but had not been since since 1846! Image, by Martin Mobberley, taken on September 28, 1995 with a 0.49-m f/4.5 Newtonian and Starlight Xpress CCD.

85P/Boethin

Of course, not all attempts to recover returning comets meet with success. A notable case was comet 85P/Boethin in 2008 which was actually scheduled to be visited by a NASA spacecraft in that year but never made an appearance! "What!," I hear you cry, "The comet never returned! Was it abducted by aliens?" Well, sorry to disappoint you but no, that was not the explanation. Some comets are very small and very fragile and little more than flying banks of icy gravel. When they heat up, as they approach the warmer inner solar system, they lose more of the volatile material on their surfaces and this heating may cause the object to break up. Sometimes, when a big chunk breaks off, there is a temporary outburst in brightness, until all the volatile material released has evaporated. All that is thought to be left of the departing chunk may be a load of boulders with no more significant icy deposits. Comets in this delicate state are often missed at a few perihelion returns that are unfavorable, such as when the comet is nowhere near the Earth. Then, maybe one orbit of the comet later, the orbital circumstances are better, or a temporary brightness surge occurs and the comet is recovered. In other cases the orbit is sometimes permanently changed if the comet has passed near to Jupiter. In the case of 85P/Boethin the comet has quite a strange history. It was first discovered by the Reverend Leo Boethin of Abra in the Philippines, on January 4, 1975 when it was 12th magnitude. He observed it over four nights before sending his discovery claim, *by post* (!), to CBAT in the USA. It was not photographically confirmed until a month later, on February 5, by Urata in Japan, but even a preliminary orbit was unavailable until early March. Ultimately it transpired that comet Boethin had an orbital period of 11.2 years. At its next return, in 1986, the comet surprised observers by becoming far brighter than expected. Astronomers were expecting it to peak at 12th magnitude but it peaked at magnitude 8 in January, some 40 times brighter than predicted. So, hopes were high that 85P/Boethin would be seen again in 1997, but no-one found it. Nevertheless, as it had been seen on the two previous returns NASA announced that it had approved a proposal by the University of Maryland to send the Deep Impact spacecraft to 85P/Boethin for a December 2008 rendezvous. This was the same spacecraft that fired a missile into periodic comet Tempel 1 during July 2005. When 85P was not recovered by late 2008 NASA was forced to change its plans. A few professional astronomers (and a few amateurs) imaged the regions along the orbit of 85P/Boethin in late 2008 and even in early 2009 in the hope they might recover the comet before an automated patrol like LINEAR, but no-one has now seen it since 1986. Nevertheless it did generate considerable interest even if no-one clinched a recovery image. One has to conclude that the comet was way below magnitude 20 and may have to be reclassified as 85D/Boethin as it appears to have become defunct!

The Remarkable Recovery of 109P/Swift–Tuttle

Probably the largest amount of prestige ever associated with any amateur comet recovery in recent times was the recovery of comet Swift–Tuttle in 1992. During the early 1980s, anticipation was high that this progenitor of the Perseid meteor shower was due to return after 130 years, but its actual return was much later. Let us look at that earlier appearance in more detail.

On the night of July 16, 1862 Lewis Swift, observing from Marathon, New York, with a 114-mm aperture refractor, spotted a bright comet (probably fifth magnitude) in the far northern constellation of Camelopardalis. Strangely, Swift assumed this was comet Schmidt, discovered 2 weeks earlier, even though the former comet was in Virgo! Because of this presumption, Swift did not claim a discovery. However, three nights later, Horace Parnell Tuttle of the Harvard Observatory at Cambridge

Massachusetts, made an independent discovery of the new comet and, ultimately, it was named comet Swift–Tuttle. Fourteen comets would ultimately carry the name Swift and six the name Tuttle. Swift–Tuttle continued to brighten steadily throughout July 1862 and was a truly superb object in the northern hemisphere skies; not comparable with the awesome comet Donati of 4 years earlier, but still a fine comet. By the first week of August it had reached +75° in Declination and it would pass 8° from the north celestial pole at mid-month. Perihelion occurred on August 23 at 144 million km from the Sun and the closest approach to the Earth (50 million km) was only a week later. By the end of August the comet was a superb second magnitude object with a 20° tail, passing through Corona Borealis: an easy naked eye spectacle.

Orbital calculations indicated a period of about 120 years and as the comet had been deduced as the parent of the Perseid meteor stream the healthy Perseid meteor showers in the early 1980s led observers to believe that the long lost comet's return was imminent. Although the first rigorous orbital solution had not been calculated until 1889 it had been realized, as early as 1867 by Giovanni Schiaparelli, that the orbit of Swift–Tuttle bore a remarkable resemblance to the orbit of the Perseid meteors. Indeed, in the 1867 *Monthly Notices* of the Royal Astronomical Society he was quoted as saying that "Swift–Tuttle is nothing more than a very large meteor of the August system." A very large meteor indeed! Whether Schiaparelli expected Swift–Tuttle to one day burn up in our atmosphere is not on record.

However, as the 1980s came to a close, many believed that the comet had come and gone and had been missed. I must admit that I was never happy with this theory as even the worst possible orbital circumstances should still have enabled the comet to be found by the world's keen amateur comet hunters, of which there were plenty. To me, it was a bit like "missing" the return of comet Halley; it just did not seem possible. Another factor was the observation (originally mentioned in this context by W.T. Lynn in 1902, in a tantalizingly brief remark) of a comet in 1737 which became known as comet Kegler.

On July 3 of that year, in Peking, China, the French missionary Ignatius Kegler had spotted a new comet that morning, near the Pisces/Aries border. The position was not far from the galaxy that would, more than 40 years later, become Messier 74. The comet was probably about fourth magnitude and Kegler observed its position over each of the next 7 days, with a "three foot long" telescope, as it drifted 15° south and into northern Cetus. Cloud prevented him making any further observations. This was, of course, a possible sighting of comet Swift–Tuttle at the apparition before 1862; however, it implied a return in 1992 and an orbital

Fig. 7.2. The prolific astronomy author and TV presenter Patrick Moore (*right*) with the meteor expert John Mason. Patrick and John famously bet a bottle of whisky on Swift–Tuttle being recovered in 1992. It was recovered, as John expected, and Patrick had to hand over the whisky! Image: Martin Mobberley.

period nearer 130 years than the accepted value of 120 years. Opinion was certainly divided on the likelihood of Swift–Tuttle returning as we entered the 1990s; so much so that two prominent British Astronomical Association members, John Mason and the legendary Patrick Moore (see Fig. 7.2), wagered a bottle of whisky on the outcome. Patrick thought we had missed it: John thought it would return.

Amongst increasing doubts that the comet would ever return, attention re-focused on the calculations of the orbit guru Dr. Brian Marsden of CBAT (see Fig. 7.3), who had, in 1973, re-examined the astrometric measurements of 1862 and produced two new possible perihelion dates for Swift–Tuttle, assuming it was the same object as Comet Kegler in 1737. One (highly publicized) took account of non-gravitational forces and the other did not. The former gave a perihelion date of November 25, 1992

Fig. 7.3. The remarkable comet and nova discoverer George Alcock (*left*) 2 weeks before his 87th birthday, together with orbit guru Dr. Brian Marsden of the Minor Planets Center in August 1999. Marsden predicted the 1992 return date of Comet Swift–Tuttle with amazing accuracy.

the latter December 11, 1992. Based on 212 individual observations of the comet's 1862 positions, Marsden's analysis highlighted a mysterious positional anomaly involving measurements made by Maclear from the southern hemisphere's Cape Observatory in October 1862. These puzzling declination errors gradually increased during the final ten 1862 astrometric measurements (as the comet headed back out of the inner solar system) and they climbed to a whopping 15 arc-sec: a huge amount in astrometric terms. Marsden called this the "Cape

Effect" and later said that it was directly responsible for the 11 years difference between his predicted 1981 and 1992 returns.

Based on those two dates of November 25, 1992 and December 11, 1992 a small number of amateur and professional observatories took photographic plates along the track to try to recover the comet in late 1991 and early 1992. However, due to a variety of reasons (such as the plate limiting magnitude being not quite deep enough for the Japanese comet discoverer Seki, or photographic flaws just in the wrong place) the comet was not recovered photographically. By mid 1992, hopes were

Fig. 7.4. Comet 109P/Swift–Tuttle on November 13, 1992. This is a manually offset-guided photograph exposed for 24 min from 1816–1840 UT using a 0.36-m f/5 Newtonian & gas hypersensitized Kodak 2415. Photograph by Martin Mobberley.

fading that the comet would arrive on time, although abnormally high Perseid rates were recorded on forward scatter radio by several observers on August 11; but surely, if the comet were returning, someone would have recovered it by now? It was therefore with some amazement that the news broke that the Japanese comet discoverer Tsuruhiko Kiuchi had recovered Swift–Tuttle visually, using 25×150 Fujinon binoculars on September 26.76, very close to the predicted position by Marsden. Indeed, the comet would reach perihelion on December 12, only one day later than Marsden's straight (no non-gravitational force component) prediction. In theory, any amateur with a 150 mm or larger reflector, dark skies and a good northern horizon could have recovered Swift–Tuttle visually but, once again, it was a Japanese observer who actually achieved it. Within a couple of days it was obvious that Kiuchi's new comet, 1992t, was comet Swift–Tuttle 1862 III which, in turn, was Kegler's comet 1737 II. The comet had been discovered with only 3 months to go to perihelion and it would be a splendid fifth magnitude object with a fine ion tail in November (see Fig. 7.4) and early December from the UK, despite far from perfect orbital circumstances. Tsuruhiko Kiuchi will go down in history as being the amateur astronomer who first spotted the parent comet progenitor of the Perseid meteor shower returning. Quite an achievement! In addition it also meant that Patrick Moore had to hand over that bottle of whisky to John Mason at the British Astronomical Association meeting of October 26, 1992.

The Mystery Remains

So, all this sounds like the orbit is resolved: Swift–Tuttle returned and the orbit was within a day of being right. What's the problem? Well, to this day, the "Cape Effect" is still a complete mystery. Those systematic errors, by an experienced observer, just make no sense. Neither does the fact that they increased steadily throughout October 1862. There are also lesser unexplained positional anomalies in July 1862 and even on the pre-recovery January 1992 photographs. Completely at odds with these anomalies is the fact that, confusingly, if treated as an asteroid, Swift–Tuttle is unbelievably well-behaved over the long term. It has now been identified on past returns over the last 2,000 years, returning like clockwork, despite being a highly active comet that sheds much material at each perihelion passage. However, one possible solution to this predictability is that it has an extremely massive nucleus. So, we have an enigma here: a comet returning every 130 years or so, exactly as theory predicts, but briefly appearing off its track when close to perihelion. Sometime ago, after all the 1990s astrometry had been analyzed, I asked Brian Marsden whether the situation was now any clearer. He e-mailed me back in minutes: "The 1862 Cape astrometry of 109P/Swift–Tuttle continues to be a puzzle, unique in my experience."

A Threat to the Earth?

So, bearing in mind that Swift–Tuttle's nucleus may be regarded as the mother of all Perseid's and its orbit intersects that of the Earth, is it a threat? Well, judging by its consistency over the past 2,000 years, the answer to that is a resounding NO: the comet is *not* a threat. Swift–Tuttle should make spectacular returns to perihelion on July 12, 2126 and August 10, 2261, passing a safe, but impressive, 20 million km from us on both occasions, and maybe reaching first magnitude at those returns. But, if the mysterious Cape effect was a real indicator of uncertainty in the comet's position, we just have to pray that its next perihelion date is not a scant 14 days later. A perihelion date of July 26, 2126, within a precise 3 min window, could cause an Earth impact with its large nucleus on August 14, 2126. Swift–Tuttle would then become a true Perseid meteor and its centuries of orbiting the Sun would be at an end. But that would be the least of our worries......

I have drifted well off the main purpose of this chapter in relating the mysterious tale of comet 109P/Swift–Tuttle but I have done so deliberately. It just goes to show how fascinating some comet recoveries can be, especially if the object in question has not been seen for a while or if it is the parent comet of a major meteor shower.

It goes without saying that the professional survey teams are very well equipped to recover comets. However, their main remit and the reason they receive funding is to discover NEOs and PHAs, thus saving the world from a potential catastrophe. They may have time to survey the entire sky searching for new objects but when the object in question is a returning comet the task facing the amateur is far less daunting because the region that needs to be searched is relatively small. Comets do not always return exactly when and where they are expected because of non-gravitational forces caused by the loss of dust and gas from the rotating cometary surface. Nevertheless, the accuracy with which Brian Marsden pinned down the return of comet Swift–Tuttle in 1992 shows that even when a comet has a 130 year period it can usually be predicted to return to perihelion with an accuracy of a couple of weeks.

For shorter period comets like 85P/Boethin the error in the return position time should only be a day or so. In general a change in the date of perihelion is the only change in the orbital elements that the potential comet recovered needs to be concerned about as any other major orbital element changes will only come about if the comet has been affected by a close encounter with Jupiter or another one of the giant outer planets. So the zone around which the potential recoverer needs to search is one that allows a shift in the perihelion date of a few days. Note that there is an important distinction here between the position a comet sits at in its orbit, a day or two either side of a given date, for a *specific* perihelion date, and a shift in the perihelion date itself. These are two similar, but slightly different things. One can have a fixed perihelion date and search along the path that the comet should move along, but that path will be subtly different to one where the search date is fixed (that is, tonight's date) and the perihelion date is changing. The terminology for the position where you look in the latter case is called the line of variation (LOV). Of course, as we saw with the Kreutz sungrazers, if you are searching for a comet that is a long way off perihelion, near to its vanishing point, where the spirals caused by the parallax from the Earth's orbit are starting, then the movement of the orbit track becomes smaller when you are hoping to discover a massive cometary chunk a full year or two from perihelion. The further away a comet is when you search for it, the less variation there will be in its position; but this will be at the expense of the comet being much fainter. One strategy here is, bearing in mind an error of only a few days in a comet's position, you can wait until a comet is at a position in its orbit where it does not appear to be moving; in other words, where the parallax effect of the Earth's motion totally cancels the comet's apparent motion in its orbit. Admittedly, with a few days of variation in the perihelion date you will not know when this will happen to better than a few days accuracy, but you can still be sure that the comet will be moving very slowly. If a comet is only moving at a few arc-seconds per hour against the starry background it may not easily betray its motion but you may only have a region less than half a degree across to search. In these situations you simply need to get as deep an image as possible and this is best achieved by simply applying the basic imaging skill rules to the maximum.

Going Faint

The larger the aperture you are using the deeper you will go, but size is not the only consideration. Sharp focusing is essential to get the lowest stellar magnitude possible and incoming comets are often very stellar in appearance. An image that is even fractionally out of focus will result in that extra light from the comet being smeared out into the sky background. Remember, when imaging to the limit of your equipment the faintest objects detectable may be less than 1% above the background sky brightness. They are literally sitting on top of a wall of sky glow. At the darkest sites in the world the background night sky will have a glow of between magnitude 21 and 22 per square arc-second. Regardless of the size of your telescope you are faced with this almost impenetrable barrier. The situation is far worse from a typical amateur observatory though. A really good amateur rural site might enjoy magnitude 20 per square arc second skies, but even in modest towns this can drop to 18, 17 or even 16 depending on how much light pollution there is (see Fig. 7.5). For comet recovery the

Fig. 7.5. Light pollution can severely hamper your ability to reach faint magnitudes: Image by Gary Poyner.

amateur really needs to be in a semi-rural location with skies darker than magnitude 19 per square arc-second. At such a site the dark adapted eye will still see the sky as grey, rather than black, but the river of the Milky Way will still be easily visible to the naked eye, along with, say, objects like the Andromeda Galaxy, M31. Having a background sky brightness of magnitude 19 per square arc second does not mean that you are limited to imaging stellar objects of magnitude 19. It is the ability of your system to extract signal from noise that is crucial. Even below magnitude 20 a steady flow of photons will be hitting your CCD chip, but will almost be drowned in the tsunami of orange photons from the sky. The key is to focus the stars precisely, track them accurately and then use long exposures so that the small signal you want gradually emerges from the much bigger random noise signal of the sky. A magnitude 20 comet, even if vaguely stellar in appearance, will be spread into a blob many arc-seconds across and so will actually have a sky brightness of the order of magnitude 23 per square arc-second; that's four magnitudes (or 40 times) fainter than a magnitude 19 per square arc-second sky. But the image should not be thought of as being totally buried under that sky background. It is slightly more reassuring to think of it as sitting atop the wall of light pollution.

Choosing nights where the sky is especially dark and haze free (such as when a cold front has passed through the region) is essential. In addition, at any one time there will be a number of comets returning to the inner solar system ready for a recovery attempt. You need to choose the comets that are closest to the zenith when you are imaging to stand the best chance of extracting them from the background murk. If you live at a high latitude of, say, 50 north or south and are imaging objects at your zenith, near 50° of declination north or 50° south, then the tracking requirements become easier by a factor of 1/(Cosine declination) which can help. Long single exposures, if tracking is perfect, are better than loads of very short exposures because for every single exposure you get readout noise from the CCD's electronics. In addition, the CCD chip's thermal noise can be just as damaging as light pollution and so it is vital to keep your camera running as cold as possible, providing you can avoid condensation in the camera housing. All cooled CCD cameras have a silica gel desiccant chamber which should be regularly baked in an oven (175°C for 4 h is typical) to keep moisture out and enable temperatures to get as cold as possible. As a general rule, thermal noise halves for every 7°C drop in CCD temperature which can mean that running a chip 30° cooler can give you a 20-fold improvement in noise.

Of course, CCD images are full of noisy artifacts and cosmic ray hits so that the merest hint of a cometary image may not be the comet at all. Comets move against the background sky and so evidence of motion will be required by the Central Bureau for Astronomical Telegrams (CBAT) as proof of the comet's reality. In addition, a second nights image, as a minimum, will be required too, especially if you are unknown to CBAT.

One powerful "trick" which a keen amateur can play in this regard is to try to book time on a large professional telescope which has an observing allocation available for amateur use. Certain telescopes, like those of the Faulkes project, are happy to consider applications for observing time from amateurs, especially if their proposal is from an experienced observer and for a goal that will enhance the reputation of the facility. A 1 or 2 m class telescope at a magnitude 21 per square arc-second site will have a much better chance of recording a faint incoming comet.

Tracking the telescope at the comet's likely motion in the sky, even if you do not know exactly where the comet hides, is another very powerful, but very tricky, technique to use. Being able to allow for a comet's motion against the starry background can make the difference between a successful recovery and a frustrating failure, but the technique needs to be mastered before it can be used successfully. In the old days a comet photographer would use a piggybacked guide telescope to observe a bright star near to the field of the comet being sought. A device known as a bifilar micrometer was then used to move fine wires in an illuminated eyepiece field at predetermined intervals and at a specific position angle such that the guide star was constantly being re-centered on the wires to simulate the motion of the comet. The technique was known as offset guiding and, as someone who carried it out in the 1980s and 1990s, I can confirm that it could be sheer mental and physical torture to successfully implement! Nowadays the situation is much easier and the amateur has two choices. The main choice is simply to use stacking software like Herbert Raab's *Astrometrica* which can stack short exposures (during which time the comet has moved by, say, only an arc-second) with reference to the motion of the comet. Stacking exposures in this way is very easy as all the work can be done after the imaging session, but it is not as deep as a single exposure would be. As a general rule if you stack lots of images together they will reach as deep as one single long exposure roughly equal to the individual short exposure length, multiplied by the square root of the number of images taken. In other words 100 short exposures of 1 min in length will go as deep as a single 10 min exposure. The second choice is to set the tracking rates of the telescope at the comet's rate using a program like Software Bisque's *The Sky* (see Fig. 7.6). However, this requires a highly accurate mounting like the Paramount ME and, even that mount will struggle to track a comet at its specific rate for more than a few minutes if the image scale is finer than 2 arc-sec per pixel, unless the field is at a very high declination.

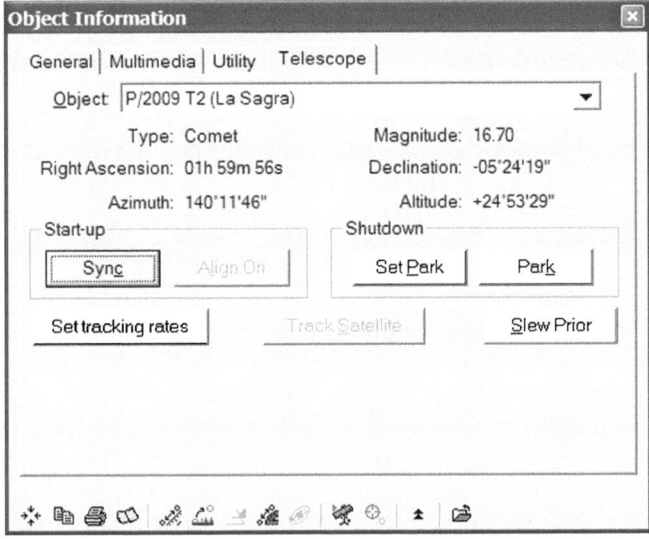

Fig. 7.6. In Software Bisque's *The Sky* package the Telescope tab in the Object Information box allows you to set the tracking rates to whatever object the telescope is pointed at, although in practice the tracking accuracy will always be limited by the quality of the worm and wheel drive.

A number of Internet sites are useful for keeping up to date with comet recovery data.

For many years Seiichi Yoshida has regularly updated his "Comets Waiting for First Observation" page at http://www.aerith.net/comet/recovery.html which is very useful for planning comet recovery projects. In addition, attempts to recover returning comets are often discussed on the Yahoo! Comets Mailing List at: http://tech.groups.yahoo.com/group/comets-ml/.

The CBAT Minor Planet Circulars have become freely available on the Internet recently and these often contain very useful data on comet orbits. These can be accessed at:

http://www.minorplanetcenter.org/mpec/RecentMPECs.html

The Central Bureau's IAU circulars and CBET (Central Bureau Electronic Telegram) are also vital sources of comet recovery information but require a subscription. Further details can be found at the CBAT website subscriptions page:

http://www.cbat.eps.harvard.edu/services/Subscriptions.html

2009 Recoveries

To illustrate some typical twenty-first century recovery circumstances we will now have a look at the details surrounding some of the comet recoveries in 2009. One of the most successful comet recovery programs is that operated by the Spacewatch team which I have already mentioned in the context of comet discoveries. Many of the 2009 comet recoveries were made by this team. Spacewatch is the oldest CCD sky survey facility and is the official name of this noted group at the University of Arizona's Lunar and Planetary Laboratory. It was founded by Professor Tom Gehrels and Dr. Robert McMillan in 1980 and Dr. McMillan leads the project in the twenty-first century. The raw power of the LINEAR and NEAT facilities, funded via military channels and using 1.0 m apertures, rather overshadowed the Spacewatch facilities in the late 1990s but the group has now changed its emphasis. They used to operate with the vintage 0.9-m f/5.3 Steward Observatory Newtonian, later upgraded to a bigger CCD system at f/3, but since 2001 they have operated a giant 1.8 m f/2.7 telescope which enables them to go much deeper and concentrate on other, more distant objects, like Transneptunian dwarf planets and incoming comets. Both telescopes are located at Kitt Peak Observatory in Arizona. Twenty nights per month, avoiding the 5 days either side of new Moon, are used for scanning the skies. The primary goal of Spacewatch in the twenty-first century is to explore the various populations of small objects in the solar system, and to study the statistics of asteroids and comets in order to investigate the dynamical evolution of the solar system. The CCD scanning programs study the Centaur, Trojan, Main-Belt, Trans-Neptunian, and Earth-approaching asteroid populations. The Spacewatch facility also finds potential targets for interplanetary spacecraft missions, provides follow-up astrometry of such targets, and finds objects that might present a hazard to the Earth.

P/2001 CV8 (LINEAR)

The first significant Spacewatch comet recovery of 2009 was that of the comet initially named as asteroid 2001 CV8 or P/2001 CV8 (LINEAR) and then renamed P/2009 D1 (LINEAR) and subsequently 216P/LINEAR on its return. It was imaged on February 19, 2009 with the 1.8-m f/2.7 reflector at Kitt Peak when it reached magnitude 20.9. The images (three taken over a 30 min time frame) showed the comet to be slightly diffuse with a 0.3 arc-min long tail at a position angle of 295°. Another set of three images were taken on the next night to confirm the motion. Compared to the initial orbit prediction the return date of perihelion had to be adjusted by −0.37 days, whereas compared to the revised prediction in the 2008/2009 Comet Handbook the correction was only −0.24 days. This was for an object with a 7.66 year period for which 130 astrometric measurements had been made by LINEAR and others at its discovery. At the time of its recovery the comet was moving at 24 arc-sec per hour, south east, in Libra, with an elongation of 111° from the Sun; it was in the morning sky and transiting at around 4.30 a.m. local time. In terms of the search zone for the returning comet it was within a few arc-minutes of its predicted position and so recovering it was largely a question of just getting to magnitude 20.9.

P/2004 K2 (McNaught)

The next significant comet recovery of 2009 was also made by Spacewatch, although it was a so-called "incidental recovery" during a routine survey on April 28. The returning comet was recovered in the course of a Spacewatch survey by T. H. Bressi, and identified as the 2004 comet by automatic software analysis at the MPC. Subsequently, independent recovery observations made on April 30 were reported on May 1 and 3, by the amateur astronomers Gustavo Muler, J. M. Ruiz, and R. Naves using a 30 cm f/6.3 Schmidt-Cassegrain and SBIG ST8XME CCD at an image scale of 1.92 arc-sec per pixel; this was achieved from the Observatorio Nazaret at Lanzarote (see Fig. 7.7). These amateurs took 200 exposures of 55 s duration to record the comet and then stacked the images to allow for the comet's motion. The cumulative exposure time was therefore 3 h and 3 min! Their magnitude estimate for the comet was 20.4. The comet appeared as a single solitary speck on the limit of each stacked image. The motion of the comet was 27 arc-sec per hour northwest and it was at 155° elongation from the Sun when recovered, in Ophiuchus. The correction to the predicted date of perihelion made prior to the comet's recovery was just −0.08 days so the comet turned up within an arc-minute of its expected position.

Fig. 7.7. Comet P/2004 K2 (McNaught) was independently recovered by Spanish amateur astronomers Gustavo Muler, J. M. Ruiz, and R. Naves using a 30 cm f/6.3 Schmidt-Cassegrain and SBIG ST8XME CCD. The cumulative exposure time to record the magnitude 20.4 comet as the merest faint speck was more than 3 h! The enlarged inset shows a DDP contrast stretched version making the comet slightly more obvious! Image by kind permission of Gustavo Muler.

P/2002 JN16 (LINEAR)

An asteroidal object discovered in 2002 by LINEAR, later re-classified as a comet when observed by the 1.2-m Palomar Schmidt, was recovered on the night of June 1, 2009. Leonid Elenin, of Lyubertsy in the Moscow region of Russia, recovered comet P/2002 JN16 as part of the ROCOT project. ROCOT stands for Recovery of periodic comets and is a project which uses the remotely operated Tzec Maun observatory telescopes and those of the Tenagra observatory. Tenagra is a worldwide facility of telescopes set up by the US amateur Michael Schwartz who discovered numerous supernovae after selling his high-tech company and investing the proceeds in various telescopes in the USA and worldwide, the largest being a massive 81 cm f/7 Ritchey–Chretien telescope. Schwartz is involved in a number of advanced astrometric and photometric projects. When Elenin obtained the June 1 recovery image he was remotely acquiring images using a 0.36-m f/3.8 Maksutov–Newtonian based in Mayhill New Mexico (USA), equipped with an SBIG ST-10XME CCD. He recorded the magnitude 20.2 comet on 16 frames and there was evidence of a 35 arc-sec long tail. The following day he requested confirmation and Michael Schwartz used his giant 0.81-m f/7 Ritchey–Chretien to confirm the recovery on June 3. The comet was more than 5 months beyond perihelion when recovered so there was a chance that it might have been missed if Elenin and Schwartz had not imaged it. It was moving at 58 arc-sec per hour in the morning sky in Pegasus when found and only elongated from the Sun by 76°. After the recovery it was designated as 2009 L1 (LINEAR) and later upgraded to 221P/LINEAR. The perihelion date (January 24) of this 6.5 year period comet needed adjusting by −0.2 days at recovery which translates to an uncertainty in its position of several arc-minutes at the time of recovery.

127P/Holt–Olmstead

On June 3, 2009 UK amateur astronomers Richard Miles and George Faillace managed to acquire observing time on the Faulkes Telescope North (FTN), a massive 2.0-m f/10 Ritchey–Chretien, based at Haleakala Observatory in Hawaii, which can be operated remotely by schools and UK amateur astronomers. Using this huge telescope they managed to image the periodic comet 127P/Holt–Olmstead at magnitude 21.3 (see Fig. 7.8). It had last been seen in 2003 (the period is 6.4 years) and was traveling at 59 arc-sec per hour east-northeast in Cetus at an elongation of 74° from the Sun. The comet was 2.5 AU from Earth and 2.4 AU from the Sun. Perihelion occurred on October 21 and by mid September 127P had brightened to 17th magnitude.

P/2004 X1 (LINEAR)

On August 2 and 3, 2009 Rob McNaught of the Siding Spring Survey noted that observations of a newly discovered magnitude 17.7 asteroid found from Siding Spring with the 52-cm f/3.4 patrol telescope, named 2009 MB9, showed a small coma. Subsequently H. Sato of Tokyo Observatory noticed that the orbital elements of this comet were very similar to those of a periodic comet found by LINEAR in 2004, namely comet P/2004 X1. It turned out that the comet was indeed the same one found 5 years earlier by LINEAR. This very small comet was passing only 0.2 AU from the Earth when discovered by McNaught and was only

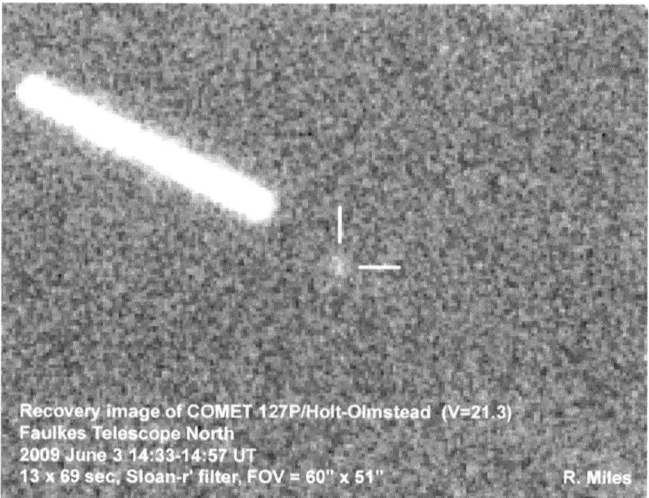

Recovery image of COMET 127P/Holt-Olmstead (V=21.3)
Faulkes Telescope North
2009 June 3 14:33-14:57 UT
13 x 69 sec, Sloan-r' filter, FOV = 60" x 51"   R. Miles

Fig. 7.8. Comet 127P/Holt–Olmstead was recovered at magnitude 21.3 on June 3, 2009. Amateur astronomers Richard Miles and George Faillace acquired observer time on the 2.0-m aperture Faulkes Telescope North. Thirteen 69 s exposures were taken and added to allow for the comet's motion. Image: Richard Miles.

a month from a small perihelion encounter with the Sun at 0.78 AU. The date of perihelion was −2.2 days in error compared to the orbital elements issued 5 years earlier and, when compounded with the close position to the Earth, meant that the comet was several degrees away from the predicted position, which is why it was not recognized as soon as it was captured. What had looked like another comet discovery from McNaught had turned out to, arguably, be a rather lucky recovery.

P/2002 S1 (Skiff)

In two observation sessions on June 15 and August 18, 2009 the Italian Remanzacco Observatory team, consisting of amateur astronomers G. Sostero, E. Guido, P. Camilleri, and E. Prosperi, recovered the comet P/2002 S1 while imaging remotely. They were using a 35-cm reflector at the Skylive-Grove Creek Observatory in Australia. On the two imaging dates the stellar looking comet had a magnitude of about 20–20.6 on the earlier date and 19.5–19.9 on the latter date. The comet has a period of 8.5 years. It was first recovered in the southern constellation of Indus about 1.7 arc-min away from the predicted position but more than a magnitude fainter than expected. It was moving at a leisurely 16 arc-sec per hour south-southwest and was 134° from the Sun. Two days later a second attempt was made to confirm the object but the comet was on top of a field star. Eventually, 2 *months* later, a confirmatory second image was obtained and the returning comet was re-designated P/2009 L18 (Skiff) and then given the permanent designation 223P/Skiff. The correction to the predicted perihelion date was a fairly modest −0.19 days.

P/2003 XD10 (LINEAR–NEAT)

From August to October 2009 the University of Arizona Lunar and Planetary Lab's veteran sky hunter Jim Scotti had an impressive run of comet recovery successes using the 1.8-m Spacewatch II telescope at Kitt Peak. His first success in that period came on the nights of August 27–29 when he recovered P/2003 XD10 (LINEAR–NEAT). The comet was at a magnitude between 21.5 and 22.0 while crawling very slowly at only 1.4 arc-sec per hour through Cetus and at an elongation of 137° from the Sun. At recovery the comet was re-designated as P/2009 Q2 (LINEAR–NEAT) and quickly given a permanent designation of 224P/ LINEAR–NEAT. A tiny correction of −0.10 days was made to the perihelion date of this 6.3 year period comet.

P/2002 T1 (LINEAR)

On the same nights that Jim Scotti was attempting recovery of P/2003 XD10 he also captured images of the returning comet P/2002 T1 at a magnitude between 20.6 and 21.7 on August 28 and 29. A tiny adjustment of the perihelion date of only +0.04 of a day was required and the object was within about 1 arc-min of where it was expected to be, moving through the constellation of Taurus at 135 arc-sec per hour northeast. It was 85° from the Sun, in the morning sky, when recovered. The comet was recorded as P/2009 Q3 (LINEAR) and swiftly upgraded to a numbered status as 225P/LINEAR.

P/2009 R2 (Kowalski) = 226P/ Pigott–LINEAR–Kowalski

Now we come to a very strange comet recovery tale. On September 10, 2009, Rich Kowalski of the Catalina Sky Survey came across a very diffuse, and possibly new, 18th magnitude comet when imaging with the 0.68-m Catalina Schmidt camera. The tireless UK amateur astronomer Peter Birtwhistle took images of the new object with his 0.4-m Schmidt-Cassegrain, which helped confirm the discovery. The comet was in the far northern hemisphere in Lynx, close to 7 h 40 m and Dec +43. The motion was a healthy 1.3 arc-min per hour east-northeast and the comet was 2.1 AU from the Sun and 2.4 AU from the Earth. It was given the designation of 2009 R2 but it soon became obvious that this was a periodic comet which had also been captured, briefly, by the LINEAR survey in January 2003. On that occasion, when the very preliminary orbit of C/2003 A1 (LINEAR) was calculated, it was noticed that the orbital elements bore a similarity to a comet first spotted by Edward Pigott of York, England as long ago as November 19, 1783, but lost since then. The 2009 re-discovery by Kowalski enabled that very long lost comet to finally be linked to the new object and the comet of 1783 and a very rare triple-barreled name was added to the periodic comet inventory! The period of the comet is currently 7.3 years and it had made 33 revolutions of the inner solar system between 1783 and 2003!

P/2004 EW38

Jim Scotti's success continued in September when, on three nights, from the 21st to the 23rd he recovered P/2004 EW38 (Catalina-LINEAR) at 22nd magnitude; it was crawling along against the background stars at 10 arcsec per hour due west and at an elongation of 135° from the Sun. Once again the 1.8-m Spacewatch II telescope was used for Scotti's recovery. The predicted perihelion date only needed an adjustment of +0.02 of a day and so the comet was found close to the expected position, even if it was extremely faint. Once the new orbit had been computed the well known comet expert Maik Meyer identified six observations of the comet from the return prior to 2004, namely that of 1997, captured by the Haleakala-NEAT survey. The comet was re-designated P/2009 S4 (Catalina–LINEAR) and swiftly upgraded to a numbered status as 227P/Catalina–LINEAR. The comet was almost a year from perihelion when recovered.

P/2001 YX127

Finally, in our look back at some of the comet recoveries of 2009, Jim Scotti notched up another success when, on October 19 and 20, he used the 1.8-m Spacewatch II reflector to recover P/2001 YX127 (LINEAR) at 21st magnitude, almost 2 years from a perihelion in August 2011 and at 3.4 AU from the Sun. The period of this comet is 8.5 years. The comet was designated P/2009 U2 (LINEAR) when recovered and quickly upgraded to 228P/LINEAR once the link with the 2001 comet was established.

Recovery Conclusions

So, to summarize the last few pages, what have we learnt? Well, comet recovery is well within the grasp of the advanced CCD equipped amateur providing he or she can image down to about 20th magnitude. We have seen that it is possible to achieve this with a relatively modest 30-cm aperture telescope with the example set by Gustavo Muler, J. M. Ruiz, and R. Naves, even if a cumulative exposure time of 3 h and careful offset-stacking of all the images, was required to achieve it. When one considers the Spacewatch II telescope's aperture of 1.8 m and its resulting 36-fold greater light grasp it is obvious that recovering comets with a 30-cm aperture is a quite remarkable feat. However, the small aperture instruments do have one advantage in that their typical 10 or 15 arc-min fields of view are just right for covering the field of an object whose perihelion date may be uncertain by a significant fraction of a day. However, we have also seen that advanced amateurs have moved with the times, using remotely operated robotic telescopes, via the Internet, to enable them to image in skies that are far darker than those at a typical amateur observatory. In addition, if a suitable proposal can be made, telescopes as large as 2.0 m in aperture can be used by amateurs, via the Internet to recover those very faint comets.

Back from the Grave!

Finally, although there are more than 20 comets which have been considered lost, or dead, and even designated with a D/rating, in recent years a few comets which were thought to have ceased to exist have been recovered after huge periods of time. Most notable amongst these are the two comets 205P/Giacobini and 206P/Barnard–Boattini. Michel Giacobini, at Nice Observatory, France discovered his comet on September 4, 1896. The orbit was calculated as 6.65 years but there were signs that it had split and the comet faded rapidly below 11th magnitude. It was then lost for 101 years and declared dead until it was rediscovered by the Japanese supernova patroller Koichi Itagaki on September 10, 2008, with his 60-cm reflector and CCD camera at magnitude 13.5. The other recently "recovered even though presumed dead" comet, had been the first comet to be discovered by photography when the astronomer Edward Emerson Barnard found it on October 13, 1892. It too was lost but then accidentally rediscovered on October 7, 2008 by the professional astronomer Andrea Boattini as part of the Catalina Sky Survey's Mt. Lemmon Survey patrol. Comet 206P had made 20 revolutions of the solar system since it had last been seen!

CHAPTER EIGHT

Discovering Comets Using SOHO

We have already seen that some of the most spectacular comets in the last two centuries have been sungrazers: those large comets that pass remarkably close to the 5,800 K solar surface. With an inverse fourth power law applying to the brightness of a comet relative to its distance from the Sun it is easy to see how even small cometary chunks can become very bright when close to the solar surface. Of course, even the most resourceful Earth-bound amateur astronomers cannot discover comets using backyard telescopes when they are within a few degrees of the Sun, even if some very bright comets have, historically, been first noticed in bright twilight before sunrise or after sunset. Under no account whatsoever should bright comets (or Mercury and Venus for that matter) be hunted for when the Sun is above the horizon. The Sun might look fairly dim when close to the horizon but it still pumps out a scary amount of heat in the infrared. Even a brief glimpse of the Sun through binoculars or a telescope will result in instant blindness and even staring at the Sun with the naked eye is bound to cause retinal scarring. The eye has no pain sensors, so when you are being blinded you will not be in agony; you will just initially be dazzled and then be walking around with a Labrador in harness for the rest of your life.

However, there is a safe way to discover comets that are close to the Sun and that is by studying images from the Solar and Heliospheric Observatory satellite, known as SOHO. This satellite was launched on December 2, 1995 and the project is a collaborative venture between

M. Mobberley, *Hunting and Imaging Comets*, Patrick Moore's Practical Astronomy Series, DOI 10.1007/978-1-4419-6905-7_8, © Springer Science+Business Media, LLC 2011

ESA and NASA. SOHO is able to permanently monitor the Sun because it orbits the Sun at the same rate as the Earth (once per year) but while sitting at a point in space 1.5 million kilometers nearer to the Sun than the Earth. At this distance and at this orbital speed the gravitational pull of the Sun, one hundred times more distant but a third of a million times more massive, cancel out, meaning the satellite can float around the so-called "First Lagrangian Point" while staring permanently at the Sun. There have been other solar satellites but these have orbited the Earth, meaning that their view of the Sun is regularly interrupted as they move into the Earth's shadow. Of course, the SOHO satellite cannot last forever and there have been a few scary periods in its history already. So, even by the time this book is published there is a chance SOHO may no longer function. Nevertheless, there have been a number of solar science satellites over the years and no doubt many more will follow, even after SOHO comes to the end of its working life. The best sungrazing comet images can be found at the home page of the SOHO/LASCO and STEREO/SECCHI comet program, based in the Solar Physics Department of the U.S. Naval Research Laboratory, Washington D.C. The web address is: http://sungrazer.nrl.navy.mil/

The SOHO spacecraft has a dozen different detectors on board for analyzing the Sun and the solar system environment but the instruments that are of greatest interest to the comet hunter are those named LASCO (the Large Angle and Spectrometric Coronagraph) and, to a much lesser extent, SWAN (Solar Wind ANisotropies). A huge number of comets have been discovered by human observers scouring the LASCO instrument's images but they are all credited with the SOHO team name. They are almost all very tiny cometary fragments which could not realistically be discovered by any other means. In addition, at discovery, they are daytime, not night-time objects.

SWAN Comets

Comets discovered by the SWAN detector are relatively rare, but even so there are now seven SWAN comets, namely C/2002 O6, C/2004 H6, C/2004 V13, C/2005 P3, C/2005 T4, C/2006 M4 and the co-discovered C/2009 F6 (Yi-SWAN). It should be noted that these comets are credited with the name SWAN, not SOHO, and behave more like normal full blown comets, because the SWAN detector looks outward at the solar system in the light of Lyman alpha ultraviolet radiation. The discovery of

C/2002 O6 (SWAN) and then the other six SWAN comets, especially the superb C/2006 M4 (SWAN) showed that SWAN can be used for the discovery of comets that are observable by amateurs in the night sky, not just cometary fragments close to the Sun. However, while a few new comets and already discovered comets (whose positions are known) can sometimes be found on SWAN images they do not stand out particularly well, especially when compared to the deep ground based sky survey images. The SWAN survey covers almost all of the sky, apart from a small region blocked by the spacecraft itself within roughly 50° of the Sun. It should be emphasized that the detector's abbreviated name SWAN is, by sheer coincidence, the same as the name of the Scottish physicist William Swan, who studied carbon spectra and whose name is attached to the Swan band filters used to observe comets by some amateur astronomers. I will mention these filters again in Chap. 11, but I just wanted to emphasize that SWAN and William Swan are not connected!

Despite the fact that comets in deep space do not immediately stand out on the SWAN detector images, arguably the best comet ever detected by instruments aboard the SOHO satellite was the comet C/2006 M4 (SWAN). It was not the brightest comet discovered by a SOHO instrument, but it did become a very photogenic fourth magnitude object. The comet was first spotted by SOHO image scourers Rob Matson of Irvine, California and Michael Mattiazzo of Adelaide, South Australia, at the start of July 2006. As well as co-discovering these comets Rob and Michael have an impressive tally of SOHO comet discoveries behind them. At the time of writing Rob Matson has 93 SOHO comet discoveries or co-discoveries to his credit and this includes two other SWAN comets, namely C/2005 T4 (SWAN) and C/2009 F6 (Yi-SWAN). He has also spotted three comets in images from NASA's STEREO (Solar TErrestrial RElations Observatory) satellites. Michael Mattiazzo has made a total of five SWAN discoveries or co-discoveries, including the excellent C/2006 M4 (SWAN); the other four were C/2004 H6 (SWAN), C/2004 V13 (SWAN), C/2005 P3 (SWAN) and C/2005 T4 (SWAN).

Matson and Mattiazzo's finest discovery, C/2006 M4 (SWAN), first appeared in SWAN images taken between June 20 and July 5. Following an alert about the potential object and because the SWAN detector images the night sky, away from the Sun, various amateurs and professionals tried to image the new object or check old survey frames. Terry Lovejoy of Thornlands, Queensland, Australia had recorded it as a 12th magnitude suspect recorded on one of his images on June 30, taken with a 200 mm-f/2.8 lens and a Canon digital SLR. On Terry's image the head of the comet was about 30 arc-sec across and green in color. The indefatigable

Robert McNaught of the Siding Spring Survey subsequently photographed the comet on July 12, with the 0.5-m Uppsala Schmidt reflector and estimated it as being magnitude 12.3 with a strongly condensed head and a short tail. Following the discovery of C/2006 M4 (SWAN) its orbit was quickly calculated and it soon became obvious that it should be a very nice binocular brightness object in October. Perihelion would occur on September 28 at only 0.78 AU from the Sun and so the weeks after perihelion, when most comets peak in activity, looked promising. The comet was already performing nicely in late October when, in the space of a few days, between October 21 and 24, it brightened from magnitude 6.0–4.3. The brightness increase was not the only factor either as in CCD images there was a dramatic increase in activity in the gas tail and, with a negligible dust tail to blur the view, some really intricate tail structure was observed in this gassy comet. Certainly C/2006 M4 (SWAN) had one of the finest gas tails this author has recorded in almost 30 years of comet photography and imaging (see Fig. 8.1a, b).

It is worth mentioning the first SWAN comet discovery in this section too. C/2002 O6 (SWAN) was spotted in that detector's images by Masayuki Suzuki of Japan on August 1, 2002. The comet was found in images captured on July 25 and 27 and it was in Eridanus at the time and so impossibly placed for most northern hemisphere observers. It appears that the comet was a very small one, with an absolute magnitude of around 10.0 at discovery, but passing relatively close to the Earth. It was seen visually by Alan Hale shortly after it was found, at around magnitude 9.5, and was predicted to reach perihelion at only 0.49 AU from the Sun at the start of September, after passing only 39 million kilometers from the earth on August 9. However, from late August it was already fading suggesting that its limited dust and gas stocks were already nearing exhaustion. Bearing in mind its feeble nature, being close to impossible to observe from the northern hemisphere, rapid motion and solar elongation it is not surprising that it evaded LINEAR and NEAT and made itself first visible on SWAN images.

The LASCO Detector

The SOHO LASCO detector is the most prolific comet discovery device ever constructed, even though it was not specifically designed for imaging cometary fragments. It consists of a set of three coronagraphs that image the solar corona from 1.1 to 32 solar radii, in other words out to a radius of 8° from the Sun. The narrowest field detector, C1, failed after a telescope shutdown period but the wider field C2 and C3 instruments

Fig. 8.1. (a) Comet C/2006 M4 (SWAN) imaged on October 24, 2006 using a Celestron 14 at f/7.7 and an SBIG ST9XE CCD. The head image consists of twenty 30 s exposures registered on the inner coma/nucleus region. In the half degree high image mosaic the tail frames consist of individual 10 min exposures with the Paramount ME tracking at the rate of the comet's motion. Image: Martin Mobberley. (**b**) Comet C/2006 M4 (SWAN) imaged on October 26, 2006 using a Celestron 14 at f/7.7 and an SBIG ST9XE CCD. Exposure details as in (**a**). Image: Martin Mobberley.

still survive and the C3 detector covers the widest field. The Sun itself is kept hidden from the detectors to prevent it saturating them but the solar corona within 22 million kilometers of the Sun can then be studied. Of course, even if a comet has already been discovered by the night-time surveys, or by amateurs, the SOHO LASCO detector can still be very useful for watching small perihelion comets as they track within 8° of the Sun and become invisible to observers on the Earth, due to the twilight and daylight glare (see Fig. 8.2). The LASCO detector works in the extreme ultraviolet, between ultraviolet and X-ray wavelengths. The blue

2003/02/18 05:54

Fig. 8.2. A SOHO LASCO image taken on February 18, 2003 showing a massive solar CME (Coronal Mass Ejection) as well as comet C/2002 V1 (NEAT) in the *top right*, which was at perihelion, only 5.7° and 14.9 million kilometers from the Sun. The SOHO/LASCO images used in this chapter are produced by a consortium of the Naval Research Laboratory (USA), Max-Planck-Institut fuer Aeronomie (Germany), Laboratoire d'Astronomie (France), and the University of Birmingham (UK). SOHO is a project of international cooperation between ESA and NASA.

images (shortest wavelengths) are centred on 171 Å. In practice objects as faint as stars of magnitude 10 can normally be seen in the images although some flare stars of magnitude 14 have been witnessed as they are X-ray bright.

The Kreutz fragments discovered mainly in the SOHO LASCO C3 images are very small by cometary standards and typically have absolute magnitudes of around 20. As I mentioned earlier a comet like Halley has an absolute magnitude of 4.5 and a typical new comet discovery has an absolute magnitude of 6 or 7. Hale–Bopp had an absolute magnitude around zero. By definition therefore when a Kreutz SOHO fragment is 1 AU from the Sun and 1 AU from the Earth it will only be magnitude 20 and magnitude 20 fuzzy objects are pretty good at avoiding night-time surveys! They are simply too faint. The perihelion distances of the SOHO Kreutz objects tend to fall into two categories, specifically, those with perihelion distances of less than 0.007 AU and those greater than this value. As 0.007 AU places a comet within one solar radius of the solar surface itself it can be seen that comets in the second category can actually impact with the solar surface and disappear forever. Indeed, the difference between a Kreutz fragment hitting the solar surface or not can end up being solely down to the position of Jupiter in the solar system when the comet made its final approach as Jupiter fractionally moves the centre of mass of the solar system away from the centre of the Sun (the heliocentre).

Up to the start of 2010 SOHO had discovered more than 1,700 comets (the vast majority using the LASCO C3 detector), of which most are members of the aforementioned Kreutz group of sungrazing comets. Roughly 85% are Kreutz group sungrazers and, of the rest, roughly 10% fit into three other groups with specific orbital elements. That leaves about 4 or 5% which are not members of the other main three sungrazing groups. The other sungrazing fragment groups, defined by there being comets with similar orbital elements in a group, are the Meyer, Kracht and Marsden groups.

Many amateur astronomers have discovered SOHO comets by analyzing the LASCO data although there can be long gaps without any discoveries. One such gap was a full 45 days long and stretched from December 21, 2001 to February 6, 2002 and a month long SOHO comet drought ran from August 4 to September 6, 2006. SOHO or SWAN comet discoveries are not named after the person who studied the images as they have been analyzed from a public source and the SOHO and SWAN names are considered the established team names of these instruments. This follows from the IAU Comet Naming Guidelines of March 2003 as adopted by the Committee on Small Body Nomenclature of Division III of the International Astronomical Union. Specifically, Guideline 3.4 (b) states:

> Comets that are discovered from data or images made public through printed publication or electronic posting (e.g., World Wide Web) are not eligible for individual names of people and generally will not be named unless there is an established program name for the origin of the images. Such discoverers are considered members of the "team."

Leading Amateur SOHO Discoverers

At the time of writing this chapter six people have discovered a hundred or more SOHO comets by scouring the LASCO detector images. The individuals concerned are Rainer Kracht of Germany, Hua Su and Bo Zhou of China, Michael Oates of the UK, the US amateur Tony Hoffman and Xavier Leprette of France. The US amateur Rob Matson is likely to reach 100 by the time this book is published. With 239 SOHO comet discoveries Rainer Kracht easily leads the table. Two specific groups of SOHO comet fragments have been named Kracht and Kracht 2. He has also discovered seven comets in images obtained by the SOLWIND (4) and STEREO (3) space probes. Rob Matson's 93 discoveries include three that were made with the SWAN detector on SOHO and he has also discovered three comets in STEREO images too. A few amateurs who have discovered comets by normal night-time patrolling, in other words, comets that bear their surnames, have also bagged SOHO comets too. Thus, Sebastian Hoenig, the discoverer of C/2002 O4 is also the discoverer of 43 SOHO comets and Kazimieras Cernis (Lithuania), the discoverer of three comets (found visually using a large Newtonian), has bagged 22 SOHO comets, including one from scouring the SWAN images. In addition, Terry Lovejoy, who discovered two comets with a 200 mm lens in 2007, has discovered eleven SOHO comets and Tao Chen of China, who found C/2008 C1 (Chen–Gao) in the night sky, with a DSLR system, has ten SOHO comets to his credit. Tao Chen's countryman and collaborator, Xing Gao, the joint discoverer of two Chinese comets in the night sky, has also discovered four SOHO comets. The German comet expert Maik Meyer has discovered 40 SOHO comets which is why a special group of SOHO comets with specific orbital elements has been named the Meyer group.

Fig. 8.3. Michael Oates has discovered 144 comets by studying SOHO LASCO images.

It was actually the UK amateur astronomer Michael Oates (see Fig. 8.3) who pioneered the technique of searching the SOHO archives from home in an attempt to discover cometary fragments in the data. Michael had no idea that comets could actually be *discovered* by browsing images posted on the Internet, until January 29, 2000. Incredibly, From January 30, 2000 to the end of July 2002 he discovered 136 comets without ever going outdoors! Doug Biesecker, a professional astronomer with the SOHO/LASCO team, used to discover the vast majority of the SOHO comets, until Michael and the other amateurs got in on the act! Doug's current SOHO discovery tally stands at 53. Michael Oates used the popular image processing package Maxim DL to process the SOHO-LASCO images and a digital imaging software package called ACDSee to compare images and look for new comets. His determination has shown that while amateurs can study professional images online the era of amateur comet discovery is NOT dead, even if the observing is carried out from indoors and you do not get the satisfaction of having the comet named after you. Figures 8.4 and 8.5 show SOHO 111 and SOHO 143 discovery images. The latter comet was discovered on an image also showing SOHO 53.

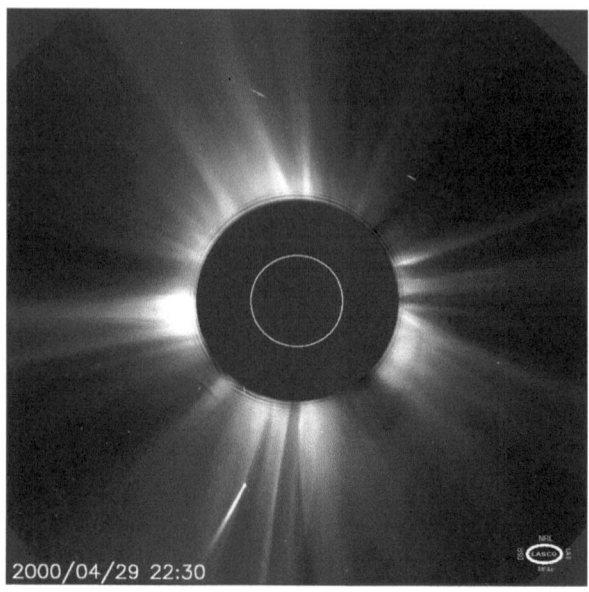

2000/04/29 22:30

Fig. 8.4. Comet SOHO 111 discovered jointly by Oates, Boschat, Lovejoy and Gorrelli. Image: ESA/NASA.

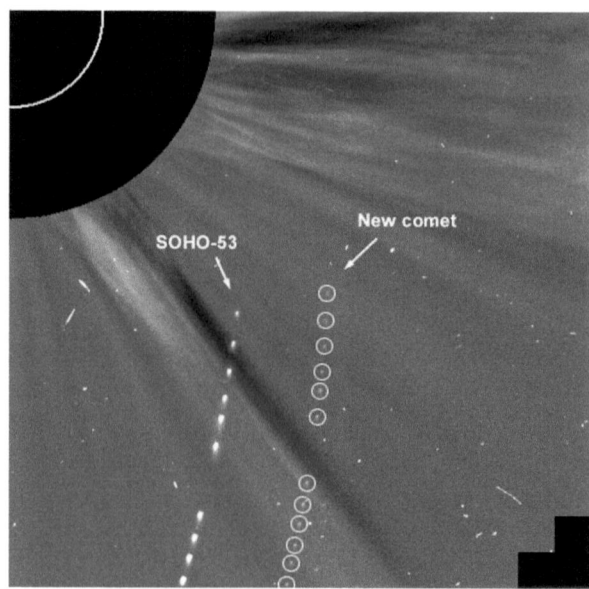

SOHO-53

New comet

Fig. 8.5. Comet SOHO 143 was discovered on images showing the brighter SOHO 53. The SOHO-LASCO images were combined by Michael Oates to produce this image. Image: courtesy Michael Oates/ESA/NASA.

C/1998 J1 (SOHO)

The brightest comet ever found in the SOHO spacecraft's LASCO images was C/1998 J1 (SOHO). The comet sneaked up from the southern hemisphere just before the LINEAR patrol system in New Mexico started full operations in the spring of 1998. However, it would have made little difference as the comet had hugged the solar twilight and stayed within 90° of the Sun for 6 months prior to its discovery on May 3, 1998. It was spotted by Shane Stezelberger in the SOHO LASCO field just 5 days prior to perihelion (see Fig. 8.6). Prior to that time, in November 1997, the comet had been below −55° declination, so well below the southern limit of LINEAR and NEAT and probably fainter than 14th magnitude as well. Its magnitude at discovery may have been as high as zero although more conservative estimates place it at magnitude 1.0. It reached perihelion on May 8 at 0.15 AU from the Sun. Needless to say, seeing such a dramatic object in the SOHO LASCO images prompted many amateurs to try to pick this comet up when it emerged from the twilight after perihelion. However, it would appear that no amateurs were successful until May 19, some 8 days after perihelion, with the comet 26° from the Sun and not far from Bellatrix in Orion. Of course, at that time of year Orion is unobservable from the far

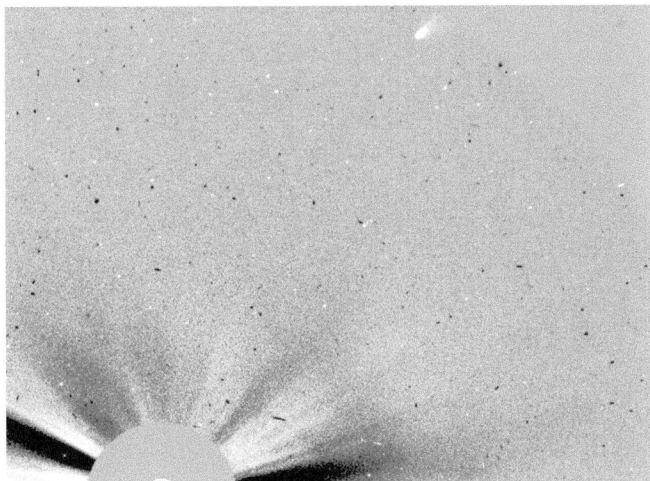

Fig. 8.6. Comet C/1998 J1 (SOHO) at perihelion on May 8, 1998. The comet is *right* at the *top* of this SOHO LASCO image frame and was discovered by Shane Stezelberger as it crept into the edge of the detector's field of view. Image: ESA/NASA.

northern hemisphere but southern hemisphere observers had better luck and managed to see and photograph the comet, now third magnitude, and its 2° tail. C/1998 J1 faded to fifth magnitude by the end of May but then surged back to third magnitude at the very start of June, eventually falling back down to seventh magnitude after mid-June; the peak weeks of the brightest ever SOHO comet were good, but very brief, and only southern hemisphere amateurs had seen it at its best.

CHAPTER NINE

Following the NEOs That Might Become Comets

I have already mentioned that it may be important in an activity like comet hunting to mix the search program with projects that can produce useful scientific data. In an "all or nothing" patrol where only a comet discovery is the aim the observer can rapidly become disillusioned with nothing to show from the search. Unlike with, say, supernova hunting, where a top hunter might average a discovery every 4,000 galaxy images there are no guarantees at all with comet hunting. An observer could spend his entire life looking for comets and get nothing. You only have to look at the example of the British discoverer Roy Panther, who spent 33 years sweeping for comets before he found one, on Christmas Day 1980, to realize how much patience can be required. So, to prevent total psychological burn out carrying out comet hunting as a secondary activity can be the key. I have already mentioned that comet *recovery*, rather than *discovery*, might be regarded as the next best thing to a genuine comet being found and I will be exploring the monitoring of outbursting comets in the Chap. 10. However, discovering that an object previously thought to be a new asteroid is actually a new comet is another option and a modest number of newly discovered fast moving objects, sometimes flagged as Near Earth Objects (NEOs), turn out to be comets on closer inspection. An image by the author of a fast moving, non-cometary, object, 2004 XP1, is shown in Fig. 9.1.

M. Mobberley, *Hunting and Imaging Comets*, Patrick Moore's Practical Astronomy Series, DOI 10.1007/978-1-4419-6905-7_9, © Springer Science+Business Media, LLC 2011

Fig. 9.1. Near Earth Object 2004 XP14 moving rapidly south-west in Draco on July 3, 2006. A 240 s exposure with a 0.35-m Celestron 14 at f/7.7 and an SBIG ST9XE CCD. The NEO moved 4 arc-min during the exposure. Image: Martin Mobberley.

Comet Naming Rules

In theory it is possible for someone who realizes an object is cometary in nature to be credited as a comet discoverer, but this is extremely rare. Let us look at the guidelines adopted in 2003 by the Committee on Small Body Nomenclature of Division III of the International Astronomical Union. The fourth Guideline, entitled "Regarding cometary nature that is not immediately noticed" is the one that is relevant here. I will quote verbatim from this guideline so that there is no risk of misinterpretation. The advice reads as follows:

4.1. It frequently happens that a comet is found by (a) discoverer(s), whether a single individual, two individuals working together, or a team, who cannot detect cometary activity with the equipment that he/she/they possess. Such an object may therefore be assumed to be a minor planet and so designated when two or more nights' worth of observations are available to the Minor Planet Center (or posted, for example, prior to being designated, on the MPC's NEO Confirmation webpage, if unusual motion is detected).

4.2. If an observer (whether an individual or a team) who is not the original discoverer of the "minor planet" finds that an "asteroidal" object has a cometary appearance, and if such cometary appearance is confirmed, both the original discoverer of the "minor planet" and the identifier of the cometary nature may be credited in the name of the comet, subject to the following prerequisites:

(a) This dual recognition in the name will occur only when there has been no prior suspicion of the unusual nature of the object; if the object was listed on the "NEO Confirmation Page" or had an unusual orbit published prior to identification of cometary appearance, the identifier of the cometary nature will not be eligible for inclusion in the name.

(b) If no prior suspicion of unusual nature is suspected, the name of the newly recognized comet can consist of two parts:
One part derived from the name of the original discoverer of the "minor planet," and the other part derived from the name of the discoverer of the cometary nature. *Each part* of the name would, however, be subject to the following conditions:

1. Only one name (either individual or team) is permitted. If the "minor planet" was credited jointly to two individuals, only one of these names can be used.
2. If the original discoverer of the "minor planet" is a team, the team cannot suggest that an individual team-member's name be used.

4.3. If follow-up observations performed by (an) observer(s) other than the original discoverer of a reportedly "asteroidal" object show the object to be a comet (that is, showing a coma and/or tail), and the provisos of 4.2 do not apply, the comet may receive a single name of the original discoverer (individual or team).

4.4. If the minor-planet designation was published before the realization is made that the object is a comet, the comet will retain the minor-planet designation. Otherwise, a new comet designation will be assigned.

4.5. If the object receives a permanent minor-planet number prior to its recognition as a comet, it shall be accorded "dual status." As such, it both retains the permanent minor-planet number and receives a new periodic-comet number:

(a) If the numbered "minor planet" has already received a name, the comet should inherit this name.
(b) If the "minor planet" has not yet received a name, a new name for the comet will be assigned according to these guidelines. The same new name will also be used for the "minor-planet" numbering, noting that minor-planet names must be unique.

The discerning reader will quickly spot that clause 4.2(a) is the critical one here because many objects initially determined to be asteroidal in nature are first chased up via the NEO Confirmation Page and many have an unusual orbit (such as a high inclination) which indicated the likelihood of the object being a comet. In many cases the discovery exposure is simply not deep enough to reveal a coma but immediate follow up by professional and amateur telescopes quickly reveals a "soft" appearance. These are cometary targets that are just too easy and if the rules were relaxed scores of amateurs would be imaging ever NEO discovery, claiming a fuzzy appearance and hoping to get a comet named after them. Obviously this would be a much easier way to get a comet named after you than checking thousands of patrol images or sweeping the sky visually for years in freezing conditions! Indeed, it would devalue the patient work of the dedicated backyard patrollers and the centuries of visual amateur comet hunting work.

Asteroids That Became Comets

It may be helpful to provide a few examples of objects that were initially designated as asteroids but were then re-defined as comets. On January 31, 1997 Syuichi Nakano of Sumoto, Japan, reported to the Minor Planet Center some observations from January 30 to 31 by his fellow countryman Takao Kobayashi, a highly advanced amateur astronomer. These were astrometric measurements of what appeared to be the discovery of an asteroid or minor planet. However, further astrometric observations showed that the orbit was more typical of a comet. In the USA another amateur astronomer, Warren Offutt of Cloudcroft, New Mexico, used his large 0.6-m f/7 Ritchey–Chrétien, to image the object and a coma was detected confirming its cometary nature. On February 3 and 4, Offutt commented that the head of the object appeared slightly diffuse and that there was a definite, very faint tail in position angle 284°, which was 17 arc-sec long on February 3 and 8 arc-sec long on February 4. J. Ticha and M. Tichy (Klet Observatory) also imaged the new object, with a 0.57-m f/5.2 reflector. When Tichy measured the images he noted that there was a suggestion of a 10 arc-sec long tail at a position angle of 300°. This comet was ultimately designated as P/1997 B1 (Kobayashi), the discoverer of the object which he initially thought was an asteroid but turned out to be a comet.

Another example was the comet P/1999 DN3 (Korlevic–Juric) which was the 89th asteroid discovered in the second half of February 1999, hence the original 1999 DN3 designation. It was found at Visnjan observatory by Korado Korlevic and Mario Juric and then they noticed its cometary nature. Their names were then added and its short period led to the P designation.

2000 WM1 is a similar case, except one where an automated system discovered the object. On December 16, 2000 a magnitude 17.8 asteroid was discovered by LINEAR. The astrometric positions allowed the Central Bureau to calculate a rough orbit and link these observations to another LINEAR object detected a month earlier on November 16 and 18. With a perihelion distance of 0.55 AU this was looking like a cometary orbit and when Tim Spahr pointed the 1.2-m Mt. Hopkins telescope towards the comet on December 20 and detected a 10 arc-sec coma and a faint tail this identity was confirmed. The comet proved to be an excellent performer and reached perihelion 13 months later in January 2002, brightening to magnitude 4.5 around mid-month with a dramatic surge to 2.9 at month's end. Its cometary designation was C/2000 WM1 (LINEAR). The periodic comet P/2001 BB50 (LINEAR-NEAT) is another object that started off as an asteroid as it was simply 2001 BB50 when first discovered and showed no sign of a coma or tail.

Perhaps the most extreme case of a solid body being renamed as a comet was when the asteroid 2060 Chiron, discovered by Charles T. Kowal in 1977 and nicknamed "the mini-planet" at the time, was found to have a weak coma 12 years later. It is generally regarded as the first of the Centaur objects and in 1977 it was the furthest known asteroid, lying between Saturn and Uranus. It mysteriously brightened in 1988 which led to the observations that determined it had a coma and even a tail (detected in 1993). As a numbered and named asteroid by then the only solution was to give it the permanent periodic comet designation 95P/Chiron, leaving it still named after the mythological centaur, instead of changing it to 95P/Kowal, after its discoverer.

Dual-Status Objects

With an estimated diameter of 230 km, 95P/Chiron = (2060) Chiron is one enormous cometary nucleus! Strictly speaking 95P now has a dual status as a comet and an asteroid and, in all truth, it is somewhere between the two.

Four other periodic comets that started life as asteroids and now have a rare dual status are 107P/Wilson–Harrington, alias asteroid (4015) Wilson–Harrington, 133P/Elst–Pizarro, alias asteroid (7968) Elst–Pizarro, 174P/Echeclus, alias (60558) Echeclus and 176P/LINEAR, alias (118401) LINEAR. The object named after Albert Wilson and Robert Harrington was discovered by those two astronomers as long ago as 1949, but when re-discovered 30 years later by Eleanor Helin at Palomar it masqueraded as an asteroid. In fact, it did such a good impersonation of an asteroid that it was given the number 4015 on its next return in 1988. By 1992 astronomers had nailed the orbit of the asteroid down so precisely that they had to accept 4015 was comet Wilson–Harrington of 1949, except the comet had died and become an asteroid. With Elst–Pizarro the situation was just the opposite with the asteroid originally designated 1979 OW7, discovered by Schelte J. Bus, being rediscovered as a comet by Eric W. Elst and Guido Pizarro in 1996. The object has the orbit of a standard asteroid, but shows cometary features. The comet 174P/Echeclus is a very similar case to 95P/Chiron in that it is a distant centaur in the outer solar system, originally classified as 2000 EC98 and then numbered and named as (60558) Echeclus. Then, in December 2005 a coma was detected. Astronomers think Echeclus is only about a third of the diameter of Chiron, but even at 85 km it is still a huge cometary nucleus. Finally, 176P/LINEAR, alias (118401) LINEAR, was provisionally designated as asteroid 1999 RE70 when found by LINEAR on September 7, 1999. However, 6 years later, in November 2005, it was discovered to have a coma by Henry Hsieh and David Jewitt as part of the Hawaii Trails project using the Gemini North 8-m telescope on Mauna Kea.

All of the above examples show that the IAU are willing to reclassify, or dual classify in rare cases, objects that are originally discovered as asteroids but then become fuzzy and vice versa. However, there is, as yet, no case of an amateur being credited with a comet discovery by just spotting a coma around an already discovered asteroid that was suspected of having a cometary type orbit (that is, an orbit with a higher eccentricity and inclination than normal). The sticking point, of course, is that clause (4.2(a)) that precludes objects with known unusual orbits or those deliberately flagged up via the NEO Confirmation Page. However, while this means that observing newly discovered objects will not get you an easy comet discovery there is still enormous respect for any amateur who regularly chases objects on the NEO confirmation page and notices cometary features. While this might seem less valuable than a straight comet discovery most amateurs would be pretty pleased if they were considered to be an equal by professional astronomers.

Nevertheless, the previously mentioned case of comet P/2002 BV (Yeung), now renamed 172P/Yeung, shows that the prior detection of an object by professional patrols (Spacewatch and LINEAR in that case) does not ban an observer from having the comet named after him. Kwong Yu Yeung independently discovered the comet as an asteroidal looking object but his re-discovery enabled the orbit to be linked, re-calculated, and determined to be cometary. Following that the professional astronomer Timothy Spahr revealed the new object's coma with the 1.2-m telescope on Mt. Hopkins. The object had not been flagged as a suspect comet prior to Yeung's observations and he had not been following the object on the basis that its orbit was uncertain: its orbit was thought to be asteroidal. So, clause 4.2(a) was not transgressed!

A Remarkable UK Amateur

One amateur astronomer who follows up NEOs, and is regarded as an equal by professionals, is the UK's Peter Birtwhistle, who lives at Great Shefford in Berkshire, England. Peter has been an active amateur astronomer for decades with a particular interest in astrometry, but in May 2002 he decided to invest in some serious equipment and purchased a 30-cm aperture fork mounted Schmidt–Cassegrain, a 2.4 m fiber glass dome and a CCD camera. Within a month an observatory code J95 was allocated to the new Great Shefford facility by the Minor Planet Center and the rest is history. Peter has become one of the world's most respected amateur astronomers and is regarded as an equal by all professionals who are aware of his work. Peter has refined his techniques to such an extent that following up NEOs posted on the NEO Confirmation Page is now a routine operation for him and every clear night astrometric results are e-mailed to the Minor Planet Center. On many occasions he has been the first to image a NEO and detect its cometary nature and his results appear on hundreds of Minor Planet Circulars and International Astronomical Union Circulars. However, it is only when some of his observing statistics are digested that his true output can be appreciated. In the

Fig. 9.2. The 0.4-m aperture Schmidt–Cassegrain of NEO chaser extraordinaire Peter Birtwhistle at Great Shefford UK, viewed through the dome slit. Image: Peter Birtwhistle.

3 years from June 2002 to June 2005 Peter measured 4,388 positions of NEOs with his 30-cm Schmidt–Cassegrain. All the results since then have been acquired with a larger 40-cm Schmidt–Cassegrain (see Figs. 9.2 and 9.3). Peter's total number of astrometric measurements exceeded 10,000 at the start of January 2009 and that total is now well in excess of 11,000. Peter uses an Apogee Alta U47 CCD on his telescope which uses a Marconi E2V 1,024 × 1,024 pixel chip (see Fig. 9.4). The pixels are 13 μm across. With a 0.63× tele-compressor at the focus of his nominally f/10 Schmidt–Cassegrain this delivers an 18.4 × 18.4 arc-min field at a scale of 1.1 arc-sec per pixel. Binning the pixels 2 × 2 to give a 512 × 512 pixel system gives 2.2 arc-sec per pixel. The Apogee camera is capable of delivering 55°C of cooling, lowering the thermal noise by a factor of several hundred, and its quantum efficiency is better than 90% from green to red wavelengths. The camera can download the images in 2 s at full resolution, or 1 s with 2 × 2 binning; these fast download speeds are essential for rapid work. Peter's observing statistics are mind boggling. The figures for 1 year alone are greater than the output of many leading amateurs over a lifetime of work. Take 2008 as one example; in that year Peter observed on an incredible 171 nights, that's 47% of the nights available in a year, from a cloudy country! His average observing session on each night used was

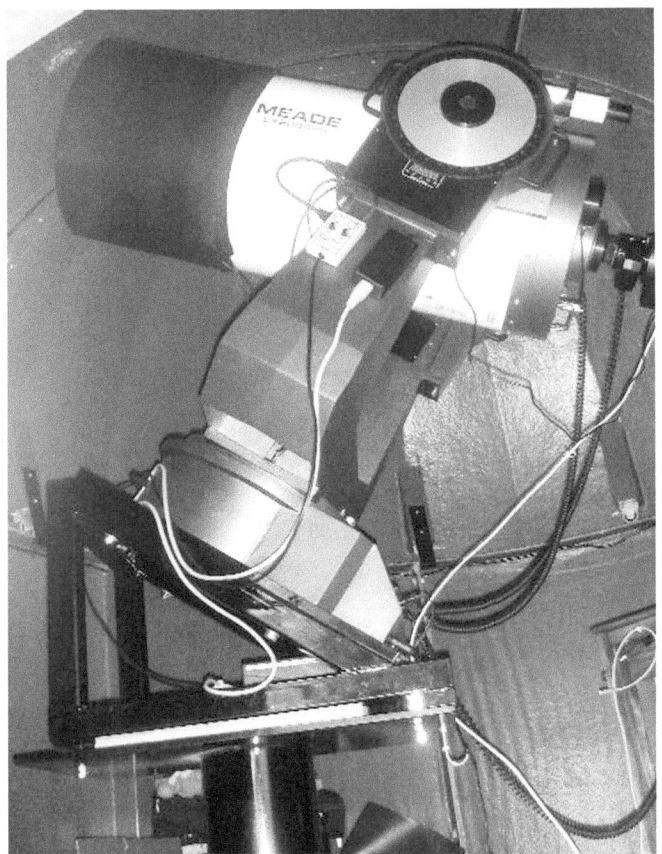

Fig. 9.3. The 0.4-m aperture Schmidt–Cassegrain of Peter Birt-whistle viewed from inside the dome. Image: Peter Birtwhistle.

4.4 h in length and 69% of those 4.4 h were spent imaging, rather than slewing or thinking! In total he obtained 91,715 images in 2008 with each image being 20 s in duration and an average of 30 s between exposures. So, that works out at 528 exposures taken on an average night. Mind-blowing! The end result is an average of about ten accurate astrometric measurements sent to the Minor Planets Center on every night used, and about 1,500 measurements per year.

Peter uses an impressive array of software tools for his production line astrometry which includes Maxim DL/CCD, Astrometrica, Focus-Max, PoleAlignMax, Pinpoint, Dimension 4, Find Orb and Sat_ID. These software packages allow him to image, measure, focus, polar align, mark his image headers, synchronize his PC clock, calculate orbits and identify

Fig. 9.4. The Apogee Alta U47 CCD mounted on Peter Birtwhistle's Schmidt–Cassegrain. Image: Peter Birtwhistle.

satellites crossing the CCD field. For the fastest moving NEOs you need to know the time of your exposure to an accuracy of better than 1 s and this not only means your PC clock has to be accurately set, via the Internet, but you may have to know how much delay there is in the mechanical shutter of your CCD camera! In addition to all these commercial tools Peter has written a number of visual basic routines himself to help speed up his observing and make it even more efficient.

So far, more than 150 new asteroids have been discovered by Peter during his analysis of more than 11,000 astrometric images from some half a million exposures taken since 2002. This discovery ratio is very similar to that experienced by the best supernova patrollers, namely, a discovery for every few thousand galaxy images. The amazing Peter Birtwhistle has not discovered any comets yet, but he has identified a number of new NEOs as being comets by being the first to image them with a deep exposure and then spotting a fuzzy coma. For example, on October 4 and 7, 2002 the NEAT and LINEAR surveys discovered an asteroidal object which was placed on the NEO Confirmation Page. On October 27, Peter was able to detect a small coma and tails with his 30-cm Schmidt–Cassegrain which were confirmed later by Tim Spahr with the 1.2-m telescope at

Fig. 9.5. Peter Birtwhistle's 330 s exposure of an object on the NEO Confirmation Page on October 27, 2002 revealed it had a coma and two tails (*centre of the image*). It was subsequently designated as a comet: P/2002 T6 (NEAT-LINEAR).

Mt. Hopkins. Peter's image showed two faint, thin, straight tails at a position angle of 107° (36″ long) and 310° (39″ long) and a nuclear condensation with a diameter of about 8″ (see Fig. 9.5). The magnitude 18.5 comet was named P/2002 T6 (NEAT-LINEAR) and reached perihelion in June 2003, at about 3.4 AU. The orbital period is about 21 years.

A similar result was achieved in July 2003 when he imaged another NEO reported by LINEAR on the 19th of the month. Peter, using his 30-cm Schmidt–Cassegrain, and professional astronomer P. Kusnirak, at Ondrejov observatory, using a 0.65-m f/3.6 reflector, found the new object had a coma and tail. Peter's description of the comet was that it showed a nuclear condensation 6 arc-sec diameter with a faint, short, broad tail about 15 arc-sec long at a position angle of 139°. He estimated the magnitude as being between 17.3 and 18.2. The NEO was declared a comet and named C/2003 O1 (LINEAR).

In May 2004 a 19th magnitude NEO, discovered by LINEAR on the 29th of the month, was found by Peter to have a 12 arc-sec coma, with a 20 arc-sec east pointing tail, when imaging with his modest 30-cm Schmidt–Cassegrain. Paolo Holvorcem and Michael Schwartz also reported a coma when the NEO was imaged with the Tenagra Observatory's 0.81-m reflector. The object was subsequently named comet C/2004 K3 (LINEAR).

In August 2006, now equipped with his 40-cm Schmidt–Cassegrain, Peter imaged a new 19th magnitude LONEOS NEO object after its details were posted on the Minor Planet Center's NEO Confirmation Page, on the 29th of that month. Peter reported that a 44 min cumulative stack of short exposures, when co-added and allowing for the object's motion, showed that the new object had a 9 arc-sec diameter coma and a faint tail about 30 arc-sec long. Exposures by J. Young with the 0.61-m f/16 Cassegrain reflector at Table Mountain and by L. Buzzi (Varese, Italy, 0.60-m f/4.64 reflector) also showed the tail. As a result the new object became the short period comet P/2006 Q2 (LONEOS). There are plenty of further examples that could be made but I think I have shown in the above that amateur astronomers like Peter Birtwhistle are every bit as respected by professional astronomers as they would be if they had discovered a comet. Peter's astrometric work gives him the satisfaction of having carried out a vital scientific contribution in addition to the fact that he has discovered 150 asteroids and helped to reveal numerous new discoveries as cometary in nature.

Comets Among the Asteroid Population

Finally, I would just like to add that it is not impossible that established asteroids, with numbers and even names, which have an unusual orbit, could first be spotted as exhibiting cometary activity by amateur visual observers. Remember, there are lots of asteroids now known (approaching half a million) and so there are plenty of unusual ones to study, especially if you have a large amateur telescope and can reach, say, 13th magnitude comets visually. In November 2009 the comet observer and Hale–Bopp co-discoverer Alan Hale drew members of the Yahoo! Comet Mailing List's attention to the favorable return to perihelion of asteroid (20898) Fountainhills which he had observed visually during its previous return in 2000/2001. No cometary activity was noted but this does not mean that the future possibility of something developing can be ruled out. Asteroids with a catalogue magnitude even as faint as eighteen can, if a coma develops, increase their brightness a 100-fold, to magnitude 13, and so asteroids with unusual orbits are well worth pointing the telescope towards, even if nothing is likely to be obvious at that location.

The Italian-based T3 project, whose website can be found at http://asteroidi.uai.it/t3.htm, was formed for the specific purpose of discovering possible cometary features among those members of the asteroid population having a Tisserand parameter with respect to Jupiter (Tj) of less than three. A Tisserand value of 3 is approximately the boundary between asteroidal and cometary orbits, according to H.F. Levison, a leading authority on comet taxonomy. It is calculated using a complex formula which compares the semi-major axis of Jupiter's orbit with the semi-major axis of the object's orbit, along with inputs for the eccentricity and inclination of the object. The T3 observing technique is relatively simple. The project coordinator, Luca Buzzi, sends a list of $T < 3$ asteroids close to perihelion to a couple of dozen top asteroid and comet imagers each week, so that all of the prime objects are covered. If any sign of cometary activity is detected this is included in the COM report line in the astrometric report to the MPC and confirmation is sought from other T3 project members of the cometary feature. Following confirmation the MPC and CBAT are e-mailed again and professional astronomers are contacted with the news. Further details can be found at the T3 website listed above.

CHAPTER TEN

Monitoring Outbursting Comets

Comets are unpredictable things and, when you think about it, this is hardly surprising. There is no design specification for a comet as it is just an object that looks fuzzy when seen from a very long distance away. Any celestial object, of any size, that vents dust and gas when it gets hot is classified as a comet. Of course, in practice there is a size limit because truly massive planet sized objects will hold onto their gases by gravitational force even when close to the Sun, but typical cometary nuclei from, say, 1–60 km in diameter will not have any significant gravitational power. The images of cometary nuclei taken by space probes show objects which look quite similar to asteroids, at least when seen at close range. Of course, when hundreds of millions of kilometers away, the nucleus becomes an infinitesimal point and we just see a sheet of dust or gas, illuminated by sunlight and projected against the vast blackness of space. Professional astronomers have complex formulae they can use to give a rough estimate of, for example, the cometary water production rate based on the intensities of strong molecular emissions detected spectroscopically. Some also have very rough estimation formulae to predict water output rates based on the magnitude of a comet imaged in different wavebands.

It goes without saying that everything in the cometary head (coma) and tail consists of dust and gas that has recently fled the surface of the comet, but precisely where it has come from and how much of the surface is covered by icy, volatile material is impossible to tell without sending a space probe to look at the object. Cometary nuclei are not perfectly spherical objects coated to an even density with a layer of ice and dust. They often appear to have various active regions on their surfaces where this type of material lives and, as a comet rotates and the icy regions come into sunlight, they heat up and evaporate. A comet may have numerous such regions on its surface and there may also be underground fissures and rifts containing icy substances too. If frozen material gets trapped in such places and then heated up it may occasionally vent small amounts of material when heated, until the pressure reaches a certain point where a geyser vents into space, causing a dramatic outburst as seen from Earth.

Many comets are known for having minor outbursts but a handful have become notorious for their outbursting ability. Comets are often discovered or recovered when in outburst.

Being the first to detect a major outburst of a hibernating comet may not put the observer in the same category as a comet discovery, but it can contribute valuable science, especially if few others are observing the object at the same time.

29P/Schwassmann–Wachmann

The most regular outbursting comet is the tongue-twistingly named 29P/ Schwassmann–Wachmann, sometimes called Schwassmann–Wachmann I as it was the first of three comets with that double-barreled name. The correct pronunciation is "Shvasman–Varkman." Arnold Schwassmann and Arno Arthur Wachmann of the Hamburg Observatory, Bergedorf, Germany, discovered this comet on photographs taken on November 15, 1927. It was approximately magnitude 13.5 at discovery but faded very rapidly and by the start of December had slumped to magnitude 16. Four years later some pre-discovery images were unearthed by Karl Reinmuth, namely some photographic plates taken in early March 1902. Schwassmann–Wachmann I appeared to be around magnitude 12 on those early plates. The comet has an orbital period of 14.7 years and the orbit is inclined to the ecliptic by 9.39°. The orbit also has a low eccentricity of 0.0441, meaning that it is quite circular compared with many other comets. At perihelion it closes to within 5.72 AU of the Sun, whereas at aphelion it only moves out to 6.25 AU. It therefore orbits the Sun just outside the orbit of Jupiter, but on a plane tilted by 9.39°. The orbital expert Kazuo Kinoshita has deduced that the initial orbital eccentricity of 29P/ Schwassmann–Wachmann was higher, at around 0.15, and the orbital period has gradually declined from around 16.0 to 14.7 years. In the twentieth century Jupiter altered the orbit of 29P twice even though there was no close encounter as such. The 1930 alteration increased the peri-helion distance by 0.05 AU and decreased the orbital period by 0.3 years. The 1974 alteration increased the perihelion distance by 0.32 AU and decreased the orbital period by 0.11 years. In 2037 Jupiter is predicted to increase the perihelion distance to 5.87 AU and increase the orbital period to 15.9 years.

29P/Schwassmann–Wachmann experiences numerous outbursts in brightness and they are observable every year too, even with amateur equipment. When the comet is in a quiet mood it can drop as low as magnitude 19 when at aphelion, or 17 when at perihelion, even when at opposition (in other words, transiting at midnight). These values are not unusual for a comet beyond the orbit of Jupiter. However, every few months or weeks it has a significant outburst. Some of these outbursts only take the comet to a magnitude of 13 or so which is rather too faint for easy visual observation with a modest amateur telescope of, say, 250-mm aperture. Remember, comets are fuzzy and their light is not concentrated to a point like a star. A magnitude 13 comet can be as tricky to see as a magnitude 16 star. Before the CCD era the number of out-bursts that 29P experienced per year was not fully appreciated as many were simply too faint for visual observers and few photographers would want to expose, develop and fix photographs of the 29P field every clear night. However, since the CCD era it has become apparent that 29P has many outbursts and images are often posted on the internet or flagged up in the *Yahoo!* Comet-Images or Comets-Mailing List groups. While a magnitude 13 fuzzy patch may sound a bit dull the comet can reach tenth magnitude in major outbursts. In terms of absolute magnitude (the magnitude of a comet if viewed at 1 AU from the Earth and 1 AU from the Sun) 29P has a quiescent value of around 7.0, fairly typical for a comet. But when in outburst it can, briefly, have an absolute magni-tude brighter than zero, that is, in the class of Hale–Bopp. Of course, at its distance, just outside the orbit of Jupiter, it only shines as brightly as tenth or eleventh magnitude at these times, not at zero magnitude, but at 6 AU from the Sun that is pretty impressive. While there are comets that have experienced more dramatic outbursts, and I will be looking at some of them later, no other comet exhibits such regular outbursts of several magnitudes than 29P. Since regular CCD imaging of comets started in the late 1990s many outbursts of 29P have been fully docu-mented. While the comet may indeed be outbursting more frequently in recent years, the continuous CCD patrols must surely mean that almost every outburst above magnitude 15 is being spotted. In early 1996 comet 29P was in outburst for several months and this behavior repeated itself in early 1998, 1999 and in July–August 2001. Several outbursts occurred in 2004 and there were further prominent ones in January 2005,

Fig. 10.1. A dramatic outburst of comet 29P/Schwassmann–Wachmann on January 24, 2009. Note the southward pointing spike. This is a composite of three 120 s exposures with an SBIG ST9XE CCD and a 0.35-m aperture Celestron 14 working at f/7.7. Field 13′ × 13′. Image: Martin Mobberley.

February 2007 and January 2009 (see Fig. 10.1). 29P's nucleus is thought to have a rotation period of around 60 days, following imaging of it by NASA's Spitzer telescope and, when in outburst, the position of a prominent jet at this time can enable the rotation period to be further scrutinized. Some astronomers think that the randomly spaced outbursts may be caused by a thermal heat wave propagating into the nucleus of the comet and thereby triggering sublimation of carbon monoxide.

Not all comet outbursts are spectacular but any case where a comet brightens by a magnitude or more in the space of 24 h is of great interest to all cometary astronomers, whether unpaid, salaried or living off research grants. There are numerous examples of this happening and one good one applies to the comet discovered by amateur astronomers Tanaka and Machholz in 1992: comet C/1992 F1. The outburst took the comet from around magnitude 8.5 to 7.0 around the full Moon period in May of that year. Full Moon is obviously a time when many comet observers take a break as the bright skies make observing comets more problematic, but it is also a time when a determined observer can get a result with less chance of anyone else spotting the outburst. On this occasion Bjorn Granslo of Norway was one of the few observers to spot this brief outburst occurring.

73P/Schwassmann–Wachmann

No, you are not seeing things! It is that tongue twisting Schwassmann–Wachmann name again. The problem is that Friedrich Karl Arnold Schwassmann (1870–1964) co-discovered three periodic comets, namely 29P, 31P and 73P, with his younger colleague Arno Arthur Wachmann (1902–1990) between 1925 and 1930 at the Hamburg Observatory in Bergedorf, Germany as well as them co-discovering the nonperiodic comet C/1930 D1 (Peltier–Schwassmann–Wachmann). Both 29P and 73P proved to be fascinating objects. As we have seen comet 29P is the most regularly outbursting comet known, whereas 73P is the comet king where breaking up into different pieces is concerned. It is also a faint comet (or faint comet shoal!) that often flies relatively close to the Earth when it returns every 5 years. Indeed, it is 73P's occasional proximity to the Earth that makes it an easy object to observe, every so often. It is debatable whether 73P should be mentioned in the same section as violently outbursting comets, because while it does outburst in magnitude when it fragments, the actual fragmentation process is its main trick. All things considered I think it probably does deserve including in this section next to its namesake 29P Schwassmann–Wachmann. 73P was first discovered on photographs exposed for the Hamburg observatory minor planet survey on May 2, 1930. The comet peaked at sixth magnitude when it passed only 0.0616 AU from Earth on May 31, 1930. It was then lost for 49 years until J. Johnston and M. Buhagiar of Perth Observatory, Australia, rediscovered it on August 13, 1979. It peaked at twelfth magnitude but its perihelion date was a huge 34 days later than would have been expected, although it had approached Jupiter within 1 AU twice in the intervening decades (to only 0.25 AU in 1953). The comet was not seen during the 1985–1986 apparition, when Halley drew all the attention, but it was seen in 1990 and peaked at ninth magnitude while passing 0.3661 AU from the Earth. During the next three apparitions the comet would start performing some party tricks! At the 1995 return the comet was not expected to brighten to more than eleventh magnitude; however, by September an outburst caused it to reach eighth magnitude and then a further brightness surge in October ramped it up to sixth magnitude. 73P was still eighth magnitude well into December when something else happened. Observers noticed that the nucleus, at high power, seemed to have split into at least three, and maybe four, tiny points. These were labeled A, B, C and D, although D was a very briefly observed phenomenon and the component now labeled as B may have a dual identity with the original "B" and a further component, labeled E at the next return, at least according to research by Zdenek Sekanina. Things were certainly getting a bit complicated by the end of that return! In 2001 comet 73P returned again. It was not favorably placed but, once again, seemed to be in outburst and in two main parts, probably those previously labeled as B and C. Not surprisingly there was much anticipation of the return in 2006 as the comet was predicted to make a very close flyby of the Earth, at a distance of only 0.0735 AU on May 13. At recovery in 2006 component C was the brightest, with B being a few magnitudes fainter. Component "B" was actually found by an amateur astronomer at magnitude 18.8–19.0, namely J. A. Farrell of Jemez Springs, New Mexico, USA, who recovered the component on January 6, 2006, while imaging with a 41-cm reflector and a cooled CCD camera. A completely new component, G, was then found almost simultaneously, by amateur Roy Tucker and professional E.J. Christensen. Then things started to get very silly indeed as ground based professional telescopes and the Hubble Space Telescope identified more than 60 individual fragments of 73P (see Fig. 10.2a, b). The brightest chunks, and those within amateur range, were B, C, G and R with C being

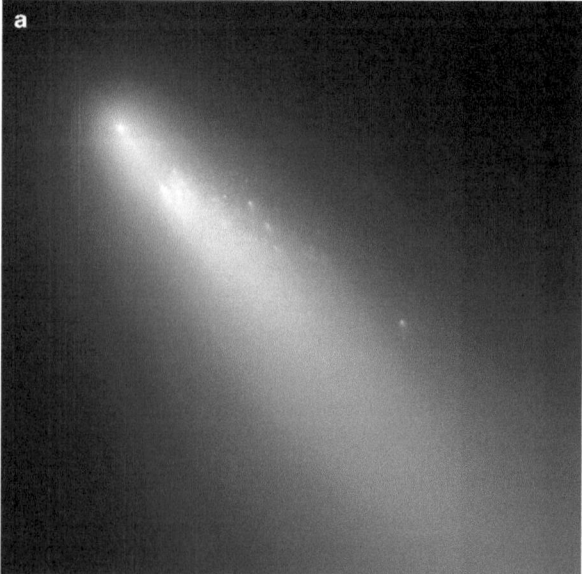

Fig. 10.2. (a) This Hubble space telescope image shows numerous fragments breaking away from component B of comet 73P/Schwassmann–Wachmann on April 18, 2006.

Fig. 10.2. (continued) (**b**) Another Hubble space telescope image taken 2 days after the image in (**a**) shows the numerous fragments from component B of comet 73P/Schwassmann–Wachmann have moved further away from the head. (**c**) This image shows the two main sections of comet 73P/Schwassmann–Wachmann, namely parts C (*left*) and B (*right*), which were 9° apart in the sky at the time of the images (individual fields are 13 arc-min wide) taken on April 28, 2006. Pictures obtained with a 0.35-m Celestron 14 and SBIG ST9XE CCD at f/7.7. Image: Martin Mobberley.

regarded as the main chunk of the original comet and with B outbursting to rival C in brightness on several occasions (see Fig. 10.2c). The comet is next due to return to perihelion in October 2011. What might we expect to see then? More fragments or nothing at all? By the time you read these words, depending when you buy this book, it may be imminent or may have already happened. Certainly, outbursts caused by fragmentation are 73P's favorite trick but will anything remain of it at its next return?

41P/Tuttle–Giacobini–Kresák

The periodic comet 41P/Tuttle–Giacobini–Kresák has something of a reputation for impressive outbursts. It was discovered by Horace Tuttle at Harvard Observatory on May 3, 1858, while in the constellation of Leo Minor; it was just 0.36 AU from the Earth and a day from perihelion. It faded rapidly and so only a few observations of it could be made; its orbit was therefore very uncertain, with the period calculated by various mathematicians ranging between 5.8 and 7.5 years. It was next "discovered" by Michel Giacobini at Nice Observatory in France in June 1907 but was only observed for 2 weeks. Twenty-one years later, the orbit expert A. C. D. Crommelin was able to mathematically link the 1858 and 1907 comets, calculating a five and a half year orbit. However, no-one saw the comet again until April 1951 when Lubos Kresák of Skalnaté Pleso Observatory re-discovered it. After a few weeks it was realized that this was the Tuttle–Giacobini comet and so its name was increased to Tuttle–Giacobini–Kresák. While the history of the comet's multiple discoveries is fascinating enough, since 1973 it has become far better known for its outbursts. In that year the returning comet was recovered by Elizabeth Roemer and J. Q. Latta of the Steward Observatory in Arizona on January 8 at magnitude 21. By mid-May 1973 it had reached fourteenth magnitude but suddenly, on May 26 and 27 Ako and Mameda, two Japanese amateurs, discovered a comet of eighth and fifth magnitude on their photographs of the region. Seiler (in Germany) also spotted a fourth magnitude comet where Tuttle–Giacobini–Kresák should be just a few hours later than Mameda, also on the 27th. In the space of a week the comet had brightened from fourteenth to fourth magnitude: a 10,000-fold increase in brightness! It was also exhibiting a 10 arc-min long tail. But a month later it was back to fourteenth magnitude. Amazing behavior! However, the story does not end there! Between July 4 and 7 the comet outburst again and by the same ten magnitude amount; it reached fourth magnitude around July 7. The veteran comet photographer

Reggie Waterfield, who had been confined to a wheelchair since contracting polio in the late 1940s, estimated its magnitude as 4.5 on the night of July 7/8 from his observatory at Woolston in England. By September the comet was at the limit of most observers' photographic equipment, at approximately magnitude 17. However, in 1978 Tuttle–Giacobini–Kresak refused to put on the same performance and was a dismal 15th magnitude object throughout its best period of visibility. Indeed, the intrinsically faint comet refused to entertain observers much from 1973 to 1995, when on August 17 and 18, 1995 L. Shotter of Union Town, Pennslyvania and H. Abe, of Yatsuka, Japan, found it had experienced another 100-fold outburst to eighth and ninth magnitude respectively.

Five years later, at the 2000/2001 apparition, it would seem that the comet was once again in the mood for a performance. In late November of 2000 comet 41P was a very tricky 15th magnitude photographic object and little more was expected from it. However, the veteran visual comet observer Alan Hale, co-discoverer of Hale–Bopp, along with the Japanese observer Yoshimoto, were monitoring the comet and found it had suddenly outburst to magnitude 10 when observed on November 27, 2000. This was not the spectacular ten magnitude double outburst of 1973 but, even so, a five magnitude surge is pretty impressive by normal cometary standards. The outburst continued into the first week of December. Comet 41P/Tuttle–Giacobini–Kresák is definitely one to keep an eye on every five and a half years. In the following year another five magnitude outburst, this time experienced by C/2001 A2 (LINEAR) 2 months before perihelion, also made the comet headlines.

51P, 52P and C/1998 U5

The faint comet 51P/Harrington was discovered with the Palomar Schmidt telescope on August 14, 1953 by Robert Harrington when it was approximately magnitude 15. The comet was never going to be a bright one but its prospects of ever being visible by amateurs again were dealt a severe blow in October 1956 when a close encounter with Jupiter changed its orbit. At the comet's return of 1960 it was a dismal magnitude 19–20 object from October to November. It was not seen again until 1980, but even then it only reached magnitude 18. However, it was seen in 1987 and in 1994 the circumstances were very good and the comet performed better than expected, reaching magnitude 12, so that it was just within amateur visual range with a decent aperture telescope. The first sign of something odd happening was on October 5, 1994 when Jim Scotti at

Kitt Peak observatory in Arizona imaged the field and discovered that as well as a twelfth magnitude primary component, two chunks of the comet of 21st and 20th magnitude seemed to have broken away and were accompanying the main object. When a splitting like this occurs it is often the cause of a surge in brightness. 51P/Harrington next passed perihelion on June 5, 2001 and 6 months later the comet was seen to split again and its magnitude temporarily increased from a single magnitude 20 image into two separate magnitude 16 or 17 chunks. Amateur astronomer Pepe Manteca of the Observatorio de Begues (near Barcelona) reported that his observations on 2001 December 6.1 UT (0.31-m f/6.3 Schmidt–Cassegrain reflector + CCD) showed the comet to have split, with the two components being separated by some 10 arc-sec on an east–west line. The components on that night were around magnitude 17.0–17.4; the western component being perhaps up to 0.4 magnitudes brighter than the eastern, which is the one that was under observation during July to November 2001. On Dec 7.0 and 7.9 Manteca gave magnitude estimates of 16.4 for the eastern component and 16.6 for the western component. The eastern component was labeled as component A at the comet's 1994 apparition, when those two much fainter components, B and C, were also recorded by Jim Scotti. The new western component was therefore denoted as component D. On December 10 and 11 the author of this book obtained further images showing the comet had split (see Fig. 10.3a, b).

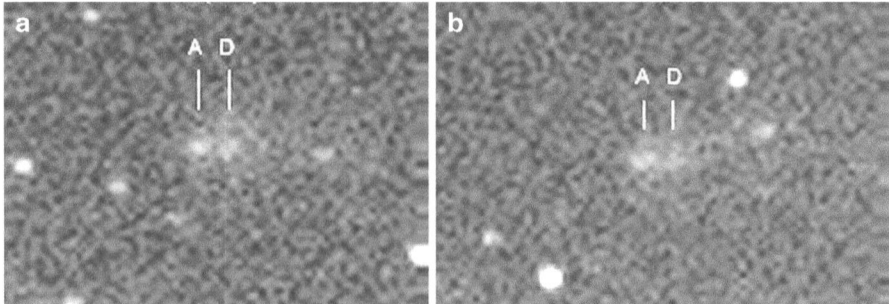

Fig. 10.3. (a) The two separate sixteenth magnitude components of comet 51P/Harrington, roughly 15 arc-sec apart, on December 10, 2001. 0.49-m f/4.5 Newtonian + Starlight Xpress MX916 CCD. 300 s exposure. Field 3′ × 2′. Image: Martin Mobberley. (b) Another image of the two separate sixteenth magnitude components of comet 51P/Harrington, roughly 15 arc-sec apart and 1 day after the image in (a), taken with the same equipment. 300 s exposure. Image: Martin Mobberley.

Of course, by any standards, a magnitude of 16 is faint, but the comet had brightened by about 40 times during the split of the nucleus.

Another comet discovered by Robert Harrington is in the records as having outburst in recent years, but this time it was a co-discovery with George Abell, also working at the Palomar Observatory. This is the comet 52P/Harrington–Abell. When discovered, on March 22, 1955, comet 52P was only seventeenth magnitude but the previous perihelion date of December 18, 1954 was soon calculated, as was its orbital period of 7 years. The comet has been seen at every single apparition since its 1955 discovery, beginning with its first recovery in January 1962 by Alan McClure. The other appearances occurred in 1969, 1976, 1983, 1991, 1998/1999 and 2006. At the 1998/1999 return the comet reached perihelion on January 27, 1999. It was first recovered on July 21, 1998 by Alain Maury (l'Observatoire de la Côte d'Azur, France) using a 0.9-m Schmidt reflector and a CCD. Maury was expecting the comet to be tricky to spot, with a magnitude of 21 or 22, but was surprised to find a comet of magnitude 12.2 on the image. This ten magnitude brightness surge was confirmed the following night by various visual observers as well as the CCD astrometrists M. Tichy and Z. Moravec at Klet Observatory, Herman Mikuz and J. Skvarc at Crni Vrh Observatory and R. Greimel and D. D. Balam at Victoria observatory. They all estimated the visual magnitude as between 10.9 and 11.8 with a coma up to 3 arc-min across. The comet's abnormal brightness continued for many months and it was still magnitude 11 around the time of perihelion. It finally dropped below 12th magnitude by the start of April. The comet had maintained a brightness of seven to ten magnitudes above that predicted from previous returns, for several months!

Comets do not have to be periodic, numbered, or seen previously on multiple returns to behave bizarrely. They can all do strange things. In October 1998 the LINEAR facility discovered an object which was soon revealed as a magnitude 14 comet. It became designated as C/1998 U5 (LINEAR) and was moving through Gemini at the time of its discovery. Orbital calculations showed that it would reach perihelion at 1.2 AU from the Sun in late December. It is quite normal for comets discovered by professionals to have their magnitudes underestimated by two magnitudes as they rarely take the whole extent of the coma into account, just the stellar core magnitude. However, the veteran comet observer John Bortle observed the comet on November 12, just 2 weeks after discovery, and found it had risen by almost six magnitudes, to magnitude 8.7, and now had a 7 arc-min diameter coma, as seen through 20×80 binoculars. When viewed with Bortle's 0.41-m reflector it showed the classic appearance of a comet in major outburst. C/1998 U5 displayed an intense central knot and a small, bright, inner coma; this was surrounded by a very extensive outer coma with no obvious boundary and an extremely faint outer halo. Guy Hurst of Basingstoke, UK, observed the comet with 15×80 binoculars 3 days later and made it as bright as magnitude 7.0 in 10×50 binoculars, exhibiting a coma one quarter of a degree in diameter. He commented that although the coma was now very large, the surface brightness was very low. An analysis of the comet's behavior indicated it was obeying an explosive magnitude law roughly corresponding to $m = 6.5 + 5\log D + 30\log r$. In the first 2 weeks after discovery the coma diameter increased from around 70,000 –300,000 km, eventually declining to 100,000 km in May of the following year.

C/1931 O1 (Nagata)

Looking back over the annals of comet observation quite a few lesser known comets have experienced major outbursts but many of the stories have largely been forgotten in the mists of time. One of these stories concerns the comet discovered by Mr Masuji Nagata on July 16, 1931. Nagata-san was born in Japan, but had moved to Brawley, California where he became a successful fruit grower. Nagata was trying to point his 75-mm refractor towards Neptune when he found a seventh magnitude comet some 3° east of the planet. He quickly informed professional astronomers Zug and Berman at Mount Wilson who confirmed the discovery and calculated that it was already 5 weeks past perihelion. By the first week in October, 3 months after its discovery and 4 months after perihelion, Comet Nagata had faded from magnitude seven to magnitude twelve and a half, but suddenly, on October sixth, it brightened by sixty times, back to magnitude eight. Professor Georges Van Biesbroeck described the comet after the outburst as having two tails, each 15 arc-min long, with nebulous matter between the two.

17P/Holmes

Without doubt 17P/Holmes has become the comet most associated with the term "outburst" since the extraordinary events of October 2007; but the comet's first massive outburst occurred 115 years earlier, in 1892. On November 6 of that year Edwin Holmes, of London, England, pointed his

32-cm Newtonian towards the Andromeda Galaxy, M31. When a fuzzy object was centered in the finder he went to the eyepiece of the main telescope and was momentarily confused at what he saw. The object did not resemble M31 at all! He soon realized that this was not M31 (although the galaxy was nearby); it was a new comet with a very bright nucleus and a 5 arc-min coma. Just after midnight, on November 7.03, Holmes worked out the position of the object and went indoors to write an urgent letter to two highly respected British astronomers, William Maw and Walter Maunder and also to his friend, Mr Kidd. The comet was quickly confirmed the next night by Kidd and Bartlett, observing from Bramley in Kent, and an independent discovery was also made the next night by the nova discoverer Dr Thomas Anderson of Edinburgh. By this time the new comet, destined to be named comet Holmes, was a third magnitude naked eye object.

Attempts were quickly made to determine the orbit of comet Holmes but early efforts floundered because they were based on the assumption that this was a long period comet with a parabolic orbit. In addition, surely this was a comet brightening as it headed towards perihelion, as how could a comet any brighter have been missed around perihelion time? Also, if it had a short period, surely it would have been seen at every return, like comet Halley? There was even speculation that this might be the lost comet 3D/Biela, last seen in two parts in 1852, tiny particles from which were seen as a meteor storm in 1872, and a small chunk of which may have survived as the modern discovery 207P/NEAT. Eventually it was realized that comet Holmes not only had a short period of 6.9 years but that it had also passed perihelion almost 5 months *before* its discovery. The truth dawned that the comet must have been discovered during a huge outburst in brightness. The renowned eagle-eyed observer Edward Emerson Barnard stated that on 9 November comet Holmes was "easily visible to the naked eye, as a small hazy star, and almost exactly as bright as the brightest part of the Andromeda galaxy." The comet slowly faded and by January 5, 1893 Barnard described the comet as "very large and very faint" even in a 30-cm refractor, indicating that it was probably only about twelfth magnitude. However, just 11 days later comet Holmes experienced another outburst which took it back to at least eighth magnitude. A photographic exposure by the pioneering British amateur astronomer Isaac Roberts showed a tiny coma some 39 arc-sec across on January 17. The comet then steadily faded once more and the last observations of it were in early April.

Comet Holmes was next seen on its return to the inner solar system on June 11, 1899, 5 weeks after perihelion. It was spotted with the 91-cm Lick Refractor by Perrine.

E.E. Barnard reported it to be at its brightest in mid August when it was only magnitude 13. Seven years later, in 1906, Barnard's rival Max Wolf, at Königstuhl Observatory in Heidelberg, Germany, recovered the comet on a photographic plate exposed on August 26, 1906. He estimated the magnitude as 15.5 and it did not get much brighter. After 1906 the comet appears to have been dead as there were no observations of it from that year until 1964; a span of 58 years. However, in 1963 Brian Marsden used a powerful computer (powerful for the 1960s) and deduced that the orbital period and perihelion distance would have altered significantly, predicting they would now be 7.35 years and 2.347 AU respectively. As a result of his calculations the comet was finally recovered by Elizabeth Roemer of the U.S. Naval Observatory at Flagstaff, Arizona on July 16, 1964 at magnitude 19.2. Since that year 17P/Holmes has been recorded at every return, but failed to repeat its spectacular 1892 style outburst until 2007.

The 2007 Explosion!

Just after local midnight on the night of October 23/24, 2007 the keen Spanish amateur astronomer Juan Antonio Henríquez Santana imaged the field of 17P/Holmes and was amazed to see a magnitude 8.0 small disc of light on his CCD image. Juan Antonio Henríquez had been using a 20-cm aperture f/9 Vixen VC200L (Visac) telescope and an SBIG ST9XE CCD to image comet 17P/Holmes in previous weeks and already had images showing it at its predicted magnitude of 17. Just as in 1892 the comet was 5 months past perihelion and then suddenly, literally in the space of 1 day, it surged in brightness by nine magnitudes prior to the outburst discovery and was continuing to brighten by 0.5 magnitudes per hour! An image taken on the previous night by the prolific Spanish CCD imager Gustavo Muler showed the comet had then been 17th magnitude. In images taken on the night of outburst 17P/Holmes had an almost stellar appearance with all the light concentrated in the inner 20 arc-sec. Between October 24.2 and 24.7 the magnitude increased further to an incredible magnitude 2.8, making it easily visible to the naked eye as an extra star in Perseus, less than 5° east of the magnitude 1.8 star Mirphak. In the first day of the outburst most of the light was coming from the core of a rapidly expanding disc, 1 arc-min across. A quick calculation told me that a sphere of debris that size (roughly 70,000 km) at the distance of Holmes (1.6 AU) would only need to reflect a 10,000th of the light falling on its surface if it were a solid globe. The thing is though, it was not a solid globe, just an explosion, and so the amount of matter being ejected from the nucleus was surely staggering for a small comet! Throughout the remaining days of October (see Figs. 10.4–10.6) and most of November Holmes' magnitude held steady, but through telescopes it started looking remarkably different, despite having a magnitude of between 2.6 and 2.8.

Whatever explosion had occurred on comet 17P/Holmes it was expanding rapidly outwards in a dramatic bubble which, through the eyepiece, looked like nothing any comet observers could ever remember seeing. It was just plain weird! By the end of October the main explosion "bubble" of comet Holmes had attained a diameter of more than 10 arc-min. Within the middle of the bubble was an intense arc-minute wide bright core and as the days passed the actual position of 17P's nucleus

Fig. 10.4. Comet 17P/Holmes on October 27, 2007, just 3 days after a massive outburst had caused it to brighten by fourteen magnitudes. The bubble of light had already expanded to 4 arc-min in diameter. Celestron 14 at f/7.7 + SBIG ST9XE CCD. This was a mere 5 s exposure! Image: Martin Mobberley.

Fig. 10.5. Comet 17P/Holmes on October 30, 2007, just 6 days after its massive outburst. The bubble of light had now expanded to 10 arc-min in diameter. Celestron 14 at f/7/7 + SBIG ST9XE CCD. This was a mere 10 s exposure! Image: Martin Mobberley.

was marked by a 30 arc-sec diameter blob slowly moving northeast. We appeared to be looking at the expanding bubble which was still centered on the explosion site while the nucleus, shrouded in dust started moving away from the densest debris. At least, that is how it looked to me! With the comet lying some 1.63 AU from Earth the radius of the bubble was expanding at some 2,000 km per hour (so 4,000 km per hour for the bubble diameter). Holmes would drift within a degree of the bright star Mirphak in mid November making it a photo-opportunity for all CCD and DSLR imagers (see Fig. 10.7). This expansion rate continued into December and by the start of the second week of that month the bubble was a degree across (see Fig. 10.8) and so, in physical terms, approaching five million kilometers in diameter! Subsequent analysis of the 17P/Holmes explosion suggested that the nucleus, with an estimated mass of some eight billion metric tons, had shed at least 150 million metric tons

Fig. 10.6. Comet 17P/Holmes recorded by the Austrian comet imaging maestro Michael Jäger on November 5, 2007, 12 days after its massive outburst. The bubble of light had now expanded to 20 arc-min in diameter, equivalent to 1.4 million kilometres! This absolutely stunning image was taken with a 20-cm aperture Astrosysteme Austria (ASA) f/2.7 astrograph and a Sigma 6303 CCD camera. The LRGB image goes very deep and so has captured a complex *blue gas tail* in addition to the expanding explosion bubble. This picture is assembled from four, deep, 10 min luminance frames with R, G and B exposures of 4, 4 and 6 min respectively. Details in the inner region have been enhanced with a Larson–Sekanina filter, as described in Chap. 11. Image: Michael Jäger.

Fig. 10.7. Comet 17P/Holmes on November 16, 2007, 23 days after its massive outburst, passing close to the bright star Mirphak in Perseus. The bubble of light had now expanded to half a degree in diameter. A Takahashi E-160 astrograph (160-mm aperture, f/3.3) and Canon 300D DSLR were used at ISO 400. Field 100' wide. 15 exposures of 30 s duration were stacked. Image: Martin Mobberley.

of dust in the outburst, or roughly two percent of its total mass! Other estimates suggested that 40 metric tons of water, per second, were being produced at the peak of the outburst. In terms of energy the explosion is estimated as involving 2,500 trillion Joules, roughly equivalent to a half-megaton nuclear blast in space.

Of course, as the bubble expanded further its surface brightness became fainter and its edge became harder to discern. The integrated magnitude of the comet stayed at third and then fourth magnitude well into 2008 but because the surface brightness was then so low it became impossible to detect the bubble easily, unless you were under jet black skies after the spring months of 2008. After having dredged up all the historical records from 1892/1893 comet fanatics were well aware that a second outburst could occur at any time. But in fact that did not happen this time, although minor photometric fluctuations in the nuclear magnitude of 17P/Holmes

Fig. 10.8. Comet 17P/Holmes on December 13, 2007, 50 days after its massive outburst. The bubble of light had now expanded to 1° in diameter. A Takahashi E-160 astrograph (160-mm aperture, f/3.3) and Canon 300D DSLR were used at ISO 400. Eight exposures of 60 s duration were stacked. Field 90′ wide. Image: Martin Mobberley.

were detected. The most violent cometary outburst ever witnessed in modern times had occurred and for dedicated comet imagers like me the October to December 2007 period had been an extraordinary time that we would never forget. A comet had outburst in brightness by more than 14 magnitudes (roughly half a million times) but no-one could quite explain why! 17P/Holmes next reaches perihelion in March 2014 and it will be well worth imaging from late 2013 onwards as it starts to heat up. The cometary expert Zdenek Sekanina has suggested that a complete disintegration of a large mass, loosely bound to 17P's surface, is the root cause of comet Holmes' behaviour. He stated in the *International Comet Quarterly* periodical that "peeling off and jettisoning of pancake-shaped layers of nucleus terrain, one by one, is an efficient fragmentation process."

The 1991 Outburst of 1P/Halley

The most famous comet of all time, 1P/Halley, was last at perihelion on February 9, 1986. Amateur astronomers across the world followed the comet throughout that year and this author managed to take a photograph of Halley as late as January 31, 1987. Professional astronomers were obviously able to follow Halley for much longer. As 1991 started this famous comet was 14 AU from the Sun, which is almost midway between the orbits of Saturn and Uranus. At this distance from the Sun the comet would have been bitterly cold and the solar heating from 5 years earlier would have radiated away, leaving the nucleus of the comet as just a ball of rock and frozen ice. At that distance the magnitude of that dark potato shaped nucleus, just 15 by 8 km in size, should have been between magnitude 25 and 26 and images taken in the previous year had confirmed this. Astronomers Hainaut and Smette working at the European Southern Observatory at La Silla, Chile, and their team member Richard West at ESO in Garching, Germany, were therefore amazed that when they imaged the field of Halley with the 1.54-m Danish telescope at La Silla, on February 12, 1991, and found that a faint coma was visible and the comet had a nuclear magnitude of 21 but a total integrated magnitude of 19, several hundred times brighter than expected. The coma was 20 arc-sec across, corresponding to a size of 200,000 km at 14.3 AU. In the frozen depths of space this event was hard to comprehend: what could possibly be the mechanism for an outburst at this huge distance? Observations with the 3.6-m New Technology Telescope at La Silla confirmed the magnitude on 15 February and Karen Meech of the University of Hawaii confirmed the outburst using the University's 2.2-m telescope on the same date. She reported a "hemispherically"-shaped coma at a position angle of 135°, extending up to a maximum of 120,000 km (projected) from the nucleus and with a diameter up to a maximum of 260,000 km.

On February 17, 1991 E. Giraud at ESO used the 2.2-m ESO/Max-Planck-Institute telescope to image Halley and obtained a V magnitude of 19.88 while A. Smette of ESO, La Silla, reported that a 30-min spectrograph with the New Technology Telescope yielded a coma that presented a solar-type spectrum with no obvious emission lines, strongly suggesting a dust-based composition to the outburst.

As is often the case with such outbursts amateur astronomers were keen to compete with the professionals now that the magnitude of the comet was just within achievable range. One of the pioneers of amateur CCD imaging, Christian Buil, along with Eric Thouvenot, C. Calvet and J.F. Touillaud, used the 0.61-m telescope at the 2,877 m altitude Pic du

Midi Observatory in the French Pyrenees to image Halley in outburst on February 18 and 19. Using hour long exposures their images recorded a coma 37 arc-sec in diameter with a total V magnitude of 19.5.

From March 12 to 18 further images were obtained with the Danish 1.54-m telescope at La Silla showing the total magnitude was still high, at about 20. The overall size of the coma was greater than 30 arc-sec with the outer contour resembling a "bow-shock" parabola, with the same general orientation as reported earlier. However, Richard West of ESO noted that there were important morphological changes from night to night. On March 13 a condensation was seen extending toward the southwest from the nucleus. On other nights, bands of enhanced surface brightness were present within the coma. It was therefore evident that the outburst was continuing and Richard West urged observers with access to large telescopes to "monitor this unique event."

So, what could have caused such a huge outburst in deep space? Various theories were put forward at the time and a paper in the prestigious journal *Nature*, in the October 1991 edition, suggested that a shock wave generated by a solar flare and propagating through the interplanetary medium could have caused the outburst. The authors (Intriligator and Dryer) argued that a solar flare on January 31, 1991 could have produced a shock wave that would have reached Halley and would have been sufficiently strong to "crack the comet's crust of fluffy ice." Needless to say, many astronomers were rather skeptical of this theory! They were also skeptical of the theory that Halley could have hit something in interplanetary space, 18° below the ecliptic plane. Interestingly, in 2008 Zdenek Sekanina, writing in the *International Comet Quarterly* periodical, drew readers' attention to a 17P/Holmes level of outburst experienced by comet 1P/Halley as long ago as January 1836.

Halley was not the only comet in outburst at the start of 1991 either. On January 7 Howard Brewington had swept up a tenth magnitude comet whose orbit was soon linked to one discovered by Metcalf in 1906 and not seen since 1907. Its initial orbital period in 1906 had been calculated as 7.8 years. Pre-discovery photographs taken by Tanaka in Japan showed that the comet had been fifteenth magnitude only 2 days before Brewington's discovery and so it had brightened a 100-fold in 2 days. Grillmair and McNaught had taken photographs of the field where the comet should have been 6 months earlier in July and it was then not detectable at all and so fainter than magnitude 19. The comet became renamed as 97P/Metcalf–Brewington but on its next return the LINEAR team found it as a distinctly asteroidal object on September 1, 2000. The magnitude was a mere 19.0 and the extremely inactive former comet was a degree and 3 days off its predicted position.

CHAPTER ELEVEN

Comet Imaging Techniques

When I first started trying my hand at deep sky photography, in the early 1980s, recording comets used to be nothing less than a battle against the insensitivity of photographic film and the inevitable arrival of cloud, on those crucial moon-free nights when a bright comet was close to perihelion. In recent years the situation has changed considerably. On the positive side modern CCDs are 20 times more light-sensitive than the best photographic emulsions and image processing is far easier than messing around for hours with revolting chemicals in a darkroom. On the negative side the modern lives of working people leave little room for learning new skills and the stress of the modern working day leaves little enthusiasm for a night-time battle with clouds and unfriendly software and hardware. I firmly believe that well thought out observatories and patient perseverance are the key to achieving success where imaging comets and deep sky objects is concerned. Basically, anyone who has learned to use a computer can learn to take good comet images; it is all a question of surmounting the various hurdles in a systematic fashion. I have always thought that it is good to have a few "hero figures" to look up to. In my early days in the British Astronomical Association I was inspired, like everyone else, by Patrick Moore as well as the lunar photographer Cdr. Henry Hatfield and the lunar and planetary photographer and telescope making genius Horace Dall, but as my interests slowly changed from the Moon and planets to comets, so I looked to other achievers.

M. Mobberley, *Hunting and Imaging Comets*, Patrick Moore's Practical Astronomy Series, DOI 10.1007/978-1-4419-6905-7_11, © Springer Science+Business Media, LLC 2011

Who could fail to be inspired by the legendary George Alcock? A discoverer of five comets and five novae and the man who memorized some 30,000 stars in patterns. Even today, if I see a partly cloudy sky and debate whether to go out, or stay in, I often think "George would have gone out." It is often the only thing that forces me over that psychological barrier to abandon the warm cozy indoors for the freezing cold observatory! Harold Ridley, who I mentioned in the Preface, is another iconic figure for me. A past president of the British Astronomical Association, Goodacre Medallist and lifelong meteor and comet observer; I often think when checking the sky "If Harold had my equipment he would definitely have gone out." Others may not need heroes to inspire them, but apparently I do! So, the first step to becoming a comet imager is to pick a few role models to spur you on!

Comets to Die For!

It goes without saying that the brightest comets will always be the ones that amateur astronomers will most want to capture, but the ultimate zero magnitude photogenic comets like Hale–Bopp and Hyakutake are very rare and such spectacles cannot be expected to appear every year. However, more modest comets like Ikeya–Zhang, which reach, say, third to fifth magnitude, are much more common; often an annual event in fact. Such comets do not require huge telescopes and long exposures to record them.

As a very rough set of "rules of thumb" I would say that any comet with an absolute magnitude of better than 7.0, a perihelion distance of less than 0.8 AU and passing within 1 AU of the Earth should make a nice target as it closes in to perihelion. One could expect, at some point, a degree long tail to develop on such an object even though, to quote David Levy once more: "Comets are like cats, they have tails and they do what they want!" Even a modest telephoto lens on a barn-door mount (a threaded rod driving two hinged plates apart) can be used to image a naked eye comet using CCD equipment. In fact, a fixed camera on a tripod, fitted with a wide angle lens, can be used for zero magnitude comets, especially in this digital era when digital images can be stacked up using software like Cor Berrevoet's Registax. With a 35-mm focal length lens individual exposures as long as 20 seconds, from a DSLR on a fixed tripod mount, can be used without stars perceptibly trailing. Extending this by a factor of ten a 350-mm focal length lens can allow two second exposures. In either case scores or even hundreds of frames can be digitally stacked with reference to the cometary head (or bright stars) to produce a very acceptable result on a zero magnitude comet. With really fast (if pricey) lenses, such as, say, a 200-mm f/2.0

model, single exposures of a few seconds duration can capture splendid images of those truly Great Comets like Hyakutake, Hale–Bopp or C/2006 P1 (McNaught). An equatorial mount is not needed with such a system.

The Internet and Handy Software

Whether you intend using a DSLR made by Canon, Nikon or another manufacturer, or a dedicated cooled CCD camera, you will almost certainly want some software to calculate where the comet is going to be in the sky. Ephemerides can also be downloaded from various web sites of course and I have already mentioned the key ones in Chap. 5.

Just to re-iterate, for convenience to the reader, the most useful websites are:

Comet Ephemerides from the IAU:
http://www.minorplanetcenter.org/iau/Ephemerides/Comets/

Minor Planet & Comet Ephemeris Service:
http://www.minorplanetcenter.org/iau/MPEph/MPEph.html

Orbital elements for popular planetarium software packages:
http://www.minorplanetcenter.org/iau/Ephemerides/Comets/SoftwareComets.html

In addition there are those two Yahoo! based User Groups for comet observers, namely the Comets Mailing List and the Comet Images Group which can be found at:

http://tech.groups.yahoo.com/group/comets-ml/
and
http://tech.groups.yahoo.com/group/Comet-Images/

Then there are those incredibly useful websites of Seiichi Yoshida, covering current comets and comets near deep sky objects:

http://www.aerith.net/comet/weekly/current.html
http://www.aerith.net/comet/rendezvous/current.html

The comet pages of the British Astronomical Association can also be found at:

http://www.ast.cam.ac.uk/~jds/

In addition the comet page of "The Astronomer" website is worth checking for recent images:

http://www.theastronomer.org/comets.html

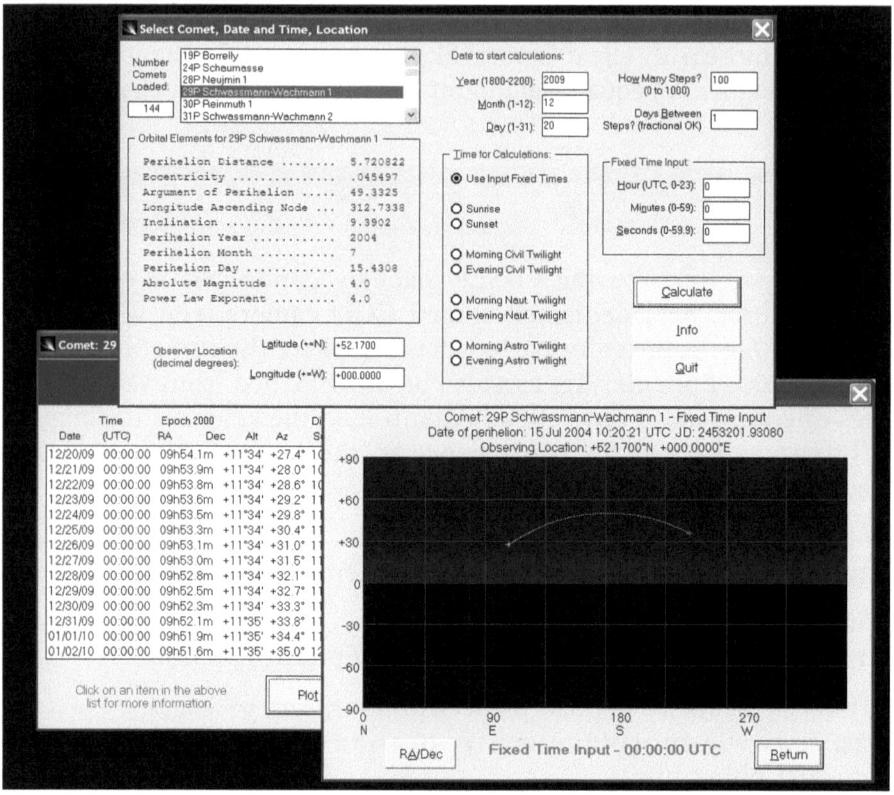

Fig. 11.1. William Schwittek's *CometWin* freeware enables the altitude and azimuth of a comet and any associated twilight conditions to be assessed with ease.

A very useful software package is William Schwittek's CMTWIN32. EXE (see Fig. 11.1) which reads a text file in which you have typed the comet's orbital elements and produces a whole host of data on the screen. Basically it tells you where the comet will be in RA/ Dec and Altitude/ Azimuth at midnight or for various degrees of twilight. It will also plot the comet's altitude and azimuth for the coming weeks and months. This is essential data for planning at what point a comet will disappear behind your local trees and houses!

At the time of writing this software can be downloaded from:

http://www.inourfamily.com/sites/cmtwin/

Richard Fleet's GraphDark software for graphically displaying comet (and other object) visibility in dark skies can also be handy:

http://www.rfleet.clara.net/graphdark/download.htm

The comet imager will often want to know how fast, and in what direction, a comet is moving. Take care: the units used by different organizations can vary. My personal preference is for arc-seconds per hour, but there are plenty of other units in use. I frequently use the excellent software Guide 8.0 published by Project Pluto: http://www.projectpluto.com

Software Bisque's The Sky, in its various forms, is a planetarium package used by many amateurs and a few professionals too. It can be used in harmony with the CCDSoft imaging package via the Orchestrate scripting tool, allowing automated observing for applications like supernova patrolling. The Sky has a slick tool for inserting new comet elements, although a few years ago there was a crisis when an accidental server directory erasure on a computer at the Harvard–Smithsonian Center for Astrophysics meant that the web addresses where new comet (and asteroid) elements for planetarium packages was stored had been erased. This meant that hundreds of users of The Sky had to alter their computer registry data (Gulp!) to point at the new comet address: http://www.minorplanetcenter.org/iau/Ephemerides/Comets/Soft06Cmt.txt.

However, once it is pointing at the correct address it is a very slick way of inserting new comet orbital elements with one key press (see Fig. 11.2). At the time of writing The Sky 6 software had just been superseded by The Sky X Serious Edition, and The Sky X Professional edition was imminent.

Guide 8.0 is a much less expensive package than The Sky X and is super-accurate too. It will also work on any PC, however old, because the graphics requirements are very simple. It contains a comprehensive list of comet elements discovered up to the time of issue and new elements can easily be added by downloading a text file from the MPC and then importing it (see Fig. 11.3). Once the comet is displayed against the background sky, right-clicking on it yields a whole host of data, including the comet's motion against the stars. Guide 8.0 also has a great feature enabling you to plot a comet's track across the sky, backwards or forwards in time (see Fig. 11.4a) and an animation dialog option allowing you to animate the comet's motion (see Fig. 11.4b).

It is often highly useful to know what comets the leading experts are imaging and how the comets are developing, especially if you have been clouded out for weeks! Seeing others images enables you to plan your own strategy. For example, if your standard set up has a 10 arc-min field and you see that the comet of interest already has a 6 arc-min head and a 2° tail you will want to use a telephoto lens with your CCD, not a telescope!

Unfortunately, the standard comet magnitude law of the form $m1 = H_0 + 5 \log \Delta + 10 \log r$ does not say much about tail length, which is

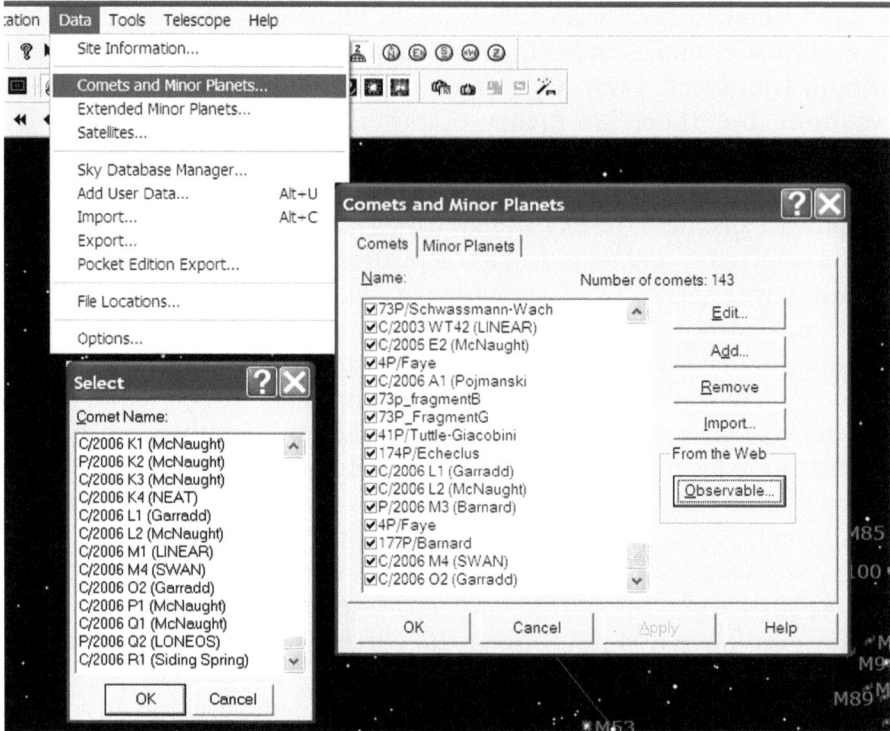

Fig. 11.2. Software Bisque's *The Sky* software allows slick insertion of new comet orbital elements from the Minor Planets Center website if the web address is correctly set.

often totally unpredictable even if the geometry of Sun, Earth and comet are taken into account. Ideally the Sun–comet–Earth geometry should form a right angle for the optimum tail geometry.

Predicting Tail and Head Sizes

So how do you know what "size" of comet to expect? Well, one method is to check out the latest comet images on the web. The Yahoo! Comet-Images group at: http://tech.groups.yahoo.com/group/Comet-Images/ always has links to recent images and Seiichi Yoshida's site has current data too, at: http://www.aerith.net/comet/weekly/current.html.

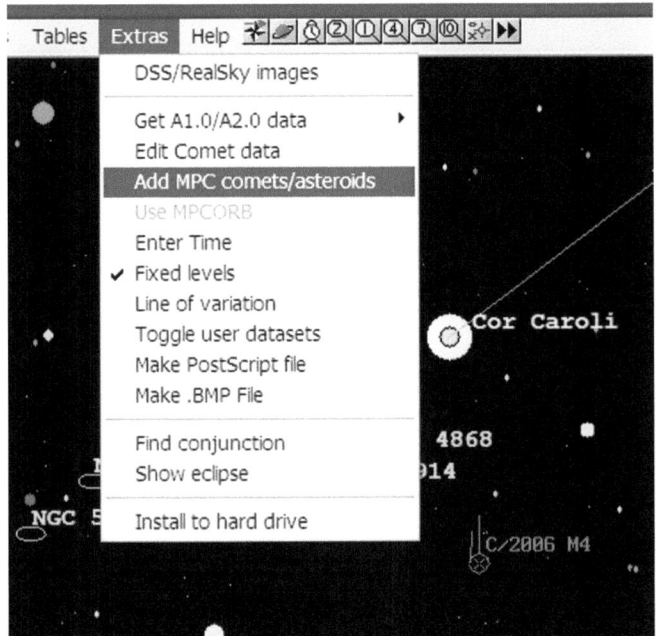

Fig. 11.3. Guide 8.0 allows new orbital elements for comets and asteroids to be added via a downloaded MPC website text file, inserted from the "Extras" option in the Menu.

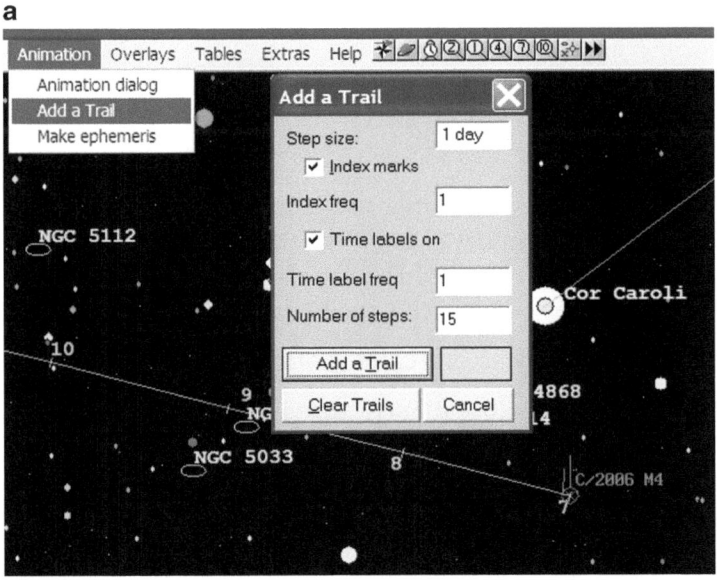

Fig. 11.4. (a) Guide 8.0's Animation/Add a Trail command allows you to plot the path of a comet against the constellations.

b

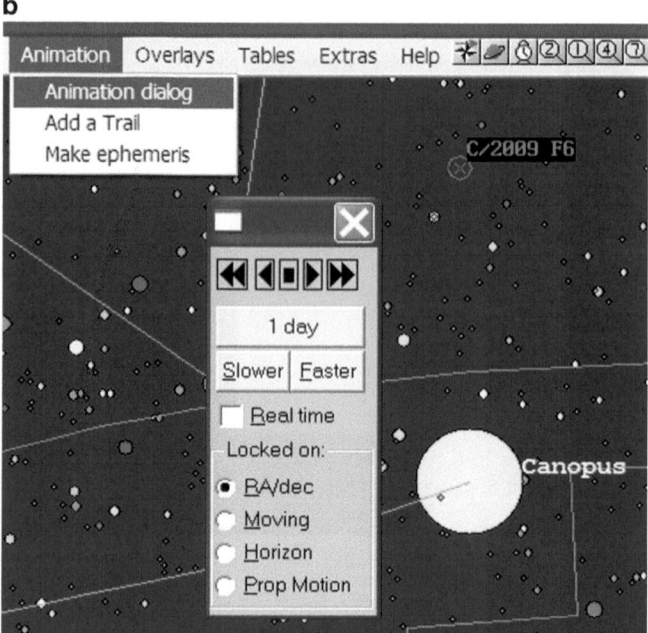

Fig. 11.4. (b) Guide 8.0's Animation/Animation dialog command allows you to animate a comet to move forward or backwards in time against the constellations.

The website of The Astronomer magazine has a comprehensive comet page which is usually updated every week or two. If a comet is within amateur imaging range there will invariably be images of it on this web site! The Astronomer comet web page is at:

http://www.theastronomer.org/comets.html.

Another method is to use planetarium software which tries to predict the size of a comet from a formula. The formula used in Guide 8.0 seems to be fairly accurate in this regard and was developed by Andreas Kammerer of Karlsruhe, Germany. It assumes that you know the absolute magnitude H_0 and the solar brightening rate K from the formula Magnitude $= H_0 + 5 \log \Delta + K \log r$ with Δ and r in AU. K corresponds to the more familiar 2.5 n value that is set to equal a value between 7 and 10 for most comets, as discussed in Chap. 1. Kammerer's approach uses a magnitude parameter independent of the Earth's distance that he calls mhelio which is simply mhelio $= H_0 + K \log r$.

This leads to an "absolute tail length" formula which Kammerer states as being:

L_0 = 10 to the power of $(-0.0075 \times \text{mhelio}^2 - 0.19 \times \text{mhelio} + 2.10)$

Finally, the actual length of the tail in millions of kilometers is given by:

$$L = Lo * (1 - 10^{-4r}) * (1 - 10^{-2r})$$

The angular size of the tail, if it were at right angles to the comet, would therefore be arctan (L/distance from Earth in millions of kilometers). This would be the formula we want except that the tail will rarely be sticking out at right angles to our line of sight and calculating the shortening factor makes the calculation even more complex!

A much simpler formula offered by Andreas Kammerer, which is far easier to use, is:

Tail length = 10 to the power of $(2.25 - 0.26 \times H_0) \times$ one million kilometers

Kammerer has proposed another formula which gives the size of a comet's coma in thousands (rather than millions) of kilometers. This uses an "absolute diameter" formula calculated by:

D_0 = 10 to the power of $(-0.0033 \times \text{mhelio}^2 - 0.07 \times \text{mhelio} + 3.25)$

From which the actual coma diameter in thousands of kilometers is calculated as:

$$D = D_0 * (1 - 10^{-2r}) * (1 - 10^{-r})$$

The angular size of the coma would therefore be arctan (D/distance from Earth in thousands of kilometers).

Kammerer's formulae are better than any other formulae but comets are still a law unto themselves and buying Guide 8.0 will show you the actual length of a cometary tail far more easily than doing the calculations step by step.

Image Scales

The fact that the tails of comets are so variable and comets can range in size between tail-less objects with a tiny coma just 10 arc-sec across and a comet like C/1996 B2 (Hyakutake) with a tail stretching across half the night sky presents a big problem. One instrument will just not suffice for all comets. Nevertheless the vast majority of comets do not have tails of even a few degrees in length and the really big ones can be captured reasonably well with a simple range of camera lenses. Nevertheless, calculating the image scale of your system is crucial to choosing an imaging telescope. If you are happy with basic trigonometry then working out the fields of view covered by a CCD chip, and even by a single pixel, can easily be accomplished using the tangent function. The inverse tangent of the CCD chip size divided by the focal length, gives you the angle covered on the sky. As an example, a CCD chip with a 20-mm width in a 2,000-mm focal length system will cover an angle of arctan $(20/2,000) = 0.57°$. An individual pixel of, say, 10 μm in size (10 μm = 0.01 mm) will span arctan $(0.01/2,000) = 0.00029°$, or 1.0 arc-sec (1° equals 60 arc-min and 1 arc-min equals 60 arc-sec). If you are not happy with trigonometry and tangents then you can use the following formula to calculate how many arc-seconds are covered by each pixel:

Arc-sec/pixel = 206 × pixel size in microns/focal length in mm.

Multiplying this result by the number of pixels across the width or height of the CCD will give you the field of view in arc-seconds, or you can use the formula:

Field of view in degs = 57.3 × chip size in mm/focal length in mm.

The number of arc-seconds that cover each pixel on your CCD is usually referred to as the image scale. For planetary imaging, where observers hope to snatch a small percentage of frames when the atmospheric seeing is very stable, image scales as fine as a tenth of an arc-second per pixel are often used. However, in deep sky and comet imaging such high resolution is definitely not required. In the vast majority of long CCD exposures the smallest star images will span 3 or 4 arc-sec and they will often be bigger than that. So, in practice, an image scale of 2 arc-sec/pixel will be more than fine enough for comet imaging, especially if you do not own an equatorial mount with near-perfect tracking. An additional consideration here is that the image scale is just as relevant to how deep you can go in a short exposure, on extended objects, as the f-ratio. If you funnel the light from two by two arc-seconds of sky into 1 pixel you will get four times the "signal to electronic noise ratio" as if you funneled the light into a two by two pixel array. An image scale that is twice as coarse always gives you four times the "signal to electronic noise" ratio and physically bigger pixels tend to be more quantum efficient too. So it is really a case of matching your focal length to your pixel size. If your drive has a poor periodic error you may well be prepared to sacrifice more resolution and go to 3 or 4 arc-sec/pixel by buying a CCD with bigger pixels and a bigger chip.

Alternatively you may wish to simply go for a shorter focal length, faster system. If your image scale is 2 arc-sec/pixel then you will get 2,000 arc-sec for every 1,000 pixels, in other words, slightly more than a half degree field of view for every 1,000 pixels. It should be stressed though that the overwhelming majority of comets within amateur imaging range have tails that are less than a quarter of a degree across. Maybe one comet per year will have a tail longer and one comet every few years will have a head larger than this! A good set up to have, for a dedicated comet imager, would be one in which the largest aperture instrument has an image scale of 2 arc-sec/pixel and a field of view of roughly half a degree, with another system with a field of view covering several degrees. For wider fields than this a standard camera lens or two, like an economical 200-mm focal length f/2.8 system, will suffice. I will have more to say on the subject of equipment in Chap. 12.

Altitude vs. Twilight

Inevitably, imaging really good small perihelion comets always seems to run into problems because of low altitude and twilight. This is hardly surprising, as such comets get close to the Sun and the Sun needs to be well below the horizon before it gets dark! The astronomical definition of real darkness is when there is no astronomical twilight, so that the Sun is 18° or more below the horizon. However, for really bright comets acceptable images can be obtained at the boundary between nautical twilight (Sun is −6 to −12°) and astronomical twilight (Sun is −12 to −18°), when the Sun is 12° below the horizon. Thus, the comet imager inevitably ends up imaging bright comets in the dusk or dawn sky. Sometimes, when a comet passes due north of the Sun, for the northern hemisphere observer, the comet imager will have the advantage of the comet being available in both dawn and dusk skies. The reverse is true for southern hemisphere observers with comets passing due south of the Sun. Schwittek's Cmtwin32.exe will show you whether this is the case with a few key-strokes and planetarium software like Guide 8.0 can easily be used to compare Sun and comet altitudes at specific times. There are various pros and cons of dawn vs. dusk imaging. Undoubtedly, getting up in the chilly pre-dawn hours is a major psychological problem for all but the hardiest individuals, especially in winter. A snug bed is far more tempting then a freezing observatory. In addition, I have personally found that there is a much greater tendency for mist and thin cloud to materialize at dawn, which is incredibly demoralizing when the equipment has been set up at such an unsociable time. The advantage of dawn observing is that the imager has had plenty of time in a dark sky to focus and check his equipment as the comet rises out of the murk. In addition, temperature changes and subsequent focus shifts are rarer at dawn then at dusk.

A lot simply depends on the observer's own horizon problems. In an ideal world all comet observers would live on top of hills with no trees or buildings above the telescope. In practice, this is never possible but a lot of planning before setting a telescope can save years of frustration. The comet observer will tend to want his eastern and western horizons to be obstruction free, but in the UK winter this will translate to southeastern/southwestern and in the UK hemisphere summer to northeastern and northwestern. Remember, bright small perihelion distance comets will generally peak where the sun rises or sets throughout the year. The author, in 1991, purchased a bungalow, specifically with comet imaging in mind. The bungalow was at the top of a south-facing slope with excellent east, west and south views. The house obstructed the north horizon, but, being a bungalow, only to an altitude of 17°: a close-to-optimum situation.

It is vital to instinctively know where your obstructions lie in an azimuth/altitude sense. Many years ago, in the film era, I prepared a long selotaped together strip-map of my whole horizon and marked the altitude and azimuth of every obstruction on it! I then knew exactly whether I could get a comet image or not, without having to go out to the telescope and see what happens. Of course, making a horizon strip map digitally is very easy in the digital era: just take loads of digital camera pictures and combine them in a package like Photoshop or Paint Shop Pro. Software like The Sky X can overlay your horizon obstructions onto the planetarium packages' horizon boundary too. Needless to say, a chain saw and powerful clippers can be a vital accessory for the comet observer who is surrounded by bushes and small trees. Of course, even if your horizon is flat, the comet will, inevitably, sink into the murk and once objects drop below about 15° altitude the star images become bloated and swollen, unless you are using a very short focal length system. However, for those sungrazing comets like the Great Comet of 2007, C/2006 P1 (McNaught), altitude obstructions are not a problem as you will probably be using a 200 or 300-mm lens and going portable, with a lightweight tripod, to find a nearby flat horizon.

Observatory "Ease-of-Use"

I have become more and more convinced, over recent years, that the most productive observers are simply those with the most user-friendly observatories. Some of the best planetary observers use Schmidt–Cassegrains on balconies, patios or roof-tops, crucially, within feet of where they sleep. The psychological issue is critical here and should not be underestimated. If all you have to do to obtain an image is roll out of bed, slide open a patio door and take a tarpaulin off, or roll a small cover off the telescope, then you will do it (see Fig. 11.5). Conversely, if you have

Fig. 11.5. A telescope housing can be simple, but effective. Karen Holland's 25-cm Schmidt–Cassegrain sits on an elevated pillar and is protected by a lightweight box when not in use. Image: Karen Holland.

to walk a hundred feet down a garden and trundle a half-ton shed off a giant Newtonian and battle with an infinity of cables and trip-wires, you will not do it. The telescope and observatory should be ready to use in minutes and permanently mounted in the observer's back garden. There is, of course, a conflict between having a telescope close to the house and the house being an obstruction, but, nevertheless, my message is that an easy-to-use telescope or observatory may be used every clear night whereas a hard-to-use telescope or observatory may be used for the first few months of its existence and then abandoned! It is a sad fact, but true. Up and down the UK there are literally tens of thousands of telescopes, but who uses them all? Like exercise bikes and home gym equipment the enthusiasm lasts for a few weeks and then the equipment is left unused.

My personal preference is for a small wheeled cover that is only just big enough to house the telescope. My main run-off shed is a pleasure to use and has dimensions of 2.0 m × 1.5 m × 2.0 m (length × width × height). The run-off sheds for my older Newtonian telescopes, in the 1980s and 1990s, were much larger and a hundred times harder to roll back. There comes a point, when a run-off shed weighs more than a few hundred kilograms or so, where it will be a real battle to run back, especially if the rails are not straight, the wheels are not lubricated or the shed is not rigid. Similarly, if a shed is more than 2 m long it will, with floor missing and door open, flex wildly when pushed back. An associated problem is that a shed with a width greater than 1.5 m cannot easily be pushed back with a hand on each side of the door-frame, so pushing on one side only is the result, with flexing and jamming being the experience! I find domes nice and cosy structures and a pleasant barrier against biting winter winds, but frustrating to the observer. I may want to look at several comets a night and not move the dome slit each time or have to automate the dome. I also want to see the night sky above me and any fireball that passes overhead; but, most importantly, I want to see any approaching cloud so I can change my strategy if I only have a few minutes to spare! My smaller run-off sheds are a joy to use and any future sheds will copy this pattern. However, if you live in an especially windy location a dome or a SkyShed Pod type of structure may be essential when contemplating long exposure work (see http://www.skyshedpod.com/).

Complete Beginners

Offering specific detailed advice on imaging comets is always very tricky as it is so dependent on what equipment the observer already owns and whether he or she really is a complete beginner or has some degree of experience.

In addition the imaging system may be a digital SLR or a dedicated cooled CCD and the software supplied with each camera system may or may not be sufficient for the observer's needs. For the complete beginner taking short exposures of bright fuzzy objects like the Orion nebula or the Andromeda galaxy has to be the first step in the learning curve. For someone at this level there will be much to learn. Simply finding and focusing the image can be tricky enough. I will assume, for simplicity, that the reader of this section has acquired either a DSLR (digital SLR) or a dedicated CCD camera and has carried out some experiments with his or her telescope or telephoto lens to verify that the basic system works and images can be acquired and downloaded at night; notice, my emphasis on the words "at night." There is a world of difference between imaging things in daytime and in darkness. Expect to encounter many nights of sheer frustration before you achieve any success at all. I am deliberately delaying specifics of taking dark frames and flat fields at this point because I know from experience that complete beginners just want to crack on and take pictures, regardless of their quality. There is a lot to learn and it can only come a bit at a time if you have no prior experience of astronomical imaging. At some point you will have to acquire software, either commercial software or freeware, to carry out the image processing, but precisely what you choose will depend a lot on what comes with your camera and your budget.

The resources section at the end of the book lists all the best image processing tools. Undoubtedly the book and software package called The Handbook of Astronomical Image Processing/AIP4Win by Richard Berry and James Burnell should be at the top of your list. You will also want to give the freeware IRIS by Christian Buil a try. Other top quality packages are Software Bisque's CCDSoft and Cyanogen's Maxim DL as well as the AstroArt package by MSB software. However, if you buy a cooled CCD camera from, for example, SBIG, you may find a package like CCDSoft is bundled with the camera purchase. For DSLRs the ever popular Photoshop or Photoshop Elements, along with Paint Shop Pro, will be a good starting point for basic brightness and contrast tweaking of the familiar jpg, bmp and tiff formats.

DSLRs

If you are keen on imaging deep sky objects containing a lot of red nebulosity you may be disappointed by DSLRs, because DSLR chips are filtered to block out the near infra-red. Thus, they will perform well on blue/green comets but not so well on objects like the Orion nebula, where the faintest red nebulosity will come out as a strange green color prior to some tweaking! However, specialist suppliers are now offering some DSLRs without the mandatory red-blocking filter and it is possible to remove the filter yourself if you only intend using the DSLR for the night sky and don't mind a slight focus shift with respect to the optical "through-the-lens" viewfinder. This is a book about comets though, and many have a blue/green cast, so even with that standard filter in place the image of a comet should not suffer and will be nearer to the visual view than an unfiltered image. But if you decide to buy a DSLR for astro-imaging, what pitfalls might you encounter in use?

For the complete novice the first hurdle is purchasing a telescope interface for your camera. Telescope draw tubes are designed for 31.7-mm (or 50.8-mm) barrel diameter eyepieces and the simplest adapters interface the 31.7-mm hole to the bayonet mount of the camera (see Fig. 11.6). Most telescope dealers can supply you with adaptors for the most popular DSLRs to interface them to the 31.7-mm size draw tube. With fast telescopes, of f/4 or f/5, "vignetting" (a darkening of the images at the edges) will start to appear when using these adaptors, although a technique known as flat field division can deal with this once you gain experience (more on this later).

As well as purchasing an adaptor I would advise you to acquire a simple remote control shutter release to minimize camera and telescope shake

Fig. 11.6. Interfaces for joining a 31.7-mm drawtube to a DSLR's bayonet mount can be obtained from most telescope dealers. Image: Martin Mobberley.

(see Fig. 11.7). Initially, you may be tempted to dial up the fastest ISO rating (ISO 1600 on many models) but this can lead to a very noisy result. Adjusting the ISO setting alters the electronic gain and reduces the dynamic range of the image but it has almost no astronomical benefit at the highest values. ISO 400 will give a far smoother result.

One of the biggest hassles can occur if you allow tiny dust specks to enter the DSLR and land on the CCD/CMOS detector surface. In normal use, with camera lenses, the detector is always covered by the shutter/reflex mirror and by the lens itself. However, in astronomy the detector is exposed for many minutes, often near the dusty, dirty environment of a telescope draw tube on breezy nights. Even the tiniest dust speck will be obvious, so taking long exposures attached to a telescope can be quite risky. However, if dust does settle on the chip, purchase a lens air blower and consult the camera manual regarding "cleaning the sensor," which may be possible via an electronic function on the camera. A few amateur astronomers that I know have, at the time of writing, purchased a device called an Arctic Butterfly SL700 DSLR CCD cleaner, which seems to be another useful weapon against dust specks on the detector.

One hassle that seems to frustrate every astro-DSLR user is the thorny issue of focusing at night. Focusing in broad daylight is never a problem,

Fig. 11.7. A hands-free remote control for your DSLR will eliminate camera shake from the user's trigger finger. Image: Martin Mobberley.

but focusing a dim fuzzy patch is near-impossible. The method I use is to find the nearest brilliant star to the object being imaged. I focus on that star first and then move the telescope to the object. Make sure the DSLR optical viewfinder is showing you a crisp focus on the ground glass screen to start with. This should be dioptre-adjusted before you go outside using the tiny thumbwheel mounted just near the viewfinder. Once adjusted to your eyes, you should be able to dispense with your glasses for focusing unless you suffer from considerable astigmatism. If you are using short focal length lenses then some distant house lights, several kilometers away, can be used to focus the camera, but do not use this method with lenses more than 100-mm in focal length as even houses kilometers away are not truly at infinity. One useful gadget in this respect is a commercial clip-on magnifier (see Fig. 11.8). Typically, these magnify 3× and clip onto the optical viewfinder. An alternative option is to tediously examine each frame exposed on the LCD viewfinder and zoom in to check the size of the exposed star images. In my case I only use a DSLR for the biggest and brightest comets when imaging with focal lengths between 300 and 550 mm. I have modified an old 3× right-angled viewfinder so that it works at 7× and I focus on a first magnitude star at 7× first, before moving to the bright comet. This technique assumes that the comet can

Fig. 11.8. A right angle magnifier fitted over the DSLR viewfinder can make visual focusing easier. Image: Martin Mobberley.

be seen through the viewfinder after you have moved from the focus star to the comet. However, quite often the comet is just too damn faint to see in the viewfinder which is when you need a small finder telescope with crosshairs perfectly aligned with the main telescope. That way you can focus the bright star through the camera and then center the comet in the finder telescope. Yet another option is taking a laptop outdoors and downloading each image as you focus and center the comet. However, this is painfully tedious unless you purchase slick software specifically for the task, such as DSLR focus. More data on this package can be found at http://www.dslrfocus.com/.

DSLR Software

DSLR images taken in warm conditions, or in bright skies, can look quite noisy, whether or not you have subtracted a dark frame (a same duration image with the lens capped). However, the simple act of digitally re-sampling the image to, say, a quarter of its original size, drastically reduces the noise. Then, in Photoshop, Paint Shop Pro, or whatever other package you have, simply applying a mild Gaussian blur, or a "texture preserving smoothing" function, can soften the noise and yet not make the image itself look soft. Most image processing packages have a range of noise reduction tools. Reducing the color saturation (to monochrome if necessary) reduces the perceived noise too. Once these initial issues have been sorted out you can experiment with other options, like saving images in Raw format, subtracting dark frames, stacking dozens of images and advanced processing. However, the starter tips above should get you over the initial DSLR astro-hurdles. Various astronomy software packages for processing digital SLR images are available. The key is to start with the basics and just take single jpeg images first. Then try stacking jpeg images with something like Cor Berrevoet's Registax software (http://www. astronomie.be/registax/) using the comet's head or a bright star to stack images with reference to. It is important to get a feel for adjusting the color balance, brightness and gamma values in Photoshop or Paintshop Pro early on in your learning curve. Once you have achieved this you can, if you wish, move on to more advanced processing. I have heard it said that it is essential to take RAW images with a DSLR so that no data is lost. For pretty pictures of comets I would totally disagree. Raw images are huge and cannot be viewed unless converted to enormous 16 bit Tiff files or the equivalent. They take up huge amounts of memory and, frankly, on most images you will not see a lot of difference between a Raw and a low compression jpeg image, as other factors like sky transparency and telescope tracking will be far more important. If you really want to do science, a proper astronomical CCD which outputs a FITS file is far more sensible than a DSLR Raw file. Most beginners will want to see some results straight away and for that the basic jpeg format will be the best first step. If the learning curve is too steep it is always a huge turn off. Jpeg or bitmap images are compatible with everything whereas Raw and FITS format images are not and can have the beginner tearing his (or her) hair out and banging their head into a wall!

Many DSLR-using amateur astronomers use the freeware IRIS developed by Christian Buil which can be found at http://www.astrosurf.com/buil/us/iris/iris.htm.

Another popular freeware package is Deep Sky Stacker:

http://deepskystacker.free.fr/english/index.html

Mike Unsold's Images Plus software is not free but can be purchased for $210 at the time of writing. More information can be found at:

http://www.mlunsold.com/

I would just like to re-iterate that it should not be assumed that you need an equatorial mount to take good pictures of the best comets with digital SLRs. For the superb C/2006 P1 (McNaught), in January 2007, I used a 300-mm focal length f/5.6 mirror lens on a fixed tripod in a howling gale, while in a farmer's field! A World War II "Pillbox" (a miniature concrete fort with tiny windows through which guns could be fired at the approaching enemy) was used as a barrier to protect me from the gales. The comet was so bright (the head as bright as Venus) that single exposures of two seconds duration were sufficient to record several degrees of tail against the backdrop of the distant trees and hedges, without the head obviously moving.

Cooled Astro-cameras

Although cooled and ultra-sensitive astro-cameras are generally more expensive than digital SLRs, and can only be used for night-time imaging, they are far more useful for taking pictures of comets and deep sky objects. They are infinitely more useful for scientific work too. There are usually four fundamental differences between a DSLR and a cooled astro-camera although it all depends on which cooled camera you are dealing with. Firstly, the cooling factor of (typically) 30°C reduces the noise by a factor of 20 or so which can be quite dramatic, especially if you are using a DSLR on a warm summer's night. Secondly, CCDS used in astro-cameras can have a very high quantum efficiency meaning you can see fainter. Thirdly, DSLR sensors, CCD or CMOS, filter the incoming light. In the standard Bayer matrix system a 2×2 pixel filter grid covers the sensor such that in each square of four filters there are two green filters, a red filter and a blue filter. The camera firmware then uses its knowledge of this system to decode the matrix and produce a resulting bmp or jpeg image with the right colors. This works fine for daylight pictures where there is plenty of illumination, but each filter causes attenuation in the incoming light resulting in the image being typically up to two magnitudes less deep than if you were using a totally unfiltered system. There is a slight improvement if you remove that infrared blocking filter I mentioned earlier, but you cannot remove the Bayer Matrix filter, which is an inherent part of the CCD or CMOS detector. I should add that there is a third type of color sensor called the Foveon X3, which uses three color layers rather than a Bayer matrix, but they do not appear to have any sensitivity advantage for astronomical work. So, to reiterate, as DSLRs give you color images you lose sensitivity and the images look very noisy when compared to an unfiltered image. Cooled astro-cameras can, of course, be used to take color images by employing a filter wheel to take images in red, green or blue light. However, when collecting such images the user will find that the color filtered frames are also incredibly noisy in comparison to an unfiltered shot. The blue filtered frames will usually be horrendously noisy unless a very fast imaging system is being used. A trick that advanced deep sky imagers use to get around this noise problem is to take a full resolution unfiltered "luminance" image, which will be nice and smooth, and then take shorter color-filtered shots to provide the color information. These filtered shots are sometimes taken at half resolution (pixels "binned" 2×2) to boost the signal-to-noise. The human eye does not notice that the color data obtained in this way is of lower resolution when it is resampled back to full size. Suitable software can be used to produce an LRGB image from the clean luminance shot and the noisier filtered R, G and B shots. However, with comets this technique becomes fiendishly complicated because they move against the background stars, so a tricolor shot will produce a star trail in three colors when allowance is made for the cometary motion. Nevertheless, not all cooled CCD cameras for astronomers are unfiltered. In recent years several manufacturers have started producing "one shot" color CCD cameras for those observers who simply cannot be bothered with the hassle of tricolor imaging and combining three (or four) images into a final picture. Of course, these "one shot" color CCD cameras are therefore not far removed from using a standard DSLR. The main difference simply boils down to cooling the chip and not having that IR-block filter in place which means that deep red nebulosity is recorded and there is less noise, although stacking hundreds of DSLR shots together is remarkably good at smoothing out the noise. I spent a lot of time going into that third difference area between DSLRs and CCDs but can now deal with the final point. The fourth difference between DSLRs and cooled CCDs is simply that in most cases the DSLRs have much larger CCD or CMOS sensors than all but the most expensive astro-cameras. I suppose a fifth difference could be pointed out, that being the ability to use some astro-cameras as autoguiders for the telescope drive, to keep it tracking on the stars. However, comets move and so things are nowhere near as simple as with straight deep sky imaging.

The great thing about using dedicated cooled astro-cameras is that the software that comes with them expects you to be carrying out astro-photography and not daylight photography. Therefore the menu exposure time options are in seconds and minutes and there is no danger of the camera trying to autofocus or auto-expose. Conversely, when using DSLRs for astronomy, the exposure and focusing options need firmly setting to manual mode. In addition, when using a cooled CCD camera controlled from a laptop PC or from a desktop PC the screen in front of you shows you a big full sized image of the view the chip is recording. There is none of this "staring down an optical viewfinder and trying to glimpse an invisible comet" nonsense and there is no staring at a tiny LCD viewfinder a few inches away when you left your damn glasses indoors! You can sit comfortably, at a table, with the PC monitor in front of view and see precisely how sharp and how deep the image is straight ahead of you.

In addition, the dedicated software that came with your astro-camera is all set up to allow you to do things like subtract dark frames, divide by a flat field and have a comparison image of the field from a planetarium software package side by side with the CCD image to check you are looking in the right place. You can even download an image of the Digitized Sky Survey while outdoors, if you have a wireless Internet connection, to check the starfield that you are imaging. It is all very civilized and precise when compared to twisting your neck through a painful angle to look at a camera viewfinder or an LCD screen. Astronomical imaging is a lot easier when you have a big astronomy-friendly graphical user interface laptop screen and keyboard confronting you, rather than tiny DSLR controls, LCDs and viewfinders.

Periodic Error

The new comet imager, whether a DSLR user or a cooled CCD camera user, soon finds out that if you can expose for 60 s or more your image becomes a lot less noisy than if you are stuck with a 10-s limit due to a very shaky mount or a poor periodic error on your telescope drive. Mass produced telescope drives are invariably built down to a price and the lower quality units often contain some very poor quality components including plastic drive wheels, tiny worm drives, toy gearboxes and drive motors right out of Scalextric cars! I kid you not (see Fig. 11.9).

All worm and wheel telescope drives have a periodic error which goes through a cycle every time the metal worm screw rotates once. By intimately understanding this error you can optimize your time outside. If there are specks of dust on the worm or on the wheel teeth these can cause non-periodic errors too. Fortunately worm and wheel sets are almost always fully sealed from the outside world these days. With many commercial systems much of the periodic error can be trained out by engaging a periodic error correction mode (PEC), during which a patient observer, with a high power illuminated reticle eyepiece, can press the keypad buttons and, with teeth gritted and fingers flying, wrestle to keep a bright star centered perfectly on a crosshair. The telescope's flash memory remembers the corrections and replays them the next time you take an exposure. This can improve tracking considerably if there is optimum contact between

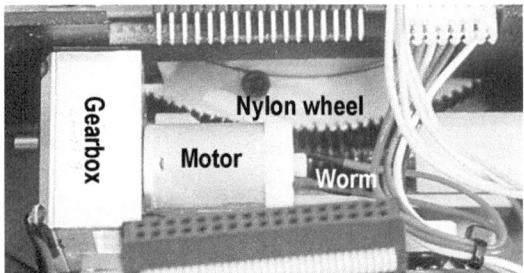

Fig. 11.9. Some mass-produced telescope drives are made from low quality junk components including toy car motors and tiny plastic components. Note how mangled the teeth are on the nylon wormwheel of this Schmidt–Cassegrain! The motor is the same as used in Scalextric toy cars and is only a few centimeters long! Image: Martin Mobberley.

worm and wheel (not loose, but not too tight either) and absolutely no dirt on the worm screw or wheel teeth.

The period of your worm rotation can be calculated by dividing the number of teeth into 1436, because there are 1,436 min in a sidereal day. For the pedantic reader, poised to complain that I have got it wrong, I will admit that the true figure is 23 h 56 min and an additional 4 s for the sky to rotate, but 1,436 min is close enough for this explanation. Many popular worm wheels have 180 teeth, resulting in a worm rotation period of almost 8 min (1,436/180). Typically, the resulting periodic error produces a sine wave deviation from the perfect tracking scenario when error is plotted against time. At the top or bottom of the sine wave the tracking error changes slowly, whereas in the middle it changes quickly. Thus, for a worm with an 8-min period you may get away with more than a minute of decent tracking near the peaks of the sine wave, but less than 15 s in the center. Knowing this information means you may choose to adopt a strategy of taking 100 1-min exposures, but just stacking the 50 or so good frames that occurred near the sine wave peaks. Periodic error is often stated as a peak to peak or RMS value. For example, the sine wave peak extremes may be ±20 arc-sec, a range of 40 arc-sec (which is the figure you want to know) but the manufacturers and dealers will often craftily quote the misleading RMS average deviation, e.g. 14 arc-sec. Fortunately for us a French amateur, Christophe Demeautis, has set up a site showing actual periodic error performance for many of the most familiar drives. See:

http://demeautis.christophe.free.fr/ep/pe.htm

Popular mounts like the SkyWatcher EQ6 Pro have a periodic error of ±30 arc-sec over an 8-min period. Go up a notch in quality and you will find mounts like the Losmandy G-11 have a period of ±7 arc-sec over a 4-min period. Then move up to the next level and the Astrophysics 1200 GTO has a periodic error of about ±2 arc-sec with a 6-min period. Finally, the Software Bisque Paramount ME (see Fig. 11.10) has a periodic error that can be impossible to easily measure with roughly ±1 arc-sec of error over a two and a half minute period.

If you are imaging a comet at a fine image scale of a couple of arcseconds per pixel then you can expose for many minutes with a Paramount ME, but with, say, a Skywatcher EQ6 Pro, you will need to be more careful. The former mount costs ten times the price of the latter mount. You can see why the Paramount's performance is better: it uses an 11.45-in. (29-cm) diameter, 576 tooth research-grade right ascension wheel, where as the EQ6 Pro uses a 3.5-in. (9-cm) diameter, 180 tooth standard-grade

Fig. 11.10. The author's Celestron 14 is mounted on a very high quality Paramount ME equatorial mounting which has a tiny periodic error. Image: Martin Mobberley.

wheel. The bearings that house the Paramount right ascension axis are also in a totally different league as well. There are several solutions to coping with a periodic error of tens of arc-seconds. One is to simply use a shorter focal length system like, say, a 200-mm f/2.8 lens. Admittedly, the pixel size needs to be remembered as well, but as a general rule a DSLR attached to a 200-mm lens on an EQ6 quality mount will be able to track for at least 60 s if polar aligned reasonably accurately. However, only the very best comets will look impressive when captured with a 200-mm lens. Another solution is to use the sledgehammer approach. Just take a hundred or so 30-s exposures, select, say, the best 20 or 30 (bin the others) and stack the good ones. Remember what I said earlier: periodic error occurs in a sine wave period; at the top or bottom of that sine wave the best exposures will be captured. The remaining solution is to use an auto-guider CCD (see Fig. 11.11a) which is attached to a guide telescope. This monitors a bright star near the object being imaged and when the star drifts from its original pixels the telescope drive is instructed to move to counter the motion. Of course, comets move and so even this is not an ideal situation, especially as there can be flexure between telescope

Fig. 11.11. **(a)** The Autoguider system on West Sussex amateur Ian Sharp's Pentax 75 refractor consists of an Orion Off-axis guider which diverts light from a small pick-off prism to a tiny Starlight Xpress Lodestar CCD autoguider (attached to the white cable in the top right of the image). An ATIK Filter Wheel and ATIK 314L CCD (bottom of the picture) form the main imaging system in Ian's set up. Image: Ian Sharp. **(b)** Some of SBIG's CCD's have a built in small autoguider chip next to the main imaging CCD, as shown in this ST9XE model. Image: Martin Mobberley.

and guide 'scope unless the main CCD has a built-in guide CCD (see Fig. 11.11b). As a result most comet imagers with high periodic error issues simply take lots of short exposures and stack them with reference to the comet's center of brightness or "false nucleus."

Active and Adaptive Optics for the Pro's

Some deep sky imagers are very keen on adaptive optics auto-guiding methods. The terms active and adaptive optics are frequently confused in the astronomical media and yet there are distinct differences between the two. A telescope with active optics is one in which the surface of the mirror can be adjusted to make it focus starlight as perfectly as technically possible. For a simple reflecting system the primary mirror needs to be parabolic to within a fraction of the wavelength of light of the perfect theoretical shape. These current generations of professional telescopes that are larger than 5 m in aperture all use frighteningly thin mirrors (or thin segmented mirrors) which, on their own, would never be able to focus the light perfectly. However with hundreds of actuators behind the thin mirrors, bending them into shape until the optics produce nice tight stars, the optical problems have been solved. This technique of bending and tilting the glass components to the right shape is known as active optics, but no amateurs yet use such sophisticated technology.

Adaptive optics is a closely related discipline, but in this case the shape of a second lightweight mirror in the light path is varied very rapidly to attempt to compensate for the turbulence (bad seeing) in the Earth's atmosphere. You may well ask "But how does the system know what the atmosphere is doing?" A good question! Well, by monitoring a bright star near to the field and rapidly adjusting the secondary mirror shape to get that star as sharp as possible the effects of atmospheric turbulence very close to the field being imaged can be compensated for. However, the star needs to be bright enough to allow very short exposures to work and in practice a star of at least magnitude 12.0 is highly desirable, even for a huge telescope. If a suitable reference star cannot be found there is another method in which a pulsed laser beam illuminates or excites particles in the atmosphere at around 13 mile altitude thereby providing an artificial source. However, the aforementioned systems are very pricey and a technical step away from amateur CCD budgets!

Amateur AO Systems

Although deformable mirrors are not available, or economically affordable, for amateur astronomers the term "adaptive optics" is still used by amateur equipment manufacturers to describe rather different rapidly adjusting "tip-tilt" systems. In these products the main aim is just to keep star images on the same pixels by compensating rapidly for telescope drive errors or for really abysmal "jerky" atmospheric seeing conditions. With a commercial AO system the image of the main field and the guide star come to their respective CCD chips via the same "tip-tilt" optical arrangement, so that when the guide star drifts a rapid adjustment of the lightweight optics re-centers the guide star on the guide chip and the imaging chip simultaneously and within a fraction of a second. Repositioning ultra lightweight optics is very rapid; by comparison, moving a heavy telescope and mounting takes considerably longer. The optical "tip-tilt" mechanism employed in the pioneering SBIG AO-7 unit was a lightweight mirror angled at 45°, but this resulted in the camera being at 90° to the telescope axis, the image being mirror-flipped and considerable back focus being eaten up. However the two main AO systems currently available to amateurs, namely SBIG's AO-8/ AO-L models and the UK manufactured Starlight Xpress SXV-AO (see Fig. 11.12) each use a much more compact "tip-tilt" transmissive (rather than reflective) plane-parallel optical window. While leading amateurs use these adaptive optics units to obtain high resolution images at scales as fine as 1 arc-sec/pixel (or sometimes less) they can also be used to improve a

Fig. 11.12. The Starlight Xpress SXV-AO adaptive optics auto-guider uses a "tip-tilt" transmissive optical window to rapidly autoguide compatible telescope drive systems.

telescope drive from mediocre to excellent. Of course, the optical windows in these models can only tilt so far before they run out of travel. In the SBIG AO-8 the range is roughly ±40 pixels (depending on the imaging camera), which equates to ±80 arc-sec at a typical imaging scale of 2 arc-sec/pixel. As most Schmidt-Cassegrain drives have periodic errors within a ±1 arc-min range an adaptive optics unit can transform a very average drive into an excellent one if a suitable guide star can be positioned on the guiding chip.

One slight fly in the ointment here is when narrowband filters are used for imaging in H-alpha and other emission bands. These filters have such incredibly narrow bandwidths that the light from potential guide stars is typically attenuated by four or five magnitudes making auto-guiding almost impossible. To get around this issue the Starlight Xpress SXV-AO provides a single 48-mm filter recess in front of the main camera chip but behind the guide chip off-axis prism. The SBIG solution is similar in design but with that product you have to purchase an additional Astrodon MOAG (Manual Off-Axis Guider) to divert the unfiltered light to a remote guide head, prior to the main CCD and filter wheel. Of course, comet imagers only image in such narrow bands if they are carrying out advanced photometry, as we shall see later. However, the biggest problem for comet imagers, yet again, is the fact that comets move and so ultra long exposures guided on stars, by whatever method, will not track the comet. In addition, if the head of the comet is being imaged with the main CCD it cannot be centered in the field of the guiding CCD where those two chips live in the same enclosure side by side, as is the case in some SBIG dual chip heads.

Also, all these adaptive optics devices take up valuable back focus distance meaning Newtonians or refractors with a limited travel may not be able to use them. However, for those who crave the very sharpest tracking or simply want to improve their existing drive accuracy a tip-tilt AO system can transform their deep sky images, at least if they can always find a bright guide star. For very slow moving comets an AO device can be used to improve a telescope drive to track for many minutes rather than for 30 s.

Anyway, I have digressed considerably here from the topic of comet imaging to one more of deep sky imaging, but, in a nutshell, if you have a telescope with a tracking capability below the level of a Paramount ME then there are many options to explore, but the simplest is just the sledgehammer approach of just stacking lots of short exposures, either with respect to the stars or with respect to the cometary coma's center of brightness. It is now time we looked at this cometary motion issue in a bit more detail.

Cometary Motion

Years ago, before the CCD era, long photographic exposures were the norm. Typically, exposures of 20 or 30 min were undertaken. Apart from the sheer torture of guiding a crummy telescope drive for this length of time, there was also the problem of comets moving against the background stars. Typically, nearby comets move at between 100 and 400 arc-sec/h, but if they are passing very close to the Earth they can move much faster. Their motion is, of course, a vector sum of the Earth's motion and the comet's motion. It does not take a genius to see that, with capturing detail of a few arc-seconds being the aim of the telescopic comet imager, if you expose for longer than it takes for the comet to drift a few arc-seconds, relative to the stars, you will lose detail. With the 30-min exposures of the photographic era there was only one solution. You have a separate guide telescope and an illuminated reticle eyepiece or a micrometer eyepiece and you try to glide a nearby bright star along a calibrated track so the telescope follows the comet. This involves a colossal amount of preparation beforehand and a period of utter misery during the actual exposure! Thankfully those days are now in the past for the CCD observer, though for 8 years, from 1985 to 1993, I was an offset-guiding comet photographer! My last offset star guided photograph was of Comet Schaumasse on February 19, 1993, though I did guide on Hale–Bopp's nucleus region in 1997. I'm certainly glad those days are over!

My own strategy with CCDs is to try to take one single image of about 120 or 180 s duration when I'm using my 35-cm Celestron 14 on its Paramount ME equatorial mounting. Now, this is a very expensive mounting and those with less wallet-bursting systems may need to keep the exposures well below 60 s maximum at image scales of 2 arc-sec/pixel or finer. I prefer this single-shot strategy with a high quality large aperture system because I vastly prefer to see the field in which the comet lies, complete with stars and maybe galaxies, as opposed to loads of straight lines. I can easily reach 20th magnitude in a single unguided 120-s exposure at the zenith and so while longer exposures will make the picture less noisy, they are not essential. The alternative, of course, is stacking lots of exposures with reference to the brightest part of the comet's head, so the stars appear trailed with the comet frozen in space. Perfectly straight star trails used to be a "badge of honor" amongst comet photographers who guided for half an hour or more, but, with modern software, there is no superhuman skill involved in this anymore. Of course, at my image scale of 1.5 arc-sec/pixel (Celestron 14 at f/7.7 and an SBIG ST9XE CCD with 20 μm pixels) if the comet is going to drift more than 3 or 4 arc-sec in 120 s (that is, a faster drift than about 100 arc-sec/h) I will usually reduce the exposure and stack the images with reference to the comet, but much depends on whether the comet is just a big fuzzy head, or if there is fine gas tail detail to be captured. A lot simply depends on experimentation for each particular comet and on other factors. For example, if the comet is rapidly sinking in the west and heading behind a house roof, it is imperative to just get an exposure quickly, of any duration, and one exposure is often quicker than taking and saving loads of shorter ones. Certainly, if a nearby comet is moving faster than 400 arc-sec/h I will take multiple short exposures of 20 or 30 s duration and then stack them, centered on the comet's center of brightness. Numerous software packages (e.g. Richard Berry's AIP4Win, Maxim DL) will do this stacking trick and there are freeware packages which are worth hunting down too.

AIP4Win Step-by-Step Comet Stacking

In AIP4Win the stacking procedure goes as follows (see Fig. 11.13). From the "Menu" select Multi-Image/Auto-Process/Deep Sky. Then "Select" all your flat-fielded (and dark frame subtracted) comet images from the sequence you captured. Go to the "Alignment" tab inside the "Auto-Process Multiple Images" window and set the "Process Type" to "Average Stack." Then pick the first frame from the sequence that you want to stack under "Select the Master Frame" and set "Alignment" to "1 star," "Track Radius" to a value of about 10 pixels or so and set "Slave Alignment Star Selection" to "Manual." The next thing to do is to click your mouse on where the center of brightness of the comet's coma is. You may wish to enlarge the "Track Radius" value at this point if the coma is large. You may also want to dim the image so you can see where the center of the coma is. This will not affect the final outcome as the stacking process uses the fixed values in the FITS images and the fact you have dimmed the displayed image should have no effect. Once you have clicked your mouse on the center of the coma, or "false nucleus" and are happy the circle is roughly centered on the brightest core region, click the "Star 1" button. The circle around the nucleus then goes yellow and clicking OK will accept this and bring up the second image to stack with a circle in the same position. For the remaining images the "Star 1" button appears at

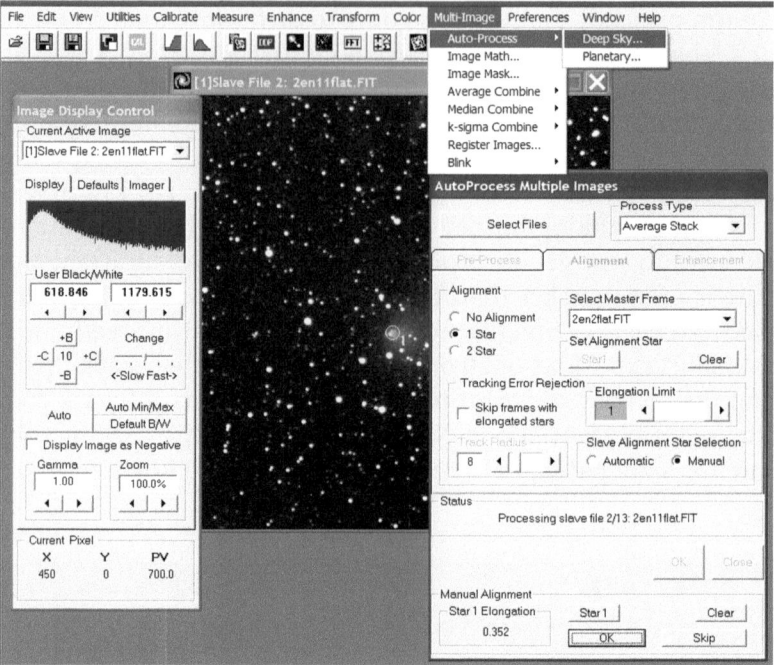

Fig. 11.13. AIP4Win's Multi-Image/Auto-Process/Deep Sky option is excellent for stacking comet images by treating the inner coma/false nucleus like a star. When a comet is moving rapidly the resultant stack freezes the comet and produces ruler-straight star trails.

the base of the "AutoProcess Multiple Images" window. You may well find that the yellow circle labeled "1" is now not centered correctly but you can drag it with the mouse and the left click button to get it centered on the nucleus and then click "Star 1" again. This process is then repeated until the point where the final image is dealt with and the software stacks all the images centered on the comet head's center of brightness. With luck you should now have a smooth stack of multiple images along with some super smooth star trails! Of course, AIP4Win is only one of many stacking programs that can be used.

There really are no hard and fast rules with regard to comet stacking and exposure duration; my advice would be to see what works for you. In general stars on deep sky images are bloated to 3 or 4 arc-sec in diameter, so allowing the comet to drift 3 or 4 arc-sec between exposures will usually let you stack the images so that the stars appear as continuous trails,

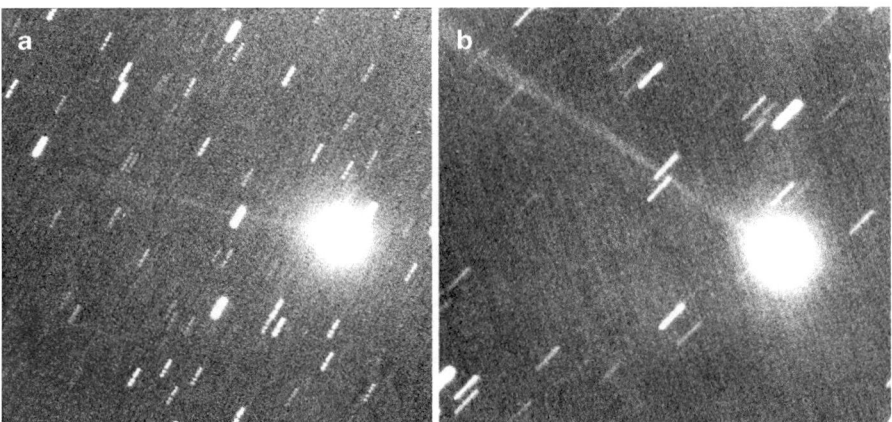

Fig. 11.14. (a) This image of comet C/2009E1 (Itagaki) taken on March 18, 2009 covers a 12 arc-min wide field and was taken with a Celestron 14 at f/7.7 using an SBIG ST9XE CCD camera. The field is 13′ wide. The comet was moving at 3 arc-sec (2 pixels) per minute NNW relative to the stars and the images is a composite of five 120-s exposures with 1 s download time between images. The blobby nature of the star trails shows that the 6 arc-sec/4 pixel motion between frames was too long to give a smooth trail: a 3 arc-sec gap, corresponding to a 60-s exposure, would have been better. Image: Martin Mobberley. **(b)** This image of comet C/2009E1 (Itagaki) taken 11 days after the image in Fig. 11.13a, but with the same equipment, is a stack of 25 exposures of 60-s duration. In this case the comet was moving at 2.1 arc-sec/min NW relative to the stars and the 2.1 arc-sec star motion between frames has resulted in very smooth star trails. Image: Martin Mobberley.

rather than a line of dots (see Fig. 11.14a, b). Observers from town sites will invariably want to stack more images than country observers, simply because their images will be noisy. Stacking tends to reduce noise proportional to the square root of the number of images stacked. Thus, a stack of sixteen images will only look about a quarter as noisy as a single image. When noise is reduced, a tail often emerges and stacked images do look nice and smooth. When images are stacked such that star trails are visible it is often far more photogenic if the trails are nice and long. Short star trails of just 10 arc-sec in length can look rather crude and blobby in my view.

Fig. 11.15. This image of comet 46P/Wirtanen, taken on February 6, 2008 using the author's C14 and SBIG ST9XE camera, is a composite of 20 exposures of 60-s duration. The image scale was 1.5 arc-sec/pixel (field 12′ wide) and the cometary motion was 2.9 arc-sec/60 s frame which delivered ruler straight star trails using the AIP4Win multi-image deep sky stacking tool. Image: Martin Mobberley.

Of course, advanced imagers may well want to stack 20 or 30 min worth of exposures to get a decent image of a faint comet (see Fig. 11.15). As long as you can see the comet, to reference the images to, simple stacking of images is easy. The difficulty comes when you cannot see any image of the comet at all, in which case you will need to offset each image by a certain number of pixels and at a certain position angle. This can be quite a mental challenge the first time you attempt it, but software packages like CCDSoft can allow you to achieve this. However, if you cannot even see the merest hint of the comet on individual exposures then the situation is usually pretty grim! A high quality mounting like the Paramount ME can allow you to move the mount at the rate the comet is drifting as opposed to the rate the stars are moving (that is, the sidereal rate caused solely by the Earth's rotation with respect to the stars). However, excellent though the Paramount ME mounting is, even that formidable piece of engineering cannot track perfectly for more than a few minutes and when objects are near to the celestial equator you will need to be using an image scale coarser than 2 arc-sec/pixel or you will see slight wiggles in the star trails after a few minutes. Wiggles of only an arc-second or so it is true, but wiggles none the less. Another way of avoiding these wiggles is if the comet is at a high declination, say, of 70° Dec or greater, when the tracking accuracy will be less critical. The improvement is proportional to the reciprocal of the cosine of the declination, so at 60° Dec (north or south) the improvement will be a factor of two as cos (60) = 0.5.

The Ultimate Conjuring Trick!

Undoubtedly the ultimate processing operation that can be performed on comets is to combine two sets of images, one set stacked with reference to the cometary head and the other set stacked with reference to the stars. The aim is to end up with a deep image of a comet, frozen against a deep image of the stars (see Fig. 11.16), something that would normally only be possible with a very fast system like a large Schmidt camera working at f/1.5 or so. At first this sounds horrendously complicated but it can be achieved, even in color, with a lot of forward planning and many hours of processing. Achieving the desired result can take a week of evenings trying different techniques until everything clicks into place.

The first step in this ultimate comet processing trick is to create a stacked image of the comet, registered, as usual, with reference to the bright inner coma's center. However, in this case we want all the stars to completely vanish from the image! Hmmm, how can we achieve that?! Well, there is a technique sometimes called Sigma combining in which images are stacked as normal but anything small that deviates from the average image at, a specific x, y position, is rejected. It helps with this technique if the individual comet images are spaced further apart in time than would normally be the case, so that in normal circumstances the stars would not appear trailed as a single line, but as a series of dots. This spacing enables the Sigma combining technique to work far better. The software CCDStack contains an inbuilt Sigma combining tool and Ray Gralak has written some freeware called, not surprisingly, Sigma, which performs this operation. These packages can be found at the following web addresses:

CCDStack: http://www.ccdware.com/products/ccdstack/
Sigma: http://www.gralak.com/Sigma/

So, after Sigma combining the images we have a nice comet image, in color or in monochrome and with all those stars and potential star trails removed. Admittedly, big stars may still need painting out manually using a clone brush, but small stars have now gone.

Fig. 11.16. This superb image of comet 2007 N3 (Lulin) was taken by Prof. Greg Parker at the New Forest Observatory in the UK on February 28, 2009 and processed by Noel Carboni in Florida. It was taken with a Celestron Nexstar 11 working at an ultra-fast f/2 with a Hyperstar III lens assembly and a Starlight Xpress SXVF-M25C one shot colour CCD camera. It is a composite of 77 frames of 180-s duration, resulting in a cumulative exposure time of 3 h 51 min! The comet moved half an arc-minute between exposures and 40 arc-min during the whole sequence. However, it appears frozen in space because two sets of images were stacked, one for the stars (with the comet erased) and one for the comet (with the stars erased) as described in the text. Image: Prof. Greg Parker and Noel Carboni.

However, that was the easy bit as we now want a deep image of the star field on its own and removing a huge object like a comet with a fat head and a long tail is far trickier than removing tiny stars. One perfect solution might be to simply wait an hour or two for a fast moving comet to drift out of the field and then expose and stack deep images of the comet-free star field. This could even be carried out on the previous night if you were 100% sure of where the comet would be on the big night! Having said that though, if it was clear you would surely be imaging the comet as the top priority and not the next night's star field! OK, maybe imaging the star field on another night might be considered cheating but, even if it was, the plan may be scuppered as the previous night could well be cloudy and the following night may be cloudy too for that matter. So, what about plan A, namely using the same images we took on the main night but removing the comet from the stack? Well, a standard mean combine function, registered to a bright star, will produce an image with fixed stars and a smeared comet with the comet's bright inner coma distinctly trailed into a line. This image will be our stellar backdrop starting point. Exactly how you remove the smeared comet, keystroke by keystroke, will depend on your software package, but the principle is the same whatever program you own. What is required is to take the "smeared comet with fixed stars" picture and subtract from it a "smeared comet with erased stars" picture with a slight offset so the background does not go black. The smeared comet is common to both shots and so disappears in the subtraction process. There may be a faint ghost of the subtracted comet remaining but this will not be a big problem once the frozen comet image is added. So, how do we create the "smeared comet with erased stars" picture? Well, it's nothing that tricky. Just applying a few aggressive noise, dust or scratch removal routines with a Gaussian blur and the clone brush (for rubbing out big stars) will do the job well enough for this stage of the process.

So, let's have a bit of a recap now. We used a Sigma stack technique registered on the comet's head to create a frozen comet image from all the sub images (which were taken with a big enough gap between exposures so the stars would not form a trail, but a set of blobs if stacked on the comet with a standard method). Then we stacked the original frames again, using a standard average stacking method, with reference to the stars to create a deep stellar background with a smeared comet. From this second stack we performed a "smeared comet with fixed stars" minus a "smeared comet with erased stars" operation (plus a slight brightness offset).

So we now have a frozen comet image, with no stars and a frozen stars image, with no comet. All we have to do now is merge the two together which can be achieved in a variety of ways. You could use Registax and average stack the two images, or you could use Paint Shop Pro's mathematical "Add" function or, just use Photoshop's Layers to merge and lighten the pictures.

The above magic trick takes some time to get your head around so do not expect to use it unless you are very patient and have plenty of imaging experience! It is most definitely not for beginners and you will certainly want to tinker with it to suit your own software tools.

For a different approach the DeepSkyStacker software at:

http://deepskystacker.free.fr/english/index.html

has a technique described in the manual that is similar and the software has a special comet stacker feature that produces "pinpoint stars" and a frozen comet.

Animations

As mentioned at the start of this book the details within the straight gas tails of comets can be quite sharp and intricate and the solar wind can occasionally cause the magnetic field lines around the comet and its ionized tail to pinch the tail off. The gas tail may then be seen to discon-nect from the head, even in the duration of a set of consecutive short exposures. The old tail can often be seen floating away from the head at the same time as a new tail sprouts forth. Normally a comet will look very similar in exposures taken even hours apart, but when crossing a solar wind boundary the tail can kink dramatically in the space of a few minutes and blinking before and after pictures of these events can be fascinating to watch (see Fig. 11.17a, b).

If multiple short exposures (say, of 60-s duration) are taken and ani-mated into a movie, the result can be quite dramatic when a tail discon-nection is recorded. The tricky part here is actually taking enough images over many nights to capture such an event, and to take a sequence lasting, say, 30 min or more, between twilight and the comet disappearing below the local horizon. Features can, occasionally, move down the tail much faster than the comet's motion against the stars, so short exposures are advisable for this kind of work. The popular package Paintshop Pro has a nifty .exe file hidden away in the Jasc Software/Animation Shop directory called Anim.exe which is very useful for creating Gif animations. Pho-toshop can also be used for creating animated Gifs using the Layers and ImageReady functions. With Photoshop Elements using Layers and then the "File/Save for Web" option is the method required. Animations for complete web pages can also be made in a Flash (swf) format using freeware like Camstudio (http://camstudio.org/) to create an AVI video first and then compress it into a compact swf file. Most animators will

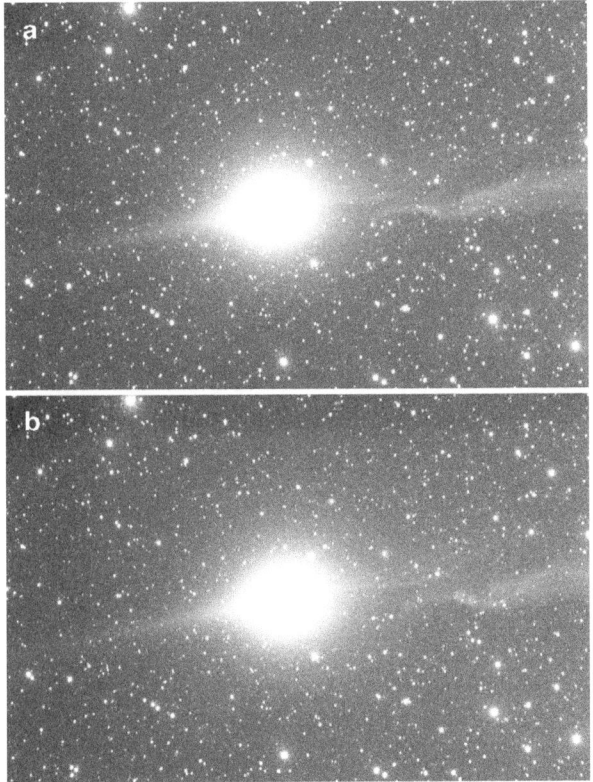

Fig. 11.17. (a) Comet C/2007 N3 (Lulin) suffered a dramatic tail disconnection event on February 4, 2009 while the author was taking images remotely over the Internet using a 0.25-m aperture f/3.4 Takahashi astrograph and an SBIG ST10XME CCD at the Global Rent-a-Scope (GRAS) New Mexico facility. The field of view in this 120-s exposure is 1° wide. Note the disconnected tail separating from the head. Image: Martin Mobberley. **(b)** Comet C/2007 N3 (Lulin) was imaged again, 27 min after the image in Fig. 11.17a, taken with the same equipment. Even in such a short time period a distinct change has occurred with the angle of the new short tail pivoting some 10° northward and sprouting subtle fan-shaped details. Image: Martin Mobberley.

probably be happier sticking with Gifs but a Flash/swf format enables a sound commentary to be added. Contrary to popular belief jpg images can be animated if called up from a webpage using JavaScript (so your browser must have Java enabled). Instead of having the entire animation encapsulated within a single huge Gif file the JavaScript simply loads jpg images rapidly to form the animation using an interval defined by the **setInterval**() function. However, having said all this animated Gif files will be more than good enough for most comet imagers.

Essential Dark Frames

I decided to deliberately delay talking about dark frames and flat fields in detail prior to this point simply because most beginners to comet imaging will simply want to crack on and take pictures of comets immediately, whatever their quality, without taking any calibration frames. With all DSLRs, and some cooled CCDs, this is perfectly possible, although the results will be less than impressive in most cases. Nevertheless, all imagers will want to get the best out of their comet pictures and so at some point understanding the very tedious facts about dark frames and flat fields is essential. It becomes critically important if you ever attempt comet photometry, although we will not tackle that particular can of worms for a while.

All CCD cameras work by changing photons into electrons. The CCD camera's "Analogue to Digital converter" measures the number of electrons in each pixel and this number is, ultimately, used to assign brightness values to each pixel on your PC screen. Unfortunately, electrons are also generated simply by the CCD camera being warm. This thermal current "noise" doubles for every 7°C increase in temperature. Fortunately, as most commercial astro-cameras can be chilled by some 30°C, this cooling gives a 20-fold decrease in noise level. At a chip temperature of 20°C many astronomical CCDs will saturate (white out) purely from thermal noise in a 10-min exposure.

A dark frame is the solution to thermal noise; it is an exposure with the camera's shutter closed, and it captures the camera noise without the image of the astronomical object (and light pollution) being present. So, it might be thought that the temperature the camera was operating at was academic; surely you can just subtract it all using the dark frame, taken at the same temperature? Unfortunately noise, by its very nature, is random. It is not a smooth level, but a choppy sea. Imaging and subtracting a dark frame with the astro-camera's CCD running as cold as possible is definitely the best approach (see Fig. 11.18a, b).

Dark frames should be taken as close to the temperature at which the main image was exposed as possible; in practice this means immediately before, or after, the main image. Some astro-software packages arrange

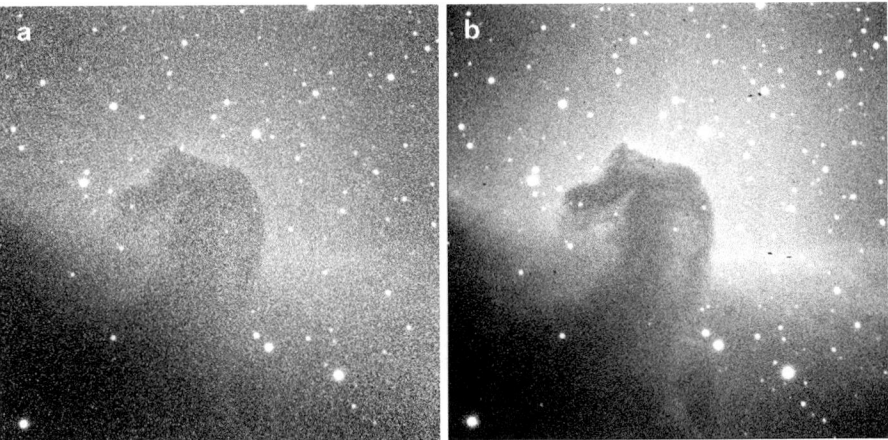

Fig. 11.18. (**a**) The famous Horsehead nebula in Orion, imaged with a Celestron 14 and SBIG ST9XE CCD and a 180-s exposure at f/7.7, but with no dark frame subtracted. Image: Martin Mobberley. (**b**) The same image as shown in Fig. 11.18a, but with the dark frame subtracted.

things so that a brand new automatic dark frame is exposed and subtracted if the exposure time has changed, or if the temperature changes by a degree or more, to ensure the best match subtraction. A few advanced amateurs build up a library of dark frames at various temperatures and exposures, along with a so-called "bias frame" (an exposure of almost zero duration, which records just the basic electronic read-out noise). Using this complex technique it is possible to create custom dark frames to avoid very lengthy time-wasting exposures on those rare crystal-clear nights. For most situations though, simply taking one dark frame (once the camera and air temperature have stabilized) and sticking to that exposure duration all night, will be the best practical solution. However, keeping an eye on the CCD chip's temperature (most astro-camera's software monitors this) is crucial throughout the imaging process.

For the experienced imager not taking a dark frame is arguably a worse crime than not cooling the camera. While some of the lowest noise Sony CCDs (and the Canon DSLR CMOS detectors) can produce acceptable images without a dark frame subtraction, the early ultrasensitive Kodak "KAF" series CCDs, used for photometry in many amateurs astro-cameras have considerable pixel-to-pixel variation; an image without a dark frame may even be unusable, whereas an image at ambient temperature with a dark frame will produce a decent, if noisy, image. The aforementioned software packages CCDSoft, IRIS and Richard Berry's AIP4Win have useful dark frame and flat field reduction routines, but you probably already have good dark frame routines with your CCD cameras software package. IRIS (powerful freeware) can be downloaded from: http://www.astrosurf.com/buil/us/iris/iris.htm

AIP4Win can be ordered from Willmann Bell at:

http://www.willbell.com/aip/index.htm

For DSLRs the previously mentioned software packages IRIS, by Christian Buil, and Images Plus by Mike Unsold, can be employed to apply dark frames as well as the flat fields mentioned in the next section.

Making the Sky Flat

So-called "Flat fields" are intended to make your final image background look a flat grey, to correct the shadows caused by parts of the telescope restricting (vignetting) the light cone, especially when tele-compressors are involved. Also, dust specks (on glass surfaces above the chip) can appear as blurry doughnut-like shadows which need removing. All good CCD astronomy packages incorporate Dark Subtract or Flat Field functions (see Fig. 11.19) and the flat field function divides every pixel in the main image by every corresponding pixel in the flat field. Once applied the background sky in your images will magically appear an even gray (flat) and those horrible dust speck doughnuts will have vanished. The flat field image itself can be produced using smooth high altitude twilight as an even source, or by using a custom built "light box" which fits over the telescope aperture.

The twilight approach to flat-field generation is very straightforward. First, you need a crystal clear post-sunset (or pre-sunrise) sky. When the sun is about 5° below the horizon, you can take very short exposures near the zenith, with your telescope, without the camera whiting out. The shorter the exposure the better, but some astronomy cameras with mechanical shutters do not let you go much below a quarter of a second. You should aim to half-saturate the CCD to get a good signal-to-noise ratio. The exposures need to be short because stars can be captured, even in twilight, with longer exposures, unless you switch the telescope tracking off. The flat field images will typically be brightest in the middle and covered with large and small rings, caused by dust specks. In addition, for a really excellent master flat field you should rattle off a few dozen flat (or rather, flat minus dark)

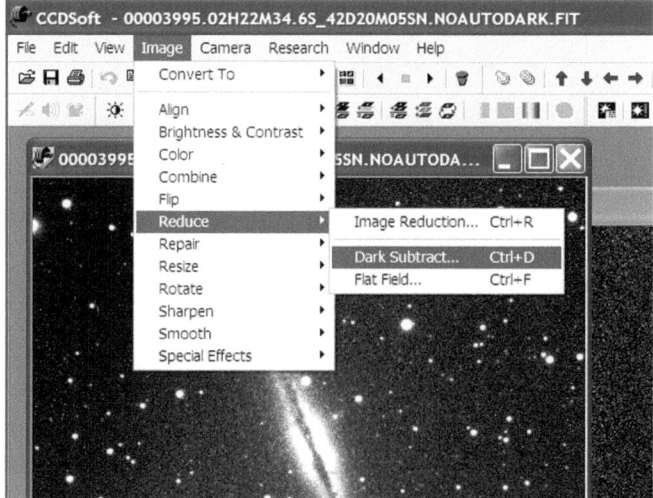

Fig. 11.19. CCD Soft's image reduction routine allows easy dark frame subtraction and flat field division.

Fig. 11.20. Stacking multiple flat fields results in a very smooth "master flat" which removes all dust specks and vignetting.

frames and stack and average them. The resulting master flat field composite will be nice and smooth (see Fig. 11.20).

As well as applying these techniques it is worth checking out, in daylight or twilight, but with the telescope aperture capped, whether your system is fully light-tight. Many uneven background field problems are actually caused by stray light (from streetlights, house lights, or observatory lights) entering the telescope or the CCD camera body at night.

The keen photometrist may want to consider making a Flat Field box for those situations where a nice even twilight sky can be elusive. Such an arrangement (see Fig. 11.21) consists of a diffusing perspex screen which lets light through to the telescope via the inside of an evenly illuminated white box. The box is constructed so that the illuminating bulbs cannot be seen from the telescope, but can illuminate the box interior. The aim is simply to present a totally "flat" uniform glow to the telescope aperture. The bulbs involved should be very low wattage as astronomical CCD cameras are very light sensitive. There are various websites about making flat field boxes and Richard Berry's AIP4Win book also shows how to construct one. Some amateurs are now using flat electroluminescent panels attached to the front of their telescopes to create flat fields. These can work well, but again the light often needs dimming to reduce their intensity as CCD cameras designed for astronomy are simply incapable

Fig. 11.21. A flat field box can produce an even illumination if a clear twilight sky is not available. Image: Martin Mobberley.

of coping with such bright commercial light sources. In this case light attenuation can be achieved by attaching sheets of white paper over the electroluminescent panel.

For the perfect comet picture flat-fielding is essential; far more so than with any type of deep sky imaging. Apart from if you are working in the precise field of photometry, there is no other area where flat-fielding is more important than in comet work. This is simply because bright comets are often imaged in twilight and the twilight illumination will enhance every horrible dust-speck and source of vignetting in your system. Specks of dust on the CCD window are a real pain for the CCD imager, but a lot depends how far away the dust mote is and what your f-ratio is. At f/3 and f/4 the fat light cone tends to suffer more from vignetting caused by the telescope draw-tube cutting the light cone or by a secondary mirror that is too small. At the typical Schmidt–Cassegrain f-ratio of 10, dust specks will be obvious appearing as small doughnut shapes on the image. It is important to maintain the same camera orientation when using a flat-field from a previous night. Some observers will stress that you need to obtain flat-fields every night because the telescope is never quite the same from night-to-night. However, there comes a point where perfectionism will simply kill new observers' enthusiasm off. Comet imaging, like all astronomy is something best learned a bit at a time. I have found that dust specks, on the glass window above the CCD, can stay in exactly the same position for months, despite others advising differently.

More on Twilight Flats and Noisy Patterns

There are a few other twists in the comet imaging tale (puns intended) which I would like to mention at this stage. The beginner will, perhaps, not want to think about them yet as just grasping what I have already covered will surely be enough for starters! However, I think these extra problems should be addressed. When a comet is really low down, and often that is when they are at their absolute best, the flat fielding requirement becomes an order of magnitude more critical. It is possible to take fairly good images of objects at high altitudes, in a dark sky, without a good flat field, or even without any flat field at all. If the sky background is dark and you keep the background dark during the processing the evidence of just how badly vignetted and dust riddled your system is will be hidden forever. However, when imaging a comet close to the Sun and therefore near to perihelion you will invariably be imaging the object against a considerable twilight glow which will show your optical system's deficiencies up in very gory detail indeed. You may well have captured the comet, but the background will look truly horrendous. Under such bright illumination, typically with the Sun only 10° below the horizon when you start a sequence of short exposures, your standard flat-field illumination will simply not be good enough.

One reason for this is that the twilight sky at 15° or even 10° altitude is measurably brighter just a few arc-minutes lower in the sky and so there will be a brightness gradient across the image; apply the usual harsh contrast stretch to the comet in the center of the image and the horizon edge whites out while the top edge goes black! Apart from this nightmare illumination scenario there is the added misery of tree branches and other obstructions complicating the picture. My own solution to this particular can of worms has been to take custom flat fields literally minutes before starting a sequence of comet images, just a degree or so away from the field where the twilight illumination gradient is very similar. If there are bright stars in the first field I try another one nearby. In these situations I am typically imaging a comet of fifth or sixth magnitude which is the brightest object in the field and exposures of more than 10-s duration will saturate the image. As the twilight gets fractionally darker the switch from flat fields to comet imaging can take place almost seamlessly and, on the face of it, there is often little obvious difference between the raw flat fields (or rather flat fields minus their dark frames) and the raw comet fields (or rather comet fields minus their dark frames), except the comet's fuzzy coma. However, when the comet images are later flat-fielded and stacked with respect to the cometary nucleus and then averaged (additive stacking would certainly saturate the image in these situations) it is amazing how much detail, including faint straight star trails, emerges. Of course, after stacking the images the tail emerges too and it can often be a battle to glimpse in which direction the tail is trailing off before serious imaging starts. Nevertheless knowing the direction of the tail before starting the main sequence is absolutely vital! There is nothing more annoying than taking a whole sequence of comet frames only to find you had set the comet's head in the bottom right corner and the tail was heading that way too – OUT OF THE FRAME! After 10 min of imaging a comet in these situations it is very common for a huge dark pattern to suddenly appear in the image meaning that the field has suddenly set behind a tree or a house and half of your telescope aperture has been obstructed.

Another issue in such dire twilight and low altitude imaging circumstances is that of fixed CCD noise patterns streaking across the image at the same position angle as the comet is moving or at a position angle which is the vector sum of the cometary motion direction and declination drift caused by the low altitude. This pattern is not normally obvious but when multiple short exposures are stacked together with a pretty much fixed offset and direction between them, and a harsh contrast stretch is applied, the pattern can literally scratch itself across the image. There is no quick fix to this problem if the usual safeguards have been applied, like cooling the chip, taking a dark frame at the start of the comet sequence and running hot and cold pixel removal routines. However, in more leisurely deep sky imaging runs amateur astronomers sometimes apply a "dither" routine such that the telescope randomly slews by, say, 10 pixels between images to break up the stacked fixed noise pattern from becoming obvious. However, imaging comets in twilight is usually a frantic enough pursuit anyway. If you still have a picture with one edge much brighter than the other after all the normal tricks have been applied you might try using AIP4Win's nice little "Sky Background Fixer" routine which appears under the Edit option in the main Menu. It allows you to cover the frame with a matrix of background sky reference points which can be deselected around the comet's head (so it does not try to dim the comet away!) and then adjusts the brightness so the background is very flat and even. The routine should not be used as a substitute for a flat field but it can solve some irritating illumination issues.

I would just like to add that with all the techniques I have mentioned here it is always essential to cool the CCD camera to a stable operating temperature well before serious imaging starts and a bright star, brighter than the twilight, should be used for focusing trials as soon as the sky becomes dark enough to image. Also, I have mainly been talking here about imaging comets in deepening twilight. Of course, the opposite happens in the morning skies, so flat fields need taking after the comet is imaged and you will have to wait for the comet to clear the trees and buildings first.

Creating a Mosaic

Comets come in all shapes and sizes, and the comet imager will often have a dilemma working out what focal length instrument is best for any one object. For the superb comet Hyakutake in March 1996, I found an 85-mm lens and 35-mm format (36×24 mm) film to be ideal, this gave a $23 \times 16°$ field. For Hale–Bopp at its best, a field of about $10°$ was close to optimum, while for the 2002 comet Ikeya–Zhang, a field of 2 or $3°$ would have been best. All of these bright comets could easily be captured on photographic film with a modest telescope, because they were bright enough that the relative insensitivity of film (it dies over long exposures due to "reciprocity failure") was not a hindrance. Indeed, they were all comets that showed distinctive colors, so color film held an advantage over a monochrome CCD. The only requirement with film was that you needed to guide the exposure for maybe 5 or 10 min. With a really bright comet and film you could guide on the false nucleus, but for a comet like Ikeya–Zhang, guiding, or (horror of horrors) offset guiding on a nearby star was necessary. Of course, DSLRs have virtually eliminated film from the equation now but the sensors in the best DSLRs are approaching, or equaling, the size of 36 mm × 24 mm film frames. However big your imaging chip, CCD or DSLR, comets still come in a range of sizes and span either a small, medium or large angle on the sky.

I suspect that most potential comet imagers will use amateur tele-scopes in the 15 to 35-cm aperture range and, when they attach a CCD camera, may have fields as small as 10, 20 or 30 arc-min to play with; but what do you do when your CCD field is, say, 20 arc-min, but the comet's tail is a degree or two long? My personal preference here is to produce a mosaic along the comet's tail (see Fig. 11.22). Positioning the telescope to take all the images may sound like a nightmare but, with a "Go-To" telescope with R.A. and Dec readout, this is not the case. The first priority is to totally familiarize yourself with which direction the N,S,E and W buttons on your hand controller move the stars on your monitor, for all equatorial mounting configurations, and by how much. It is essential to know this precisely as, in the heat of the imaging moment, it is easy to get confused with directions, especially if you are frozen rigid at 3 a.m. Also, bear in mind that, for example, at 60° Dec, a minute of right ascension on the celestial sphere will correspond to 7.5 min of arc, not the 15 min of arc you get at 0° Dec! Remember to apply the correction factor pro-duced by the cosine of the declination. The next step is to be prepared for which direction the comet's tail is pointing, either by looking at your own image from the previous night, or other images on the web. You can then do a little sketch before going out to estimate the amount you need to move the telescope between frames to follow the tail, but with plenty of overlap. Do not confuse north and south either! This is all too easy to do with a German equatorial mounting when swapping from pointing west to pointing east and vice versa: north and south swap positions on the image. If you get the planning right, taking a mosaic of frames only requires moving the scope in RA and Dec by a set amount for each frame. For most "Go-To" telescopes a slow creeping of the telescope to offset by the right amount is controllable to an accuracy of ±1 arc-min, which is enough accuracy if you allow a few arc-minutes of overlap. If you are

Fig. 11.22. A mosaic of images along the tail of comet C/2002 V1 (NEAT) taken on January 25, 2003. Each exposure was 80 s with a 30-cm Schmidt–Cassegrain at f/6.3 and an SBIG ST9XE. The field is a degree wide. Image: Martin Mobberley.

really super-confident of the RA and Dec offsets required to take pictures along the tail, or if the tail is just streaking away due north, south, east or west, I suppose you could use the Software Bisque Orchestrate program to run a prepared script file. The Orchestrate software controls Software Bisque's The Sky X and CCDSoft so they work in harmony, slewing the telescope and imaging with the CCD camera, all according to your script. However, for only half a dozen frames in a mosaic I think most imagers would be happy to control the process manually "on the fly."

The next step, once indoors, is to equalize the background sky contrast and brightness of each image where it interfaces with the next image. There may well be ways of doing this automatically, but I prefer to judge the degree of "grayness" by eye. Unfortunately, if the comet is in twilight, which changes in brightness every minute, it will be very difficult to achieve a seamless mosaic, as even if the sky "grayness" of adjacent images is matched, the comet's tail brightness will alter as it crosses the image boundary! Mosaic-ing is best achieved in a dark sky. Once each image's "grayness" has been adjusted they need to be aligned. My preference here is to put the comet's head image (image 1) into Paint Shop Pro and then enlarge the canvas size ready for the subsequent images. I prefer a black canvas background, but that is just my personal choice. The next step is to call up image 2 and then, using Paint Shop Pro's "clone brush" (and the preferred brush tip size), right-click on a star on the boundary of both pictures in image 2 and then left-click on the same star in image 1, and commence painting a clone of image 2 into image 1. This may not make much sense if you don't use Paint Shop Pro, but many people do and other packages, like Photoshop, have a similar clone brush feature too. The process is continued for the image 2/image 3 boundary, image 3/image 4 boundary and so on.

It should be stressed here that taking multiple images of each comet frame in a mosaic, to go deeper, will not work with a fast moving comet, but this may only hit you after you have tried it. As well as the fact that it is advisable to complete a mosaic quickly, before the comet and its long tail have drifted too much against the stars, if you think about it you will realize that you can only stack sub-frames of the first image in a mosaic; this is simply because only the first frame, containing the bright comet's head and false nucleus, can be used to register the sub-frames! There will be no other concentrated light source within the comet's tail and so nothing to register the comet tail images with respect to!

Coping with the Weather!

On average I find that on about one night in three, from deepest Suffolk on the eastern side of the UK, I can do some useful work between gaps in cloud banks, gaps in quite serious cloud or, in a totally clear sky. Anyone familiar with living in the UK knows that the trend is for a lot of active weather fronts coming from the west, often with few cloud-breaks; but when clear nights do arrive, they invariably come in a row. In a cloudy country like ours it is highly frustrating being the sort of observer who specializes in observing time-critical phenomena. We all know that one-off solar eclipses, lunar eclipses, asteroid occultations and the peak nights of meteor showers are usually clouded out. Fortunately most bright comets are at their peak of activity for several weeks and so the chance of being clouded out for all of that time is low, although not impossible. With CCDs the half, gibbous, or even Full Moon is no longer an easy excuse for staying indoors. Image processing can perform miracles on a bright sky. The same applies to broken cloud: it really is simply no longer an excuse! While 30 min offset guided exposures on photographic film were thwarted by any amount of cloud, 60-s exposures can be clinched between gaps in cloud. Indeed, since the early 1990s this has been my main strategy and enables me to average an image on as many as one night in three, when a good comet is about. Hanging around a telescope staring at broken cloud is about as exciting as watching paint dry, but it is worth it if a break of a few minutes enables you to clinch a spectacular picture! In addition, with a remotely controlled system you don't have to hang about outside as you can just keep taking frames from indoors until stars appear on the PC screen.

Satellite pictures are a great way of checking what the weather is doing of course but some care is needed. The visual wavelength picture is the best indicator, but at night, for obvious reasons, it is as useful as an ashtray on a motorbike. The infra-red satellite picture is available all day and night via the Internet but some care is needed. The infra-red view is a record of the temperature of cloud tops or of the ground. It does not show fog or thin cloud and frozen ground often looks like cloud, so caution is necessary! The best type of weather for the comet imager occurs following the passage of a cold front from north to south across a northern hemisphere country. From my observing site the cold northern air is always dry and crystal clear and, by day, a superb pure blue sky at the zenith will fade into a lighter shade of blue at the horizon, but without a trace of the milky haze that ruins transparency. In such a sky good CCD images of naked eye comets can be taken, even at Full Moon, provided the comet and the Moon are on opposite horizons. Remember, many comets peak in astronomical or even nautical twilight, when the sky is as bright as if it were moonlit anyway! Hazy skies are the worst for the comet imager. They often coincide with a high pressure system that has been around for too long, or humid air coming up from the south. I prefer total cloud cover to a hazy sky; at least with total cloud cover you know it is not worth bothering. With a hazy sky I usually make the effort, but then wonder why I did!

In passing, it should go without saying that for a good image the telescope should be critically focused. With older Schmidt–Cassegrains, this can be a tiresome procedure as their mirrors can "flop" when the telescope changes position and tubes can expand or contract with temperature. I have also found that dew heater bands, while essential for long observing periods in the UK climate, can easily alter the telescope focus as they heat up. If you are going to use a dew heater it should be switched on half an hour before you try focusing the telescope, so everything reaches equilibrium early on!

Like anything in astronomy, or in life itself, practice makes perfect. Over the past 30 years I have taken more mediocre comet pictures than I have taken good ones, but these have all been part of the learning process. Ultimately, the results one gets depend on a number of factors, not least (for most observers) distractions from tedious work and family commitments which always coincide with the clearest of nights. Possibly the most important factor is simply a stubborn but patient single-mindedness to surmount all the obstacles placed in the comet imagers way in the form of physical tiredness, other commitments, trees, light pollution, cloud, cold, dew, equipment failures and financial restrictions. Easy! Well, maybe not. However, with modern equipment, imaging comets is far less of a battle than it used to be and hopefully, this chapter might inspire others to take up the challenge, however much cloud hovers over his or her observatory when a bright comet appears.

Imaging at Swan Band Wavelengths

The Scottish physicist William Swan (1818–1894) is responsible for the term "Swan bands" as he was the person who first studied in detail the spectral bands of carbon and the carbon radical C_2 in 1856. These bands are seen in the spectra of gassy comets, carbon stars and hydrocarbon fuels and in total there are many Swan bands associated with carbon across the 400–600 nm region. Sir William Huggins first identified these bands in a comet in 1868 when he obtained a long exposure spectrum of comet 1868 II Winnecke and compared it with a carbon spectrum produced simultaneously by decomposing olive oil with the spark from an induction coil!

The US Company Lumicon manufactures an interference band filter for amateur astronomers which provides a narrow, 25 nm wide, window through which the dominant Swan bands at 511 and 514 nm can be viewed, along with the astronomically significant Oxygen OIII lines at 496 and 501 nm. As I pointed out some time back, the name Swan is all too easily confused with the abbreviation SWAN used for the SOHO satellite's Solar Wind ANisotropies detector but there is no connection with the Swan filter. The big advantage of any narrow band filter like this is that it increases contrast and lets through light from a comet's ion tail, while blocking the light from its dust tail, thus greatly enhancing any gassy tail emissions in the 511 to 514-nm waveband; this is not too far from the region of the spectrum where the human retina (even a dark adapted one) is most sensitive. The manufacturer's data for this filter includes a claim that Howard Brewington discovered Comet Aarseth–Brewington, 1989a1, using the Lumicon Comet Band Filter. Howard is quoted as saying that the filter doubled the contrast, making the comet very easy to spot. Sounds impressive, but can the filter be used for any benefit by comet imagers? Well, with very bright comets, yes, especially when you want to peer through the dust and enhance any gas activity. The problem is that many comets are very dusty and the dust simply reflects the light from the Sun. Using a 25-nm bandwidth filter on a comet will attenuate most of the visual band light output by as much as three magnitudes making the image very dim and attenuating background stars by a similar amount. Swan band filters are definitely worth experimenting with but in many cases the comet may look rather dull and sad unless it is a very bright one or one with a substantial gas tail, gas activity near the nucleus, or one with low dust output.

Special Processing Techniques

It is perhaps time to remind ourselves at this point that comets and deep sky objects are faint, and I mean really faint! The faintest comet tail details visible in a typical amateur image can glimmer at less than 1% of the background sky brightness caused by light pollution and natural sky glow, and yet even that background sky brightness is feeble compared to broad daylight. If a CCD or DSLR image of a comet is left untouched and just displayed on your PC screen without any kind of processing the sky will appear black, with just the head of even a bright comet becoming visible as little more than a dim smudge. For this reason it is always necessary to carry out a basic contrast stretch if nothing else. The contrast stretch algorithm simply looks at the narrow brightness region from the sky background value to the "sky background plus cometary detail value" and stretches that range of values to fill the whole dynamic range of the image. At a stroke this brings out all the faintest detail but can also give a very "soot and whitewash" effect where the head of the comet is just an overexposed white blob. The contrast stretch can be applied in numerous ways in numerous different packages. In some, like Paint Shop Pro, simply playing with the brightness and contrast options will produce an acceptable, if crude, result. However, for more advanced processing the transfer function between input image value and output image value can be adjusted more carefully so that all significant values above the sky background do not simply saturate to white. The gamma function is useful here because it allows the user to just alter the mid-range brightness values without plunging the dark values to black or forcing the light values to white. Playing with the menu options that allow you to individually alter the highlight, mid-tone and shadow values is highly recommended and, in some packages, the histogram adjustment or "curves" feature may be useful too. It should be remembered that a basic PC screen has 256 brightness levels corresponding to eight brightness level "bits" of data. When all these bits are set to the binary number 00000000 the screen is black and has a decimal value of 0, but when they are all set to 11111111 the screen is white and has a decimal value of 255. If the input values of the unprocessed image are untouched then every pixel brightness value in the initial image, when scaled down to an 8 bit system, maps onto every pixel brightness value in the final image. However, by subtly adjusting the way in which brightness values map across from original to final image the maximum detail can be coaxed out. To further complicate matters the original raw FITS images from astronomical CCD cameras are often 16 bit images and so, strictly speaking, they have values from 0 to 65,535, but even the best PC screens cannot reproduce such a wide range of values.

Dedicated astronomy packages, such as Richard Berry's AIP4Win, offer a massive range of processing options which are aimed at bringing the maximum amount of detail out of the images. The problem with comets is that while you want the maximum contrast stretch to bring out the faintest tail details, the same algorithm will usually turn the head into a huge bleached blob with no features. A logarithmic stretch is often worth using but my favorite comet processing tool is AIP's gammalog stretch (see Fig. 11.23) which manages to bring out faint details while still avoiding saturating the head. Another option is the old favorite the "unsharp mask." This may seem a curious name for a processing technique, as who on Earth wants to make an image "unsharp?" Well, the term dates from the photographic era when photographers were frequently frustrated by the fact that they found it impossible to transfer all the detail in a film negative onto the much narrower dynamic range of a photographic print. Details in the mid range were captured but bright regions white out and dim regions went to black as soon as any contrasty photographic paper was used. To prevent this from happening photographers created a blurred negative image of their original film frame which could be aligned with the original image. This was then used to attenuate large scale brightness extremes while preserving small scale variations, simply because the mask was blurred. The same principle can now be applied digitally with the "unsharp" mask feature being available in most image processing packages.

An alternative approach is to try the favorite tool of the galaxy or deep sky imager, namely the Digital Development Processing or DDP method. A "DDP tool" may well be present in your preferred astro-software package

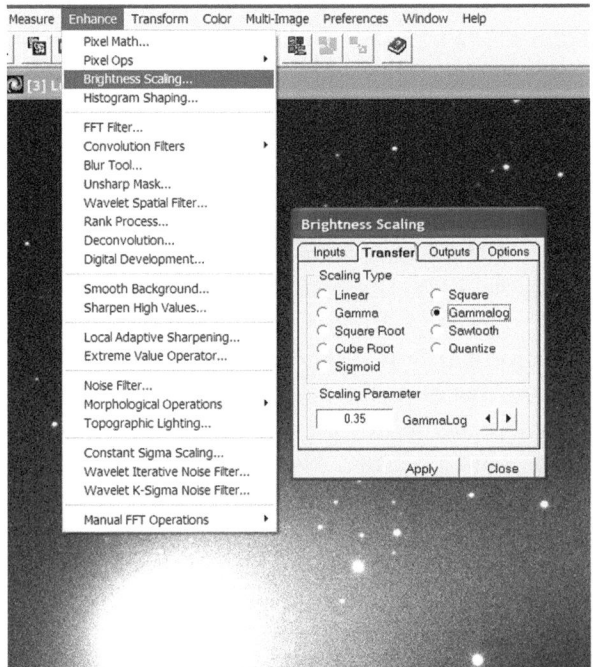

Fig. 11.23. AIP4Win's Gammalog function, as well as eight other useful brightness modifying tools, are available under the Enhance/Brightness Scaling menu. The Gammalog function is especially useful for enhancing faint comet tail details without saturating the head of the comet.

and can work even better than the "unsharp" mask technique. Photographic film, although very insensitive in long exposures, used to have a nonlinear response to light (a nonlinear "gamma curve") which worked well in many astronomical situations, as well as a natural "edge sharpening" response which worked particularly well on comets and galaxies. The DDP algorithm effectively adjusts the relationship between brightness levels in the raw image (that 16 bit or 65,536 gray level image) and the displayed image on your 8 bit/256 gray level PC so that the contrast of faint features is enhanced but not at the expense of the brighter parts. In addition, DDP also throws in an edge sharpening effect similar to an unsharp mask which sharpens up fine details and makes stars or star trails look sharp, rather than bloated. Some comet imagers also like the tools that show the brightness contours in an image by producing a contour map or a false color map.

Larson–Sekanina and Radial Filters

In some ways comet images are like total solar eclipse images in that they benefit from an image processing routine that brings out radial detail, that is, detail that streaks away from a central point. Deep within the center of the coma of a comet you will find the star-like nucleus, sometimes referred to as the false nucleus because the actual nucleus is far too small to see or image. Everything that emerges from the comet, namely the gas and the dust, emerges from this tiny dot. But the nucleus is rotating and only specific parts of it will be releasing volatile material when directly heated from the Sun. Therefore jets of material will be 'geysering out' from the surface of the nucleus on the sunlit side only to be blown back by the solar wind pressure. Normally these tiny features are simply whited out by the bright inner coma in CCD images, but by using very short exposures (a few seconds), or averaged stacks of many short exposures, the region near to the nucleus containing these jets can be revealed. However, the details are still very subtle and so an extra tool can be applied. Sometimes these tools are referred to as rotational gradient filters or even rotational "unsharp" masks. I know that some beginners can become confused by the term "filter" in the context of processing routines as most new observers will interpret the word "filter" as meaning a colored piece of glass in the light path. However, "filter" just means "image processing routine" in this context.

The most popular radial processing routine for comets is often called the Larson–Sekanina filter which was invented for the very purpose of enhancing jets and dust shells near the nucleus (see Fig. 11.24). Not every astronomical image processing software package contains a Larson–Sekanina routine though and it has to be used with care because, as with all powerful image processing routines it can be scientifically dangerous and lead to false artifacts being created. The method was devised by Steve Larson of the Lunar and Planetary Laboratory in Arizona and Zdenek Sekanina of the Jet Propulsion Laboratory in California. Any image in a digital form consists of x axis and y axis columns and rows comprised of pixels. This system is usually referred to as a Cartesian coordinate system. However, the image can also be represented in a polar coordinate system where an origin is selected and the position of any point in the image can be located using a radius r, and an angle theta. Of course no CCD of this type exists, but the mathematical formulae applied to the images do not care about that! The key point here is that if the apparent nucleus of a comet is selected as the origin, powerful processing routines can be applied which bring out detail that is radial to the nucleus, in other words the jets given off by the comet. In practice the precise

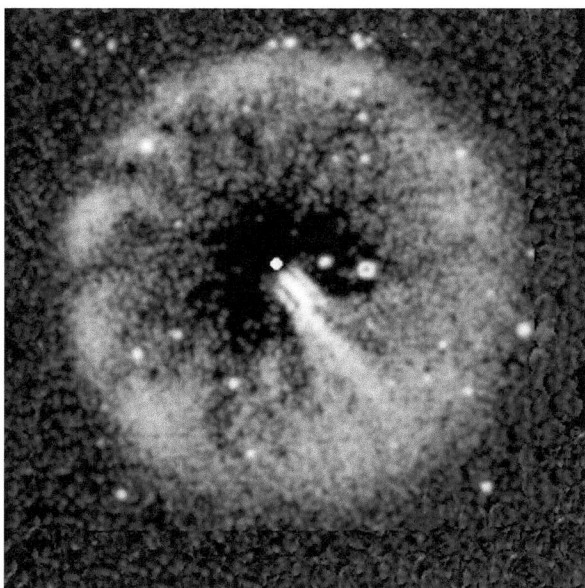

Fig. 11.24. A Larson–Sekanina filter applied to an image of the expanding head of comet 17P/Holmes on October 28, 2007. Image: Martin Mobberley.

technique simply requires the user to input the x and y coordinates of the nucleus on the CCD image (by pointing the mouse cursor at it) and then to specify a radial shift (often set to zero for jets) and a rotational shift (often set to zero for dust shell enhancement). Once these values have been entered the software creates a new image that has been shifted and rotated by the values specified and subtracts it from the original image.

Of course, as the operation involves the subtraction of two similar images the result can end up being very dark, which is why there is usually an option of providing a background offset brightness on top of which the processed image sits. As I have hinted above, if jets streaking out from the nucleus are the main interest then a radial shift away from that nucleus, followed by a subtraction, may not be required at all. However, if dust shells that are concentric (rather than radial) to the nucleus are the main target of interest then the user of the Larson–Sekanina filter may well not want a rotational subtraction. In practice though the user will always play around with the values until the maximum jet and dust shell activity is revealed. Any stars in an image where a Larson–Sekanina filter has been applied will tend to appear twice, as the original white star and as a subtracted dark star some distance away. Therefore the image is not so much an accurate image of how the comet actually looks but more a tool to show where maximum jet and dust activity is occurring. It is dangerous to assume that everything in such a strongly processed image is genuine. There will definitely be artifacts and those black stars are the most obvious example, so the Larson–Sekanina images must be treated with care. Not all packages refer to this type of processing as a Larson–Sekanina filter, some just call it a "rotational gradient" filter, but Larson and Sekanina were the pioneers in the days when images were mainly photographic and computers were very slow to carry out the transformation. Their original paper on the technique was published in 1984 in the Astronomical Journal.

The popular IRIS freeware by Christian Buil has a Larson–Sekanina type routine which can be applied by going to the Menu option "Processing" and selecting the last item, namely "Rotational Gradient" (see Fig. 11.25).

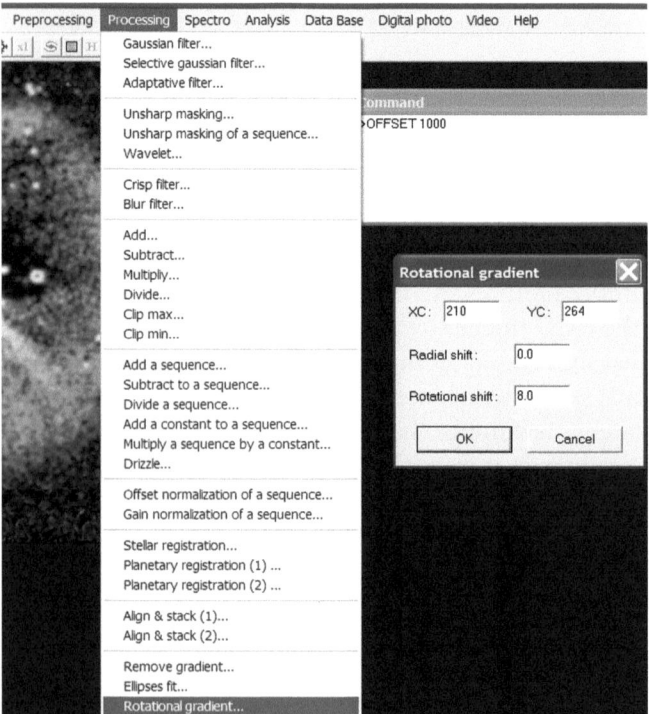

Fig. 11.25. The rotational gradient filter in the Iris software can be found as the last item under the "Processing" option on the main Menu.

In this package a brightness offset value can be added by clicking the command line icon and typing, e.g. OFFSET 1000. Other software packages allow similar processing and the popular AstroArt package has a rotational gradient filter as does Cyanogen's Maxim DL.

Other powerful jet and dust shell enhancing routines exist and the two techniques most often used by professional astronomers, or advanced amateurs, are known as azimuthal median subtraction and the radial weighted method. Azimuthal median subtraction measures the median values of pixels around circles of different radius from the nucleus. When these average values are subtracted from the original image an enhanced concentric map of subtle brightness differences in the inner coma emerges. The radial weighted method is similar in function except rather than using median pixel values at a distance "r" from the nucleus to subtract from the original, theoretical pixel values are used. The theory is simple and assumes that dust brightness will fade in direct proportion to its distance from the nucleus r, in other words brightness is directly proportional to 1/r.

The two techniques of azimuthal median subtraction and the radial weighted method can produce very similar plots in well-behaved comets and can be used to double check the validity of features revealed by the Larson-Sekanina tool. However, all these processing methods need a strong signal-to-noise inner coma image to work upon and so only the brightest comets produce good results with these powerful techniques.

CHAPTER TWELVE

Lenses, Telescopes, Astrographs, and Mountings

The f-ratio of a lens, or a telescope, is calculated by dividing the instrument's focal length by the aperture. For the comet or nebula imager the instrument's f-ratio (smaller means faster) is of vital importance, along with the sensitivity of the detector and the darkness and transparency of your sky. For big comets that move rapidly across the sky imaging with a very fast wide-field system has huge advantages, especially if imaging in color, whether using filtered RGB methods, LRGB methods or using a one-shot color CCD detector.

Schmidts and CCDs

Traditionally, and especially in the days of photography, the super-fast (typically f/2) Schmidt camera was the professional astrophotographer's biggest weapon. Designed by Bernhard Schmidt in 1930 the system used a spherical primary mirror and an aspherical correcting plate to counteract the severe optical aberrations from such a fast f-ratio. In addition, the focal surface ended up being curved so that the film itself had to sit on a curved plate inside the camera. Because of the obvious difficulties of focusing such a fast device Schmidt cameras were made from extremely low expansion materials and the focus was set at the factory. Thus, in theory,

M. Mobberley, *Hunting and Imaging Comets*, Patrick Moore's Practical Astronomy Series, DOI 10.1007/978-1-4419-6905-7_12, © Springer Science+Business Media, LLC 2011

it would never need focusing in use. Professional astronomers made bigger and bigger Schmidt cameras to photograph large chunks of the sky, culminating in the huge Tautenberg Schmidt at the Karl Schwarzschild Observatory near Jena in Germany, which has a 1.34 m aperture and a 2 m mirror. Eugene and Carolyn Shoemaker used a much smaller (45 cm f/2) Schmidt at Mt. Palomar to discover 32 comets in the 1980s and 1990s.

Remarkably, Celestron manufactured three sizes of f/1.5 Schmidt cameras for amateur astronomers for many years and, despite being extremely fiddly to use, many top comet and deep sky photographers of the 1970s and 1980s purchased them. They mirrored the main 5.5, 8 and 14-in. aperture range of Celestron's Schmidt-Cassegrains (140, 203 and 356 mm in metric dimensions) but with longer tubes and an access door into which the user could, in the dark, fit the curved film holder, fully loaded with a sensitive emulsion. The 14-in. model was discontinued after a while, but the 8-in. proved very popular amongst amateur astronomers. From the mid 1980s, until 2001, Celestron ceased production of the 5.5 and 8-in. Schmidt cameras but still supplied a small company called Epoch Instruments with the optical components which they then assembled into working cameras.

The advent of CCDs has changed the need for optics as fast as f/1.5 but deep sky and comet imagers still crave short f-ratio instruments. With a fast astrograph (astronomical camera) CCD exposure times can be shortened to 60 s or less such that the periodic error of amateur drives becomes almost irrelevant. The best modern cooled CCDs are more than three astronomical magnitudes more sensitive than the best astrophotographic emulsions of the 1990s, which is equivalent to an f-ratio more than four times faster. In other words a high quality f/6 instrument with a cooled CCD can match an f/1.5 Schmidt camera of the same focal length loaded with film. However, a fast f-ratio is still desirable to reduce exposure times when using amateur equatorial mountings and brings with it a wide field of view, as well as a deep image.

If you have a very quantum efficient camera and an average telescope drive and are prepared to stack dozens (or hundreds) of images together then a slow f-ratio system can always be used for imaging. However, in general, fast wide-field systems are a dream to use by comparison, provided they do not introduce too many optical aberrations, which is their Achilles heel. Most "exotic glass" doublet lens refractors of f/6 or so, labeled as semi-apochromats, give pin point stars in the field center, but these start becoming distinct stellar tadpoles outside of a 15 mm diameter field. With a triplet apo the stars are normally sharp out to a 20 or 25 mm diameter film. A quadruplet, or a triplet with a field corrector, can give a pin-sharp star field covering a 36 × 24 mm "full frame" DSLR sensor. With field flattening and compressing lenses some apochromat level refractors can work as fast as f/4.5 or so.

Commercial Fast F-Ratio Solutions

Perhaps the ultimate example of a fast apo is Takahashi's FSQ 106 f/5 refractor; a much prized ($5K) instrument in deep sky and comet circles. Another solution, for those brave enough to gingerly remove their Schmidt-Cassegrain's secondary mirror (Gulp!), is Starizona's Hyperstar. This enables the f/2 focus of an SCT to be used for DSLR imaging and is very powerful. However, you can't use the telescope visually at the same time and it is definitely not a solution for the clumsy "butterfingers" amateur. Of course, a Schmidt-Cassegrain is, as the name suggests, a modified Schmidt camera. It is modified by using a 5× secondary mirror to boost that fast f/2 light cone to f/10. But unscrew the secondary mirror and put a correcting lens and CCD at the f/2 focus and you have a potent fast imaging camera (see Fig. 12.1).

If you like those huge trademark white or black telephoto lenses all the sports photographers use, and have thousands of dollars burning a hole in your pocket there are some very nice 300–500-mm f/2.8–f/4 lenses

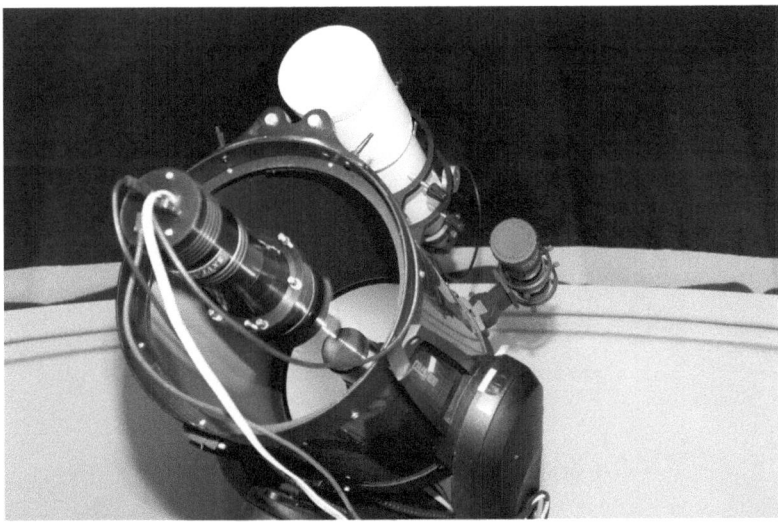

Fig. 12.1. The 11-in. (280-mm) Celestron of Prof. Greg Parker, fitted with a Hyperstar III to give an f/2 system at his New Forest Observatory. A Starlight Xpress SXVF-M25C is attached to the new f/2 focus. This system was used to take the Comet Lulin picture processed by Noel Carboni and shown in Fig. 11.16. Image: Prof. Greg Parker.

out there from Canon and Nikon which work well with bright comets. However, they do contain many lens elements to cope with the f-ratio and also with focusing at near and far objects, hence their cost. Canon has recently introduced an ultra-fast 200-mm f/2.0 lens, the EF200 mm f/2L IS USM which has great potential for astrophotography but is frighteningly expensive. As I mentioned earlier in the book the best value lenses of this type are invariably 200-mm f/2.8 systems.

Perhaps the best system for ultra-fast DSLR astronomy work again comes from the Japanese company Takahashi, in the form of their Epsilon series (see Fig. 12.2). These use a hyperbolic primary mirror, and a four element lens field corrector to give 10 μm sized star images across a DSLR field. Their 180 mm f/2.8 model is very tempting, but very pricey too, at more than $5,000. A budget solution would be to seek out a Schmidt–Newtonian optical tube. Celestron made a 5.5-in. (14-cm) aperture f/3.6 "Comet Catcher" in the 1980s and Meade have produced f/4 Schmidt–Newtonian tubes more recently. Failing that, standard f/4.5 Newtonians fitted with coma correctors are cheap and fast instruments which work well with comets.

Fig. 12.2. A fast astrograph like Takahashi's 160-mm aperture f/3.3 Epsilon E-160 is ideal for comets, covering a degree or more of the sky. Image: Martin Mobberley.

Fig. 12.3. Orion Optics' AG series are 20–40 cm aperture systems and all work at a fast f/3.8 ratio. Image: Orion Optics.

The telescope company Orion Optics in the UK has recently introduced a splendid range of fast astrographs in the form of their AG range which have apertures from 20 to 40 cm and all work at a fast f/3.8 ratio (see Fig. 12.3). The AG telescope tubes are constructed from a carbon fiber sandwich design which means they are very strong, to avoid flexing. They also have zero expansion to ensure the focal plane does not change over long photographic exposures. To assist cool down of the telescope, each model is fitted with three fans, all positioned behind the primary mirror, and precise focusing is achieved by the application of a 76-mm Crayford focuser on all models. Large bearings are placed in optimum positions to spread the load of the drawtube, corrector and any camera fitted to the telescope. A 10:1 focuser drive is included as standard and electronic focusers are available too.

To produce the remarkably flat photographic field, Orion Optics have designed and built a Field Flattener/Corrector based on the Wynne Corrector design, often used in similar telescopes. Orion have taken this design a stage forward and increased the performance over a wider field. Spot sizes across the field are therefore incredibly small and regular across a very wide field. The use of specialized ED glass types ensure color correction, coma and spherical aberration are optimized to achieve an almost perfect, and very wide, photographic field. In addition, 79 mm of back focus allows virtually all modern CCD cameras to be used with the system along with their filter wheels. The corrected field flattener is coated with low loss, multi layer, anti-reflection coatings to maintain bright images. A CNC machined mono construction secondary spider and trim ensures stability in collimation and the main mirror is held in a computer optimized, nine point suspension cell which balances the mass of the mirror over a large area and provides sharper, better defined images. Not surprisingly these AG series astrographs do not come cheap and range from £4,000 to £10,000 for just an optical tube assembly. Nevertheless they are an excellent option for the comet and deep sky imager with cash to spare! The Belgian comet enthusiast Erik Bryssink, mentioned in the final chapter, uses a 40-cm f/3.8 AG astrograph for his imaging work.

Some very similar fast reflectors to those AG systems made by Orion Optics are made by ASA (Astrosysteme Austria). Their N-Series Astrographs can be used for three different focal lengths, by using a quality range of tele-compressor and tele-extender lenses. The basic system focal ratio is f/3.8 (as with the final Orion Optics AG series f-ratio), but it becomes even faster, at f/3.6, using the 76-mm diameter Wynne corrector lens. An alternate corrector can make the system even faster, at f/2.7 and, if a longer f-ratio is required a 1.8× tele-extender lens can take the system to f/6.8. The ASA N series range is available in apertures of 200, 250, 300 and 400 mm (see Fig. 12.4). The new ASA H-series, introduced in 2010 uses hyperbolic primary mirrors to give a fast f/3.0 ratio. An initial 200-mm aperture f/3.0 model is available with 250 and 300 mm apertures available on request. Once again, you need deep pockets for these instruments as they range from 4,500 to 13,500 Euros.

The Astro Optik company (www.astrooptik.com/) whose telescopes are designed by the renowned optical expert Philipp Keller have gained a reputation for supplying observatory class equipment to advanced amateur astronomers. Their main products are large Cassegrains or "Hypergraphs" with primary mirrors of f/3 or so and a final f-ratio of f/8 or f/9. However they also supply systems where the f/3 prime focus can be used for very fast wide field imaging.

All of the systems I have described above are designed for wide fields and fast f-ratios, but many amateurs will want rather longer f-ratio

Fig. 12.4. Astro Systeme Austria make a range of superb astrographs suitable for wide field deep sky or comet imaging. They are designed by Philipp Keller of Germany and work at f/3.6 to f/2.75. Image: Astro Systeme Austria.

systems because big comets with tails a degree or two in length are relatively rare. You may well want a telescope that can be used for longer focal length work so that you can image faint comets, small galaxies and planetary nebulae. It might be thought that moving away from ultra-fast systems would mean that all those optical aberrations that cause fast astrographs to be so expensive would vanish. Sadly, while they are less of a problem they do not vanish completely and for fields of view greater than a quarter of a degree wide we still encounter the basic problem of coping with coma. When each pixel is covering an arc-second or two at long focal lengths coma will still be obvious in an uncorrected system.

Coping with Coma

In a perfect deep sky imaging system all star images would focus to a diffraction limited point, even at the edge of the field. In practice, with all normal optical systems, as you travel away from the centre of the optical field, stars rapidly resemble tadpoles and the further you move from the field center, the worse things get. With Newtonian reflectors the off-axis coma is the main cause of aberrations and becomes exponentially better as the focal ratio becomes longer. Move from an f/5 to an f/10 Newtonian and the diffraction limited "sweet spot" doesn't just double in size, it improves by a factor of eight, from about 3–24 mm in diameter! In angular terms a fourfold improvement is seen. However, in practice, for long exposure imaging you do not need diffraction limited performance. In an f/5 Newtonian stars will still look fairly sharp over a 10 mm diameter circle at the focus. But for comet work you often want a faster f-ratio and so then the coma problem becomes serious.

Spot Diagrams

Manufacturers often demonstrate wide field optical performance using a spot diagram which shows the physical size of the tadpole-like stars at the edge of the field of view, in microns (thousandths of a millimeter). Most full frame CCD detectors have pixel sizes in the seven to nine micron range and so keeping those tadpoles below about 20 µm at the CCD edge has become the astrographic instrument designer's goal. In practice this corresponds to <2 arc-sec on a 2 m focal length system. Each spot in a spot diagram corresponds to the light from a star hitting a specific part of the primary mirror. The entire spot diagram shows where all the light from a single star, hitting every part of the mirror, ends up at the focal plane.

In recent years CCD detectors have been getting bigger and broadband internet access to deep sky websites has become widespread. These advances mean that amateur imagers are now looking more towards optical systems which can deliver pinpoint star images across a 24 × 36-mm mega-pixel CCD diagonal (43 mm). One solution to these wider fields is to use a coma corrector like TeleVue's Parracorr (see Fig. 12.5). This substantially reduces coma on Newtonian systems between f/3.5 and f/8 and also increases the focal length by about 15%. Many modern imagers use Schmidt-Cassegrain Telescopes (SCTs) because of their compactness and portability. A standard f/10 SCT of 25 cm aperture will only deliver acceptable star images across a quarter degree wide CCD field, similar to an f/6 Newtonian of that aperture. However, the popular 0.63× tele-compressor/field flatteners available for f/10 SCTs do tend to sharpen the stars noticeably at the field edge. The 0.33× tele-compressors work OK, but at that very fast f-ratio it is quite common to get serious aberrations developing on any CCD detector outside a 10-mm diameter circle.

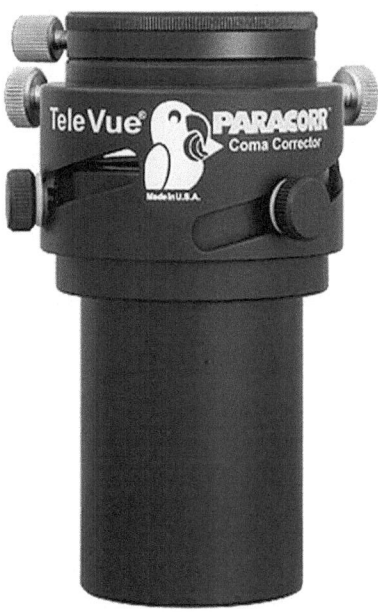

Fig. 12.5. TeleVue's Paracorr is designed for optimum coma reduction on f/4.5 Newtonians but reduces coma on reflectors as fast as f/3.5. Image: TeleVue.

Coma-Free Designs

Over the last decade more and more of the world's keenest (and wealthiest!) amateurs have been turning to the Ritchey Chrétien telescope design to fully exploit the larger CCD detectors available, while still using compact Cassegrain style tubes and apertures of 25-cm and above. Typically these telescopes work at about f/8 or f/9 with an f/3 hyperbolic primary mirror and a hyperbolic secondary providing three times amplification. Optically, an f/10 Newtonian would have a similar performance, but with an extremely unwieldy tube length. The Arizona company RCOS is arguably the best known supplier of quality Ritchey Chrétiens, but with prices ranging from $9,400 for a 25 cm astrograph, to $54,000 (!!) for a 50 cm model these are not instruments for the beginner (see Fig. 12.6). Genuine Ritchey Chrétien's typically feature a very large secondary mirror, typically 40% of the diameter of the primary mirror,

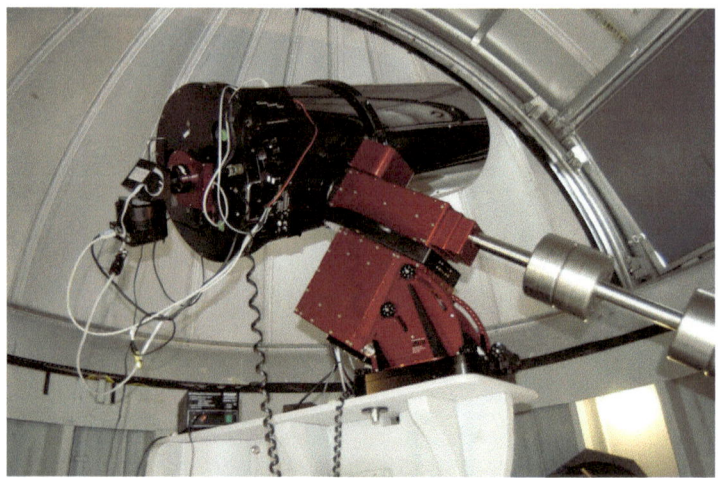

Fig. 12.6. The 0.4-m RCOS Ritchey-Chrétien of Gordon Rogers, a well known UK deep sky imager. Image: Gordon Rogers.

which means the contrast is reduced if they are used for planetary work. Optically the Ritchey-Chrétien design is a beautiful solution as only two reflecting surfaces are involved and therefore there is less chance of any ghost images as can occur in some fast systems using multi-lens correctors. The RCOS website is at www.rcopticalsystems.com/

In 2002 Meade brought out its RCX400 line of telescopes, followed by the LX200R version of its LX200 SCT. Both were advertised as employing Ritchey-Chrétien optics, even though they both had Schmidt style corrector plates like an SCT. While the usable fields of these instruments were sharper and wider than a conventional SCT they were not, technically, Ritchey-Chrétiens and after a legal ruling against Meade on January 17, 2008 (in favor of Star Instruments and RC Optical systems) Meade were forced to rename their RCX400 and LX200R ranges as LX400ACF and LX200ACF models respectively. The ACF stands for "Advanced Coma Free" because the Meade design is basically a coma free Schmidt-Cassegrain, first designed by Ronald Willey in the 1960s and described shortly afterwards in Sky & Telescope. The Meade models are cheaper than genuine Ritchey–Chrétiens and have tempted many astro-imagers who want better wide field performance than offered by a standard SCT, without having to re-mortgage their homes!

There has been an additional development in the quest for aberration-free optics in recent years too. Two former engineers from the Celestron Company, Richard Hendrick and Joseph Haberman set up a new company called PlaneWave Instruments. They currently offer three apertures of their CDK (Corrected Dall–Kirkham) telescope, a 12.5-in. (318-mm) f/8 instrument, a 17-in. (432-mm) and a 20-in. (508-mm) f/6.8 instrument. At $9,900, $22,000 and $32,500 for the optical tubes these are serious pieces of kit too. If you have just won the lottery they now make a complete alt-azimuth mounted 27.6-in. (0.7-m) CDK system for $185,000! PlaneWave claim their instruments outperform Ritchey–Chrétiens by minimising off-axis astigmatism and field curvature as well as eliminating coma. Their website can be found at: www.planewaveinstruments.com/index.php

Celestron themselves have not been idle and in 2009 started marketing their Edge HD range which boast a wider, flatter, field than their traditional Schmidt-Cassegrains and at considerably less cost than the CDK or Ritchey Chrétien ranges! There are many other Ritchey–Chrétien manufacturers I have not mentioned, a high percentage of which are supplied with optics by Eastern European companies; but, do you actually need to part with enormous amounts of money to take great images? Frankly, NO! In practice Ritchey–Chrétien's are only for the dedicated astro-imager with deep pockets. For fields of view less than a quarter of a degree the standard SCT with a telecompressor/field flattener will do a great job. Apart from bright comets, the only Messier galaxies that cannot be shoehorned into a $15 \times 15'$ field are M31, M33 and M101. An f/4.5 Newtonian equipped with a coma corrector will also deliver fine images.

There is another issue here too, which is very important to amateur astronomers. Most amateurs do not own hermetically sealable, temperature controlled observatories like the pro's and so their telescopes can get very damp inside sheds and under tarpaulins. They will, therefore, always tend to go for a sealed tube system, like a Schmidt-Cassegrain, because they really do not like to see their aluminum mirror coatings corroding after 6 months! Ritchey Chrétien's and other open tube designs are best suited to solid observatory buildings where the optics will be better protected against dew and dust.

Nevertheless, if you are rich and you have already perfected autoguiding and image processing and plan on using full format CCDs to image half degree and wider fields, with decent apertures, a Ritchey–Chrétien, or a similar coma-free system, may be the next step. You may need your bank manager's approval though! However, most comet imagers work on a tight budget and may well prefer Celestron's Edge HD telescopes, a Meade ACF tube, or a modest Newtonian with a coma corrector to a more exotic ultra-wide, ultra-sharp, or ultra-fast system.

John Broughton and His 0.51-m Newtonian

A 0.51-m aperture Newtonian is large by any amateur observatory standard. But when the telescope is home-made, works at f/2.7, and has been used to discover hundreds of asteroids, a few NEOs and two comets it is worth describing in more detail. I am very grateful to John Broughton for sending me lots of information about himself and his telescope for this section (see Fig. 12.7). John's observatory is located at Reedy Creek on the Gold Coast in Queensland, Australia. Remarkably he made the 0.51-m mirror for his Newtonian in just 5 weeks, at the end of 2002. The mirror was made to an ultra-fast f/3.7 but John admits that he simply did not know that it was possible to get it much faster when he made the mirror. John made a custom focal reducer to correct coma and reduce the focal ratio from 3.7 to 2.7, for imaging a full square degree per frame with his Apogee AP6 camera. He designed the focal reducer to incorporate two stock 75-mm plano-convex lenses, so that he only had to make a single plano-concave lens from a specific glass by grinding against an 80-mm steel ball.

After the mirror was completed John made the "English Yoke" style telescope mount and a semi-Serrurier truss style tube framework in 2003. The yoke is attached at both ends to the concrete observatory building with minimal support structure being necessary. Inexpensive 1 in. (25.4-mm) bearings on each axis are sufficient to carry the weight of the yoke, frame and mirror. When it comes to the telescope drive system, stepper motors drive threaded rods meshed to large-radius worm wheels; these are comprised of pairs of curved threaded rods. According to John the telescope acts as an LX200 clone by way of Mel Bartels stepper drive system, which is further controlled by John's own software to automate the pointing and imaging sequences during his sky patrols. Accurate RA tracking for imaging is one absolutely crucial requirement and with six variables in the adjustment of John's unique worm and worm wheel system, it wasn't until the second year of operation (2005) that he had reduced the periodic error to almost zero.

The Planetary Society provided a grant for John's Apogee AP6 camera which sits at the prime-focus, so there is no secondary mirror. He estimates that the whole telescope project cost no more than $2,500, although hundreds of hours of construction work were involved with no guarantee of success. John's 2.3 m diameter observatory was not originally intended to house such a large telescope, so to drastically reduce telescope bulk

Fig. 12.7. One of the world's leading amateur astronomers, John Broughton, inside his Reedy Creek Observatory in Queensland, Australia. John's home-made 0.51-m Newtonian uses his own design of tele-compressor to bring the f-ratio down from f/3.7 to f/2.7 and the telescope sits in an English Yoke design of equatorial mounting. Image: John Broughton.

above the declination axis, he designed a radical tube comprising a triangular spar on one side of the mirror and six struts. The 24-kg of steel at the top of the spar acts as a counterweight to the primary mirror. The whole spider assembly moves up and down the spar for focusing and incorporates a PVC pipe whose thermal expansion negates that of the spar by pulling the assembly by an equal amount but in the opposite direction. Ingenious!

John's incredible success with his home-made system goes to show that while spending tens of thousands of dollars on a commercial system may get you an impressive looking telescope, it is sheer energy and determination that will bring discoveries.

Although John has now discovered more than 800 asteroids, far exceeding the combined output of all other amateur sites in the southern hemisphere, his comet discoveries actually came out of an 8,000 square-degree survey for a trans-Neptunian planet south of −20° declination. As mentioned earlier in the book the comets John discovered were P/2005 T5 (Broughton) and C/2006 OF2 (Broughton). By the start of 2010 several NEOs had also been discovered with the 0.51-m f/2.7 instrument. Two of them are potentially hazardous and one has been numbered after John recovered it in 2008. Numbered PHAs are pretty rare. When John last looked in early 2010, even the Siding Spring Survey, with all of their discoveries, did not have one!

John's main competitor, from his southern hemisphere site, is the Siding Spring arm of the Catalina Sky Survey, manned by Rob McNaught and Gordon Garradd. John has provided some data (see Table 12.1) to allow his telescope and the Siding Spring Uppsala Southern Schmidt system to be compared. While John's aperture and f/ratio are very similar to that of the Siding Spring Survey equipment his CCD and sky quality means he does not get as deep or have as wide a field in each frame as the professional survey. Specifically, for the CCD the Siding Spring detector is a 4,096×4,096 pixel device cooled to −90°C whereas John's 1,024×1,024 Apogee AP6 CCD is only cooled to −20°C and, despite bigger pixels (24 μm compared to 15 μm) still only covers a field half the width of the Uppsala Southern Schmidt.

Table 12.1. Siding Spring and John Broughton's Reedy Creek facilities compared.

Parameter	Siding Spring	Reedy Creek
Aperture/speed	0.50-m f/3.4	0.51-m f/2.7
Sky	Very dark	Mod. light pollution
CCD QE	90%	68%
CCD cooling	Liquid N. −90°C	TEC −20°C
Pixel scale	1.8"	3.6"
Typical exposure	30 s	120 s
Limiting mag	20.5	19.2
Field of view	4.20 square degree	1.05 square degree

John has supplied the following fascinating information on his patrols. In order for him to compete at all he has to follow what he calls "certain strategies and loopholes."

First, the week between last quarter and new moon is a valuable opportunity for John to find new objects in areas not scanned for at least 10 days and before the surveys have had time to once again blanket the sky. John's first PHA and second comet were found during such periods.

Second, John points out that the current surveys avoid following up on objects that do not have unusual motion or do not have a non-stellar appearance. However, he observes for at least two nights on anything new, to get a provisional designation, and there's always a small possibility that the object is either a NEO or a comet (with a very small coma) and its instantaneous motion happens simply to be indistinguishable from a main belt asteroid at that time. John's first comet, his first NEO, and a few high eccentricity asteroids he has been credited with all had such characteristics.

Finally, John comments that the Siding Spring Survey's object detection software cannot handle crowded fields, so that is where he is often "scavenging" for anything on the move.

John's second PHA was spotted close to the galactic centre, crossing a dark nebula.

As all the subsequent observations over the 12 day arc were targeted ones this was one important object that otherwise would have escaped detection. John Broughton is living proof that with a large aperture system, covering only a square degree field, a dedicated amateur can still discover comets, as long as he has a complete knowledge of his professional rivals' weaknesses.

Build Your Own Schmidt Camera

Many amateur astronomers used to build their own telescopes, although it seems to be much rarer in the credit card (and credit crunch) driven twenty-first century when no one has the time to actually build equipment. However, as we have just seen, some amateurs still do build telescopes and some even build Schmidt cameras for comet imaging, although they are mainly amateurs who have considerable experience at mirror making anyway. It might be of interest to the reader if I describe a system made by a friend of mine which he uses to image comets on a low budget.

Mike Harlow is the chap in question and he is a member of the Orwell Astronomical Society of Ipswich in Suffolk, England. He recently built a Schmidt camera which has been extremely successful when imaging comets. In the photographic era Mike ground and polished an ultra-fast 220-mm aperture primary mirror until it had a focal length of just 400-mm, and then started on a 160-mm corrector plate which defined the aperture and f-ratio, i.e. 160-mm f/2.5. The oversized primary mirror is normal in Schmidt cameras and guarantees an even illumination over a wide field. Mike then ground another glass disc to act as a mold for a plastic resin film holder with the same curve as the focal plane, in other words a curved surface to clip the 35 mm film too; just like the one in Celestron's 1970s and 1980s Schmidt cameras. Despite the

Fig. 12.8. The home made 160 mm f/2.5 Schmidt camera of Mike Harlow, as described in the text. Image: Martin Mobberley.

camera's photographic success, fiddling with small bits of film was still a big hassle. However, in recent years Mike has revived his Schmidt camera, placing an early Starlight Xpress MX916 CCD at the focal plane instead of the curved film. Because the CCD chip is only 8.7 × 6.5-mm across, compared to 36 × 24-mm for the curved film, the field curvature can be totally ignored: it is just not a problem, even in the corner of the image. Of course, this reduces the field of view by a factor of 4, to about 1°, but it is still a very fast f/2.5 system and wide enough for most comets. Figure 12.8 shows Mike's home-made Schmidt camera on its home-made friction drive fork mount.

Telescope Mountings

I have already covered the subject of telescope drive periodic error in Chap. 11. The bottom line is that, in general, you get what you pay for with telescope mountings and the larger the worm-wheel the better the telescope will track. At the very top of the quality league table for telescope mountings is Software Bisque's Paramount ME (www.bisque.com/sc/pages/Paramount-ME.aspx).

At $14,500 this is not an inexpensive mounting and is totally outside most amateur astronomers' price range. I purchased a Paramount ME in 2003 when they were priced at $8,500. I have used it on about one hundred nights per year over a 7 year period and have taken thousands of images with it. On every night of use it has tracked perfectly for 2 or 3 min at the celestial equator and far longer at higher declinations, at an image scale of 1.5 arc-sec per pixel. It always puts the object I slewed to onto the 13 arc-min wide CCD chip too: mostly near the middle. So, it was money well spent. I have a Celestron 14 mounted on top of the mount. In contrast the 30-cm fork mounted Schmidt-Cassegrain I purchased in 1997 has been nothing but trouble and the plastic toy car parts inside it failed totally at least once per year. If you buy something of high quality it is generally worth the extra cash and the Paramount ME is definitely the best astronomical investment I have ever made. Nevertheless there are other good mounts out there too. Astrophysics (www.astro-physics.com/) have a fine reputation for making the highest quality telescopes and telescope mountings, even if there is usually quite a waiting list to purchase one. Their Mach1GTO, 900GTO, 1200GTO, 3600GTO and 3600GTOPE mountings cost between $6,000 and $27,000 but every model is made to the highest standard. In addition to the Software Bisque Paramount ME and Astrophysics mounts the mountings made by Losmandy, the Gemini G42 mount, the mounts made by the Italian company 10 μm and Celestron's CGE Pro are all heavy duty high quality systems. The Japanese manufacturers Takahashi and Vixen also have a reputation for manufacturing reliable mountings with a low periodic error. However, in recent years the best selling mounts have been those made by SkyWatcher, specifically the EQ6 Pro (as it is marketed in Europe). While the SkyWatcher mountings cannot compete with the tracking accuracy of the most expensive mountings their ability to carry heavy payloads (up to 20 kg for the EQ6 pro) for a price tag of $1,000 and below has proved very attractive.

There is also a completely new form of drive on the market too: the direct drive DDM 85. In a direct drive system there are no gears at all, just one massive motor with an impressive amount of torque, capable of tracking at low speeds which are orders of magnitude below the stalling speed of a normal motor. Of course, without a worm or high gearing to hold an imperfectly balanced telescope in position, what happens in a power cut? Well, according to Astrosysteme Austria (ASA) who manufactures the DDM 85 direct drive telescope mounts, mechanical brakes avoid uncontrolled movements of the mount in such a scary situation. With such huge amounts of torque available the DDM 85 can easily slew at up to 10° (or optionally 20°) per second even if it does draw 20 A when doing so. Also, with no worm and wheel, there is no backlash either and the claimed tracking accuracy of <0.25 arc-sec over 5 min is staggering. However, before you decide that a very pricey direct drive telescope is the way to go, some caution is needed. Firstly, however accurately a mount tracks, unless account is made of polar misalignment and telescope flexure, errors of arc-second level and greater will still occur. Secondly, and this is far more important, telescope performance and reliability over months and years and in a range of temperature and humidity conditions is the acid test of a good mount, something where the Software Bisque Paramount and Astrophysics mounts have a proven track record. It will be interesting to see, in the coming years, whether more telescope manufacturers switch from traditional worm and wheel designs to direct drives. Customer satisfaction, as always, will be the decider.

I have mentioned a lot of pricey tracking gear in the last page or two but I certainly would not want to give the impression that an outlay of $10,000 was necessary for comet imaging. The best comets can be captured with the simplest equipment. If you have a budget of, say $1,000, you could look no further than the AstroTrac TT320X. This is a portable (450×75×40-mm) system weighing just 1.5 kg which has a micro-controller regulated tangent arm drive. The system attaches between a tripod and a DSLR system (or a small telescope). It will track for 20 min and, depending how well it is polar-aligned, can track perfectly with 200 or 300 mm focal length lenses. It is ideal as a portable system for imaging those bright naked eye comets that come along every few years. See www.astrotrac.com/ for more information.

CHAPTER THIRTEEN

Imaging Comets Remotely and via the Internet

One question this author is regularly asked by potential deep sky and comet imagers is: "What, precisely, must I do to set up an observatory where I sit comfortably indoors in the warm, while my telescope slews around flawlessly in the cold?" Freezing weather can sap even the keenest amateur astronomer's enthusiasm and so making a telescope and CCD camera controllable from indoors can be extremely productive.

There are five essential requirements, namely:

- A data link from the PC indoors to the "GO TO" telescope
- A fast data link from the CCD camera to the PC's USB port
- A reliable link to a motorized focuser
- A good dew heater for the corrector plate of any Schmidt–Cassegrain
- A field of view much larger than your telescope's slewing accuracy

The first three criteria are achieved by the system in Fig. 13.1. Of course, you might not use a Schmidt–Cassegrain, but even if you do not a dew heater may be needed at some point, such as behind the secondary mirror on a Newtonian, a well known attracting point for dew. To control your telescope remotely you only need an old fashioned RS-232 serial connection. Essentially RS-232 is a trustworthy but slow (typically 9,600 bits/s) communications system using a minimum of three wires (Transmit/ Receive/Ground) which sends the information from your computer's

M. Mobberley, *Hunting and Imaging Comets*, Patrick Moore's Practical Astronomy Series, DOI 10.1007/978-1-4419-6905-7_13, © Springer Science+Business Media, LLC 2011

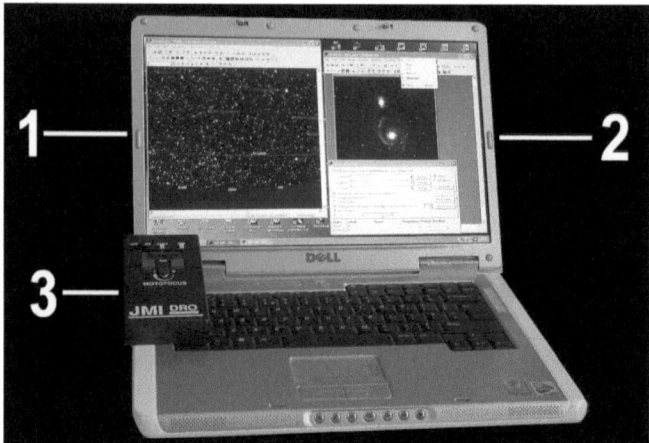

Fig. 13.1. A basic system for controlling a telescope remotely should enable control of the telescope (*1*), control of the CCD camera (*2*) and control over the focusing (*3*). This is the author's own system and consists of Software Bisque's The Sky (*1*) and CCD Soft (*2*) as well as a JMI motorized focuser. The telescope is controlled from indoors via 35 m of high quality cable.

planetarium package (e.g., The Sky X or Guide 8.0) to your telescope's "Go-To" serial port. Using quality "Category 5" cables, the signals can reliably travel over 50 m or more. If the modern PC you own has no serial ports, but your old mount has an old RS-232 port, do not despair: USB-Serial port replicators can be purchased and many telescopes can now accept USB control signals too. USB is fast becoming the telescope control system interface as well as being the system via which images are transmitted, despite the fact that only a small number of bytes are needed to control a telescope.

The data link to the CCD camera itself must be fast, namely a high speed USB 2.0 connection, because there is a huge amount of information to collect in a megapixel image. This might create a slight problem as your telescope is probably more than five meters away. USB is not guaranteed over such distances, so for communication lengths of more than five or ten meters you will need a quality USB extender system. The Icron range of Extreme USB units are used by many US amateurs. Alternatively, a cheaper solution is a series of USB repeater cables. Yet another alternative is to set up a local area network with Microsoft Windows or, for example, Symantec's PCanywhere with one PC close to the telescope controlling slewing and imaging, while relaying its status to a remote PC indoors. A reliable motorized focuser (such as a JMI unit) is absolutely essential. Some are available with RS-232 control but most will work fine if you just extend the handset leads by tens of meters!

There is little point in having a remote observing system if you have to keep bounding down the garden with a dew clearing hairdryer, every 5 min. Without a decent length dew cap and a powerful dew heater system an SCT's corrector plate will dew over in minutes on a typical British night. So buy them! Finally, you need to ask yourself how accurately your own, "Go To" telescope really does slew, in practice, ignoring the manufacturer's glossy hype! For typical 20–30-cm f/10 SCTs you would be well advised to have a system with a comfortable field of view at least 15–20 arc-min wide. With smaller CCD chips this can mean purchasing a 0.63× or 0.33× telecompressor. I would also strongly recommend reducing the slew rate on any low priced equatorial mounting carrying heavy equipment to about 1 or 2°/s. This will greatly extend the life of your mount or SCT fork's tiny plastic gearbox!

Armchair Observers

The term "armchair observer" used to mean someone who may have never used a telescope but was simply a student of astronomy, or of its history. However, in the twenty-first century some of the world's keenest amateurs observe, like professionals, from indoors and many sit in an armchair too! Moreover, some observers are not even using their own telescopes as, via the Internet, instruments at far-flung locations can be employed. This is a big advantage from a country like the UK where most nights are cloudy, some whole months are cloudy, and most of the southern hemisphere is totally inaccessible. Another potential advantage of using a distant telescope via the Internet is that from, say, a UK apartment building in a major city, light pollution and the lack of anywhere to put a telescope (except, perhaps a balcony) prevents conventional observing anyway. These days, with a few sweeps of a mouse you can be controlling a telescope at a jet black sky location, and imaging a faint comet with equipment way beyond your financial reach. In addition, if you are controlling a telescope in, say, Australia, you can do all your observing at a civilized time; it might be the dead of night at the telescope's location but it can be your lunch break in the UK. Of course there are big disadvantages too. Remote observing means you can never look through an eyepiece and actually see faint objects with your own eyes, so you can never feel the thrill of actually gazing at an object first hand. In addition, remote observing is of little use for the planetary imaging specialist who will want to continuously download gigabytes of video data covering tiny fields of view at huge focal lengths. Some amateurs are now willing to pay serious money for regular telescope time, either by taking a holiday at a very dark sky site, with loads of clear skies, or via the Internet.

Global Rent-a-Scope

Undoubtedly the best facility for the advanced amateur is operated by GRAS (Global Rent-a-Scope) at www.global-rent-a-scope.com/who currently operate a network of thirteen telescopes based in New Mexico (six telescopes) and Australia (seven telescopes). The instruments range in size from 106-mm aperture Takahashi FSQ 106 apochromats to a 400-mm RCOS Ritchey-Chrétien and a 400-mm Astrosysteme Austria corrected field Newtonian. Many of the telescopes are Takahashi or RCOS instruments, in other words they are top quality pieces of equipment (see Fig. 13.2). All are mounted on Paramount equatorial mountings: the only ones reliable enough to guarantee such a service, night after night. The CCD detectors are mainly high quality units from SBIG. Of course, with many of the telescopes based in Australia this offers the far northern hemisphere comet enthusiast the chance to image objects that are totally unobservable.

It might be thought that remotely operating a telescope over the Internet would be fraught with hassles and highly expensive. For example, how do you focus a camera more than ten thousand kilometers away and will the telescope always hit the target. How do you know if the sky at the site is cloudy and even if it is nighttime there? In addition, how do you know which telescopes in the network are available for hire and are dark frames and flat fields available for each instrument? Also, surely, someone else might be using the telescope you want? All these are valid points, but after having used the GRAS system on many occasions I can confirm that the GRAS website control environment makes things pretty clear. Yes, there is a learning curve, but some useful video tutorials are available which can put you at ease before you log on and start using a specific telescope. In addition the GRAS founder & owner (Arnie Rosner) seemed to be permanently contactable while online with almost instantaneous feedback, which a beginner will find very useful.

Fig. 13.2. One of Global Rent-a-Scope's telescopes based in New Mexico and controlled via the Internet. Image: Arnie Rosner.

Rental Costs

So how much does all this cost? Well, there is a complex pricing structure which takes quite a while to work out and it depends exactly what facilities you want. But, the cheapest option costs $39.95 per month. This buys you a number of telescope user points but, in essence, it will get you about 2 h of telescope time on the cheapest instrument. The more expensive pricing plans buy you a lot more time and at a better time per dollar rate too. So, one could envisage an occasional UK user choosing to spend the minimum $39.95 (roughly £25) every month (the finances are handled via PayPal) for an annual cost of £300 or so. Once you have taken your images the full resolution files are placed on the Internet for collection (via ftp) within 2 h. Obviously you need a reliable Broadband connection for it all to work. A one off trial membership is available for $19.95 and buys you 1 h of telescope time, at least on the cheapest instrument. When you consider each imaging system represents a $20,000–$30,000 investment, and is at a really dark site, with mainly clear skies, it makes a lot of financial sense, apart from the fact you can never look through it visually or impress the neighbors with it!

GRAS is not the only online observing system. The SLOOH (a corruption of the words Slew and Ooh!) system is aimed more at beginners and for educational purposes. Membership of SLOOH allows you to watch a space show where the telescope chosen goes to various popular objects (chosen by SLOOH members) each night in a sort of guided tour "live show." The Moon and planets can be targeted by SLOOH, but not at the sort of resolutions a planetary expert like Damian Peach would want! No doubt, in this broadband era, many more online facilities will spring up in the coming years.

Educational Robotic Telescopes

In recent years there has been a surge in the use of robotic telescopes as educational tools for schools and colleges, or simply for use by University Departments wishing to access darker and clearer skies at more favorable latitudes. In the UK there are three major Robotic Telescope Projects and these are the Bradford Robotic telescope (BRT), run by Bradford University, the Faulkes Telescopes Project and the Liverpool John Moore's University National Schools Observatory project. There are also similar educational projects in the USA. The BRT consists of that proven optical package, a Celestron C14 (355 mm f/11 Schmidt–Cassegrain), mounted on that proven robotic mount, the Paramount ME. The telescope (www.telescope.org) is sited on the Observatorio del Teide site of the Instituto De Astrofisica De Canarias, in Tenerife, Canary Islands, Spain. The Teide Observatory is the best in Europe and is situated at an altitude of 2,400 m (7,874 ft.) on the northern part of the Teide volcanic caldera. At the time of writing the telescope had completed more than 15,000 image requests and in excess of 20,000 users were registered with the telescope site. The two Faulkes telescopes (http://faulkes-telescope.com/) are a considerably bigger technological investment as they are substantial 2 m aperture instruments based in Hawaii and Siding Spring Observatory, Australia. The Faulkes Telescope Project was launched in March 2004, with 10 million UK pounds of funding from the Dill Faulkes Educational Trust, established by business entrepreneur Dr. Martin "Dill" Faulkes. The Faulkes telescopes were designed and built by Telescope Technologies Limited (TTL), a spin-off company from Liverpool John Moores University in northwest England (*TTL* is now a wholly-owned subsidiary of the Las Cumbres Observatory Global Telescope Network or LCOGTN). The 2.0 m Liverpool Telescope (http://telescope.livjm.ac.uk/), based at the Observatorio del Roque de Los Muchachos on the Canary island of La Palma, Spain was TTL's first 2.0 m instrument, and became actively robotic in 2004, paving the way for TTL's Faulke's instruments, an online educational facility for schools, and a future "ROBONET" project to build up to seven 2.0 m telescopes around the world for nonprofit educational research. It goes without saying that access to observing time on 2.0 m telescopes is not that easy for amateurs to acquire, but, if a sensible proposal is put forward, anything can happen. The main advantage of all these University-based robotic telescopes is that they operate on a zero profit basis with schools and educational use a top priority.

Robotic and in the Backyard!

Robotic observing does not necessarily mean using a telescope on the other side of the world. The well known UK amateur and former BAA President Richard Miles is an expert at photometry (the precise measurement of variable star and comet magnitudes) and has a very neat housing for his robotic photometry system. A hinged, counterweighted box opens up his observatory containing a Celestron 11 and two Takahashi FS60c refractors on a Vixen Atlux mounting. Because the box lid is precisely balanced by the weights (note the cube on the lid side) it takes little effort to open the tiny observatory up. The telescope and camera can then be controlled from indoors (see Fig. 13.3a, b).

Fig. 13.3. (a) The remotely controlled observatory of the UK asteroid and comet observer Richard Miles in Dorset, UK. The instrument is a Celestron 11 mounted on a Vixen Atlux. Note the cleverly designed and counterweighted box lid arrangement which enables the observatory to be opened quickly and with little effort. Image: Richard Miles. (b) The observatory featured in (a), except with the lid fully opened, revealing the telescope. Image: Richard Miles.

CHAPTER FOURTEEN

Comet Photometry

Any amateur astronomer who takes images of comets will wonder, at some point, if they can be used for a scientific purpose. We have already seen that astrometry, the measurement of an object's precise position relative to the background stars, is a very useful contribution, especially for newly discovered comets and NEOs. However, comet photometry, the measurement of the brightness of comets, is also valuable. The problem is that while comet astrometry is a straightforward and very precise science, with little chance of major errors once you have mastered the technique, comet photometry is not. I have seen respected astronomers becoming very hot under the collar about comet photometry on a number of occasions, largely because if they are wrong in their approach to the subject years of work may be regarded as distinctly dubious. The bottom line is that comets are fuzzy objects that fade away into the light polluted background sky and whose images are peppered with background stars. Measuring their precise brightness is incredibly difficult and potentially subject to huge errors. In many ways the visual observers have a similar set of problems to face, but I think most of them would accept that their estimates are, at best, just that: estimates. Conversely, CCD photometry is supposed to be a precise science. However, at the present time it is more like the proverbial can of worms! Before we can even start to consider comet photometry we should understand the basics of stellar photometry, in other words, measuring the magnitudes of stars.

M. Mobberley, *Hunting and Imaging Comets*, Patrick Moore's Practical Astronomy Series, DOI 10.1007/978-1-4419-6905-7_14, © Springer Science+Business Media, LLC 2011

Color and Wavebands

Various factors need addressing before a CCD image can be used for an accurate photometric measurement, stellar or cometary. The first factor is one of color. A CCD detector has a far wider spectral range than the human eye. It can see deep into the near infrared and into the ultraviolet too. For astronomers this is a problem as stars can behave differently in, for example, the infrared compared to their performance in the normal human visual range. Gassy comets can be bright in various emission bands too. Professional astronomers take measurements in specific wavebands called U, B, V, R and I (Ultra-Violet, Blue, Visual, Red and Infrared) which enables them to precisely define a star's performance at each color and to calibrate their photometry accurately (see Fig. 14.1). The V band corresponds to the center of the human visual range, centered on the color green (the peak response of the eye in darkness is slightly different but we will ignore that subtlety). At first the use of filters may seem a backward step. A CCD is so much more sensitive than the eye or film, but then we have to make it less sensitive by slapping filters in front of it. However, it is all in a good cause: the quest for scientific accuracy; having

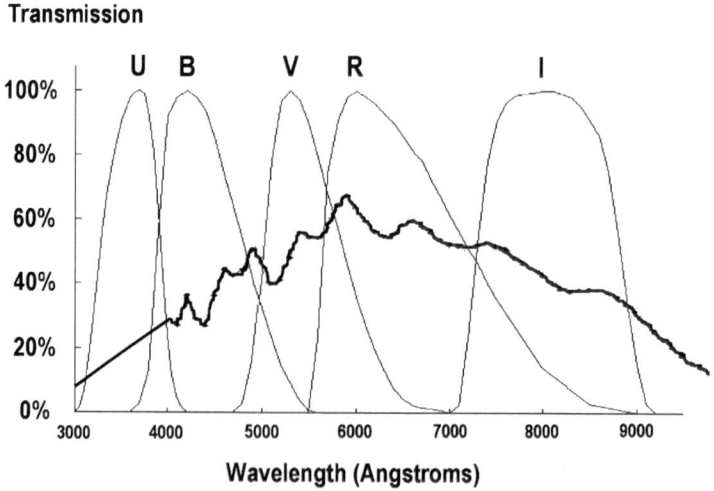

Fig. 14.1. The transmission properties of the standard U, B, V, R and I photometric wavebands used by professional astronomers and advanced amateurs. The response of the Kodak KAF 0261E chip in the author's ST9XE CCD camera is also plotted as a *line*, peaking at around 70% Quantum efficiency at 6,000 Å.

said this, where comets fainter than eleventh magnitude, or stars fainter than fifteenth magnitude are concerned filters are frequently abandoned. The simple fact is that when faint objects are being monitored for subtle fluctuations, a filter, causing as much as a two magnitude loss in sensitivity can push those fluctuations into the noise.

When reporting unfiltered CCD observations a "C" symbol is usually used after the magnitude estimate. V, B and U are used for CCD measurements with Johnson V, B or U Band filters. R and I filter measurements are usually suffixed with a small j or a small c depending whether those filters correspond to the Johnson or Kron–Cousins photometric system. The abbreviation CR is sometimes seen and refers to an unfiltered CCD measurement that has been "zero point adjusted" to conform to a Kron/Cousins R band filter measurement. The second critical factor relates to the linearity and usable range of a CCD. Much CCD photometry is based on comparison with a single star. Both the comparison star and the star being measured should be well above the noise floor of the image and well below the, potentially non-linear, saturation point. In general CCDs are very linear devices with the charge collected in each pixel being accurately related to the number of photons collected. However, in devices with anti-blooming gates (ABG) appreciable non-linearities creep in when the pixel "well" becomes more than half-full, that is more than half way to saturation and white-out. Anti-blooming gates are generally not found on purely scientific CCD detectors. Their sole role is to drain away excess charge, much as a storm drain copes with a surge of water. This draining activity prevents any likelihood of a very bright star "bleeding" down the image and so is vital for commercial applications where a pretty picture is the aim and not a scientific measurement. When an anti-blooming gate first starts to work it is siphoning vital data away from your measurement of a bright star in the field, which is NOT what we want. Rather predictably, Non-anti-blooming gate detectors are designated by the abbreviation NABG. In short, ABG CCDs should only be used for photometry when the comparison and target star (or the comet) are each filling no more than 50% of the available range of the detector. It is also imperative that no image processing routines (except basic dark frame subtraction and flat-fielding) are carried out prior to the photometric analysis.

Counting the Photons

We have just seen that color and linearity are major factors that affect the accuracy of a CCD magnitude estimate. These are not the only issues though. Measuring the brightness of a star or a cometary nucleus with a CCD full of pixels is not dissimilar to measuring the amount of rainwater in a field full of buckets. If some of the buckets (pixels) are literally overflowing (with a bright star's charge) then we have an inaccurate measurement. If there are only a few drops of rain in the bottom of the bucket the measurement will be almost impossible to make accurately (equivalent to measuring a faint star) especially if the bucket was damp (noisy) to start with. Before an image of a variable star or a cometary nucleus is exposed through the appropriate filter a few preparations need to be made. Firstly, the object you are measuring, cometary nuclear region or a star, needs to ideally fall in the mid-range of the image. For a typical 16 bit ($2^{16} = 65,536$) measurement, the maximum range will actually be 65,535 ADUs (Analogue to Digital Units), on an NABG camera. If the target object and comparison star are, say, giving readings of between 20,000 and 40,000 ADUs that is excellent: well above the noise and well below saturating the detector. In practice, things will never be quite that easy and, where a faint star or cometary nucleus is well below the faintest star on the photometric sequence, it is quite possible that the faintest comparison star may end up registering 50,000 or more ADUs, with the target down in the murk at the 5,000 ADU level, even in a deep exposure. In other words, a cometary nucleus might be 2.5 magnitudes fainter than the comparison star, so the accuracy will suffer. Some prior knowledge of the number of ADUs produced by a given exposure through a specific filter on a specific star is needed and this can only come with experience.

With my own system, a Celestron 14 at f/7.7 plus SBIG ST9XE camera (KAF-0261E detector), a 2 min unfiltered exposure is just safe for stars, or stellar looking comets, of magnitude 12.5, but too close to saturation for comfort with stars of magnitude 12.0. Down at the noisy end of the range, by the time I reach nineteenth magnitude stellar objects, they are just an extra 10% addition on top of my typical noise wall and random noise is badly eating into my confidence of the measurement. Nineteenth magnitude stellar objects can still be measured, yes, but with a poor accuracy, e.g. 19.3 ± 0.5, compared to, say, a star or cometary nucleus of magnitude 14 which might be measured fairly easily as 14.1 ± 0.05, with some care and the same equipment. Essentially, with my unfiltered system, at 1.5 arc-sec per pixel, the noise ripple can be thought of as being similar to that of an underlying carpet of random twentieth or twenty-first magnitude stars, with everything else sitting on the top. Of course, every observer's situation is unique and, in my case, I am fortunate to have an observatory in a dark country location and a Paramount ME mount which can track for 2 min or more without guiding.

With filtered work we really need to knock at least 1.5 magnitudes of sensitivity from the above calculations. As a rule of thumb, accurate, filtered photometry on stellar objects down to magnitude 17 could be undertaken with my equipment and the same exposure time. Obviously if I want to go fainter, a longer exposure time will work in our favor as although the sky background and thermal noise will then increase, the amount by which the faint stars rise above the random noise will increase too, in other words, we will have more ADUs to safely play with.

Photometric Terminology

Even the stars, let alone cometary cores, are not point sources and their light spreads out over many pixels on a CCD detector attached to a modest astronomical telescope; in fact the eye, studying a CCD image, often misses the fainter outer halo of a star, just seeing the bright core. For a really accurate photometric measurement we need to collect all of the star's light for measurement, but we do NOT want to collect light from nearby stars that might be very close to the star being measured. Often one encounters a situation where there are some very faint, almost invisible stars close to the star being measured. If their light is collected the resulting magnitude estimate could be inaccurate. Note, I am only really talking about stars at this point, because the situation descends into utter misery where a comet lives in a dense (or not so dense) star field.

Some technical jargon needs explaining at this point because terminology does have the power to turn beginners completely off the subject of photometry. The jargon I would like to cover here is that covered by the terms "aperture" and "FWHM". Now you may think you know all about aperture. It is the diameter of the telescope's mirror or lens, right? In the photometric context no, you are absolutely wrong. In photometry the aperture is the diameter of the ring within which the starlight is being collected. A further term which is sometimes encountered is "annulus", that is, a measurement ring surrounding the star in question, solely collecting background sky brightness and thermal noise. This is used to calculate the height of the noise wall on which the star is sitting. So, as well as being very careful about deciding just how big our photometric aperture is, to collect, say, 99% of a star's light or a cometary nucleus' light, we have to devise an accurate way of measuring the background brightness in units equating to magnitudes/square arc-sec. We do not want faint stars to confuse this background measurement either. Before this starts to sound too daunting, don't panic, most software packages do all this for you, at least for stellar photometry, but they still require a bit of intelligent user input. The sky background does not have to be measured by sampling the area just outside the aperture either, especially not in the case of comets. If this annulus is polluted by other stars (inevitable with comets) another nearby patch of barren sky can be used for a "sky aperture" measurement, provided the background sky brightness is flat. Typically the annulus radius for stellar photometry is about twice the aperture radius. The bigger the annulus the greater the statistical accuracy of the sky background measurement, BUT, there is also a greater risk of background stars polluting the field, so nothing is fixed. You frequently need to intelligently adapt your techniques to the star field in question. I will repeat myself again here: I am discussing the basics for precise stellar photometry here, not the much knottier problem of cometary photometry.

The Gaussian-like curve of a star's profile can be defined in various ways which help the photometrist. The abbreviation FWHM is often seen in this context and stands for Full Width Half Maximum. This is the width of the star's profile at half its brightness maximum. In turbulent air, the effects of optical diffraction will be swamped by the spreading of the light due to atmospheric seeing, especially in large amateur telescopes. However, the profile will still have a roughly Gaussian shape. As a rough rule of thumb, many amateurs use a photometric "aperture" of roughly four times FWHM. In other words, take the star's apparent diameter at half the maximum brightness and multiply it by four. In practice, with a telescope like mine, at a low altitude site, even the faintest stars leave obvious disks several arc-sec across in long exposures and a sensible "aperture" to set is one of around 20 arc-sec. With my 1.5 arc-sec/pixel scale this translates into approximately 13 pixels diameter or an area of about 140 square pixels. However, the best plan is to acquire a star chart of a photometric test region and, by experimentation, deduce what is the best aperture for your system by measuring stars of precisely known magnitudes. In cluttered Milky Way fields you may wish to use a slightly smaller photometric aperture. Of course, for the most accurate work photometric sequences for the filter band in use should be used. In many cases amateurs take images that are unfiltered and which roughly correspond to an "R" (Red) filtered photometric sequence. Unfiltered measurements with the same CCD camera are still valuable, but to be scientific they should be labeled "Unfiltered" so that any analysis takes this into account. Yet again, this topic of filtered versus unfiltered becomes horribly murky when we try comet photometry.

Images used for photometric purposes should never, ever, be subjected to the kind of advanced image processing techniques that amateur astronomers use to produce spectacular comet, galaxy and nebula images. The *only* processes that should ever be used when photometry is being considered are dark frame subtraction and flat-fielding which I covered fully in the imaging chapter. Both of these processes make an image far more appropriate for photometry, whereas operations that alter image contrast, gamma, and the sizes of star disks (such as unsharp mask routines) totally wreck the suitability of the image for photometry.

In addition, after a good imaging session every astronomer should treat his or her original images like gold dust. Typically, these files are in FITS format or the camera's own "Raw" format. These files are much larger than the compressed jpeg images which we are all familiar with, and an image carrying out photometry every minute for several hours can easily have hundreds of megabytes of data. Nevertheless, the temptation to delete the raw files should be resisted when the images are scientific ones. CD, DVD, memory stick and archival memory storage is cheap these days, and those really clear nights are pretty rare events. In addition the FITS image headers contain lots of useful exposure data which you will want to keep.

Photometric Wavebands

As mentioned earlier, professional astronomers measure magnitudes in accordance with internationally agreed standards, in other words the sensitivity of their equipment is accurately defined at specific wavelengths. This is absolutely essential, as for precision stellar photometry you need to know that your CCD detector/filter combination is measuring exactly the same waveband as every other advanced amateur or professional. Once again, the issue is far more complex where fuzzy comets are concerned, but we will deal with that minefield very shortly now! For stars at least it is all a matter of calibrating your system across a wide range of colors. In the 1950s, Harold Johnson of the Yerkes and Macdonald observatories created filtered measurement bands for the visual (V), blue (B) and ultraviolet (U) regions to suit the rather primitive blue sensitive photomultiplier detector he was using. Later, he added red (R) and infrared (I) bands as he then had an enhanced, red sensitive, photo-multiplier. Twenty years later Cousins and Menzies (South African Astronomical Observatory) recreated the Johnson measurement bands using different filters on more sensitive detectors. Then, in the 1980s Kron and Cousins modified the R and I system to better match the much more red sensitive CCD detectors. In 1990 Bessell (Mt. Stromlo and Siding Spring Observatories) defined new filter bands that would recreate the entire UBVRI system for CCD detectors.

Essentially, filter manufacturers, working with CCD manufacturers now produce scientific grade Johnson–Kron–Cousins UBVRI filter sets using the prescription defined by Bessell in 1990.

The V or visual band is of the most interest to amateurs measuring variable stars. It approximates the visual band of the human eye so that CCD magnitudes through a V-filter are reasonably comparable with visual magnitude estimates. But the B and R bands are also of interest. Some variable stars vary considerably in the blue end of the spectrum, even more than they do in the visual band. CCDs are at their most sensitive in the R band, so this region is also of interest.

The I band is of interest to specialist professional astronomers but the U band is close to the limit of most CCDs spectral range. When the photometric magnitude of a star is determined in both V and B passbands the B−V (B minus V) color index can be calculated. This color index can tell us a lot about the star; it can also be used to calibrate a photometric CCD system. When choosing filters for a CCD camera and variable star work it is important to understand that the Johnson–Kron–Cousins passband boundaries define an ideal system and it is often not possible to perfectly match a given set of filters to a CCD. The CCD will have its own response at specific wavelengths and to derive the response of the system it is necessary to *convolve* the spectral response of the chosen filters with the spectral response of the CCD chip. By convolve, we mean multiplying each point on the curve of the filters' spectral response, for every wavelength, with each point on the curve of the CCD's spectral response; but don't panic, because most CCD manufacturers will sell you, or recommend third party vendors for, appropriate Bessell prescription filter sets, to match their CCD cameras. As an example, SBIG can sell you specific photometric filters for their CFW-8 filter wheel, or a complete photometric filter wheel matched to their cameras sensitivity. Even if the filter-CCD match is not perfect, calibration correction factors can be applied to make it near-perfect if perfection is required. These correction factors can be verified by images of known test sequences in the sky, for each passband. However, in practice, many amateurs tend to carry out unfiltered photometry on all but the brightest targets because a V filter can easily knock 1.5 magnitudes or more off your CCD camera's magnitude limit.

Photometry Packages

Increasingly there seem to be two main photometry packages that are routinely used by amateur photometrists, namely: Software Bisque's *CCD-Soft* and Richard Berry/James Burnell's *AIP4Win*. However, for comets some observers are now also using Herbert Raab's *Astrometrica* which I mentioned in Chap. 5 and many Spanish observers use *Focas II* which I will mention again very shortly. The tools in the professional package IRAF were the *de facto* standard at major observatories and a decade ago it was often suggested that it would be a good idea if amateur observers used the same techniques by installing IRAF on a PC that was running the LINUX operating system. However that simply has not happened as most amateurs prefer to stick with a *Microsoft Windows* based operating system even if most have given *Windows Vista* a wide berth and either stuck to *Windows XP* or crossed over to *Windows 7*.

Making photometric measurements of stars in Software Bisque's *CCDSoft* is relatively painless if you have a good photometric star chart for the object to hand. Simply click on the "photometry set up" icon and enter your telescope aperture, f-ratio and pixel size. You then have to select whether the seeing was excellent, good, fair or poor. You then click the photometry "reference magnitude" icon and when you click on a star of known magnitude, simply enter the magnitude in the box. Then the third photometry icon (labeled "determine magnitude") can be selected. Clicking on your target star then gives you the magnitude of that star relative to the reference magnitude star. If you have Software Bisque's planetarium package *The Sky X* installed on your PC other options are available to you, as both packages working in harmony can identify your CCD image star field and call up the magnitude data on all the stars in the field. However, some caution is necessary here because you will want to know that the stars you are using are not variable and have been catalogued accurately. In the case of the default Hubble Guide Star Catalogue this is notoriously inaccurate as it was not intended as a photometric reference, merely as a source of guide stars. The best catalogue to use for stellar astrometry is the US Naval Observatory's USNO UCAC3. The USNO CCD Astrograph Catalog (UCAC) is an astrometric, observational program, which was started in 1998 and has recently been completed. The goal was to compile a precise star catalogue for fainter stars, extending the precise reference frame provided by the ESA Hipparcos and Tycho catalogues down to 16th magnitude. This is a very accurate astrometric catalogue and it can be used for photometry too, although for comets we, once again, enter a can of worms on that subject. Yes, I will say it again: we will learn more about this shortly! Unlike some of the larger catalogues the UCAC 2 fitted onto a modern hard disc (taking up roughly two Gigabytes) and its successor the UCAC3 was issued in 2009 on a DVD. Other US Naval Observatory catalogues, like the 80 Gigabyte USNO-B1.0 can be accessed via the web at http://www.vizier.u-strasbg.fr/viz-bin/VizieR/.

So, to get back to the plot, *CCDSoft* can be used to measure star magnitudes, and is very powerful when combined with *The Sky X*, once you have set up all the required parameters and accessed the Research/Comparison/Star Chart menu. Full details are given in the *CCDSoft* user manual of course.

The impressive book and software package *AIP*, which has now evolved into a second generation product called *AIP4Win* has become a firm favorite amongst many variable star photometrists. It has a slick photometry tool as well as a highly comprehensive photometry chapter which is well worth reading. The photometry tool is easily accessed under the "Measure-Photometry" menu (see Fig. 14.2) and you are then presented with a choice of options including: "Single Star", "Single Image" and "Multiple Image" photometry. These terms are fairly self explanatory. The "Single Star" photometry option simply gives you a result on a single

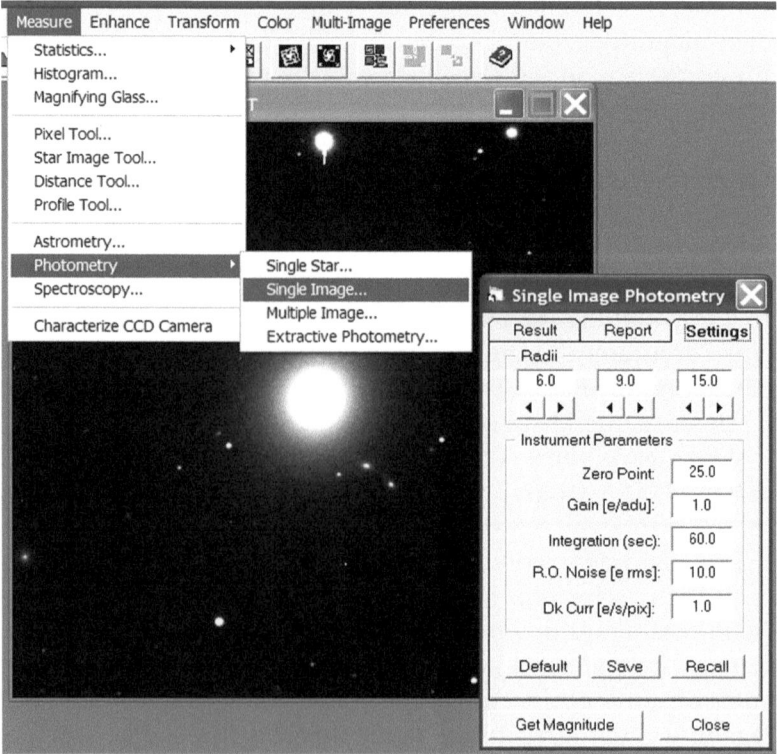

Fig. 14.2. AIP's measure/photometry/single image tool is excellent for measuring stellar magnitudes.

star by adding up its light contribution and deducting the background contribution. Using this tool gives you a raw instrumental magnitude value, but one which will have little bearing on the actual magnitude of a star unless all of the image data has been entered meticulously. This data consists of some highly technical stuff like Zero-Point Magnitude (the magnitude of the sky background), readout noise, gain and dark current, as well as the integration time. In practice to obtain a single good magnitude result the observer will choose the "Single Image" photometry option. This option enables you to do differential photometry, i.e. comparing the different magnitudes of two or more stars. This is what virtually all amateur photometrists will want to do as they will have star charts with stars of a precisely known magnitude on them (determined by professional astronomers, or advanced amateurs). In most cases they can have complete confidence that the comparison stars have been accurately measured and are not themselves variable. Knowing this they can make a differential measurement with confidence using *AIP4Win*'s "Single Image" photometry tool.

The great thing about the *AIP4Win* tool is that you can see the aperture and inner/outer annulus rings on the screen surrounding the stars or cometary nuclei/inner coma you are interested in (see Fig. 14.3). This helps greatly in choosing an appropriate aperture to surround the stars and an appropriate annulus for the background sky, and it helps avoid any faint stars which might complicate the measurement. The first object you click on is the variable object (V) and the second star is the comparison star (C1). If you click on a second comparison star then that can serve as extra security just in case the first comparison star turns out to be variable.

The photometry tool has various controls which are accessed via three tabs labeled Result, Details and Settings. The Result tab, not surprisingly, displays the results from the measurements in various ways. The Details tab provides information on factors such as the number of pixels in the aperture and annulus, the signal-to-noise ratio, the number of comparison stars and the standard deviation in the measurements. The Settings tab allows you to actually set the aperture and sky background radii, as well as set various instrumental factors which may be needed to enable the absolute magnitude value initially offered to appear in a sensible range. For example, the gain of the CCD camera can be entered at this point (expressed in electrons per Analogue-Digital Unit or ADU) as well as the exposure time, read out noise and dark current. As with any comprehensive scientific package there is a lot to digest to get precise photometric results, but in terms of comprehensiveness and value for money *AIP4Win* is hard to beat.

Fig. 14.3. AIP's photometry tool shows the photometric aperture and background annulus clearly.

Photometry Applied to Extended Objects

Now, at last, and as promised, we cross over from the precise world of variable star photometry to the murky can of worms called comet photometry! Estimating the magnitudes of comets has long been the most vital scientific contribution of the *visual* comet observer. In essence the technique is very similar to estimating the magnitudes of variable stars. The target object is compared with stars of known and similar magnitudes that are not highly colored and are not themselves variable stars. Of course, visual magnitude estimates of any kind, stellar or cometary, are just that, they are *estimates*. The human eye and brain is not a photometer and so while there are experienced observers who can make magnitude estimates of stars well above their visual threshold, to an accuracy of 0.2, or sometimes 0.1 magnitude, many observers will struggle to get anywhere near this degree of accuracy. Where the star is close to the limit of detection the magnitude estimate can be very approximate, with an accuracy of little better than 0.5 in extreme cases. With comets the problems are far greater, because they are not point sources, but extended sources. Therefore the standard visual technique is to defocus a nearby star until it looks as bright as the focused comet image (there is more on this "black art" at the end of the chapter). This is a very tricky skill to master, especially if you are halfway up a ladder in freezing cold weather with your neck twisted at such an angle that the blood supply to your brain is almost cut off. So, can the comet imager hope to achieve any greater photometric accuracy? Well, hopefully, yes. However, even when digital images are analyzed and used to make a magnitude estimate there are still huge hurdles to be surmounted and comet photometry experts often disagree amongst themselves about how measurements should be made. Needless to say comet imagers do *not* defocus their images to make a photometric measurement, they simply use software to measure the charge collected in each pixel relative to the background sky. However, the very nature of comets makes the techniques used to digitally measure their magnitudes quite a challenge, even for the experienced variable star photometrist.

Standardizing the Method

The British comet enthusiast Mark Kidger is both an amateur and a professional astronomer and has worked with Spanish amateur astronomers and comet observers since the 1980s, as he has been based both at Tenerife and in mainland Spain. In the early 1990s he noticed that there were hundreds of thousands of comet magnitude estimates hidden in the Minor Planet Center's astrometric database as a data spin-off from software based astrometry, where the software used had made a crude estimate of the magnitude of the object whose position was being measured. However, a student of Mark's noticed the data featured an enormous four magnitude scatter, equivalent to a factor of 40 times in the measurement! Mark soon realized that the scatter was caused by a total lack of standardization. The magnitude measurements were being made both with and without filters and the photometric aperture (the angular region being measured) was not detailed in the results. In addition, the precise way the measurement was reduced from the raw data to a magnitude value was not explained. When Mark started receiving photometric comet data from Spanish observers with CCD imaging equipment, from the mid-1990s, he decided a standard method should be applied to his contributors' data. However, this had to take into account the practical limitations of amateur observations made using relatively small telescopes from light polluted sites and it also needed to be simple and enjoyable, or amateurs would not participate for long. The approach adopted was to standardize on a specific photometric aperture, the same reduction software, and the same reference star catalogue. Also, the CCDs used were to be unfiltered and the assumption would be that this roughly corresponded to a photometric "R" band measurement, as unfiltered CCDs have more sensitivity at the red end of the spectrum. Most of the comets being measured were small and faint and a photometric aperture of 10 arc-sec, or however many pixels that corresponded to with the system being used, was chosen. This decision was made on the basis that the typical professional aperture of 30 arc-sec, for a nuclear magnitude estimate, would allow too much background light pollution to affect the data collected from a typical amateur's site. Going narrower than 10 arc-sec would cause other problems because star images from backyard sites are rarely less than 4 arc-sec across. Later, larger photometric aperture options, up to 60 arc-sec in 10 arc-sec steps, were permitted in the analysis.

Astrometrica and Focas II

The Spanish team also standardized on using Herbert Raab's software *Astrometrica*, even though it is primarily a package for carrying out astrometry, not photometry. They also decided to subtract the sky background using the median value for the whole image rather than using a small annulus around the comet, as is used in standard variable star measuring software. This was achieved using software called *FOCAS*, now upgraded to *FOCAS II*, which can be found on the Astrosurf web pages at: http://www.astrosurf.com/cometas-obs/_Articulos/Focas/Focas.htm.

The Spanish abbreviation *Focas* stands for: Fotometría de cometas, asteroides y variables. There is a page, in Spanish, on the astrosurf site, describing *Focas II* and there is also a Spanish *Yahoo!* site for Spanish comet observers and photometrists at: http://www.es.groups.yahoo.com/group/Cometas_Obs/.

However, unless you speak Spanish this is probably not going to fill you with optimism, even with an online translation service, but it may be worth describing briefly how *Focas II* works as it is used by a few non-Spanish amateur astronomers too. Essentially *Focas II* uses the log file produced by Herbert Raab's *Astrometrica* software and derives photometry by using it in conjunction with either the USNO-A 2.0 star catalogue or, restricted to declinations between +50 and −30, with the fourteenth Carlsberg Meridian Catalogue.

Mark Kidger and the Spanish observers decided to use R magnitudes from the USNO-A2.0 catalogue for the measurement of comparison stars as that catalogue contains high numbers of stars measured in the R band (so there will be suitable stars on every CCD image), whereas the Tycho 2 catalogue only contained B and V measurements. However, there is a risk in using the USNO-A 2.0 catalogue as some 6% of the R measurements have a measurement error of about 0.5 magnitudes.

Af(rho)

The Spanish group's magnitudes can be converted into what is known as an *Af(rho)* dust production rate measurement, originally devised around 1980 by Mike A'Hearn (see the Spanish group's 9P results in Fig. 14.4). The *Af(rho)* result is often expressed in centimeters or meters and relates

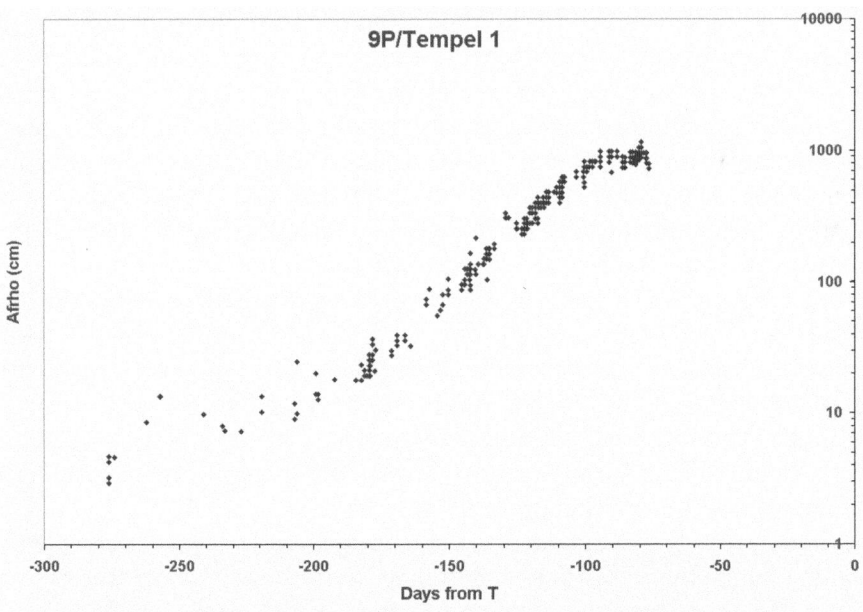

Fig. 14.4. Mark Kidger's Spanish group's Afrho dust measurements for comet 9P/Tempel for 280 to 60 days before perihelion. Image: Mark Kidger.

to the total cross section of the grains in the coma; it is often used as an indicator of dust production rate. Values of tens of centimeters for Af(rho) are typical for small comets and values in meters are typical for brighter comets. In extraordinary cases, as with Hale-Bopp, an Af(rho) figure of many kilometers is possible! It can be difficult to visualize precisely what Af(rho) means, but to repeat myself, it simply relates mathematically to the *total* cross section of the grains in the observed coma. The term *A* in *Af(rho)* stands for the albedo of the cometary dust grains, which typically reflect only a few percent of the light. The term *f* (for filling factor) represents the quantity of the line of sight filled with dust grains. The term *rho* is the radius of the photometric aperture in kilometers, at the comet's coma. The term *rho* is, not surprisingly, a substitute for the Greek symbol ρ. The other parameters involved in calculating *Af(rho)* are: Δ and *r* (the geocentric and heliocentric distances in AU); *Solflux*, my term here for the solar 1 AU flux (equivalent to absolute magnitude) in the observed band; and *Comflux*, my term here for the observed comet light flux in an aperture defined by *rho*. However, the eventual resulting measured dust production rate, often called *Qd*, relies on major assumptions about the density, the size, and the velocity distribution of dust grains. The density is always a crude guess and 1 g/cm^3 is often used. Similar production values, typically around the hundreds of kilograms per second value for a cometary nucleus, can be conjured up for water, or rather the OH radical gas production, but just how seriously any of these values are taken varies considerably from astronomer to astronomer. One leading comet astronomer told me: "I hurl all the CCD and visual data into a big *Excel* file and calculate using that. It works like magic. In fact, it's a pretty simple calculation to convert a carefully measured CCD magnitude to dust production in tons per second and also very simple to convert a total visual magnitude estimate to water production in tons per second." Hmmm! Well, I remain rather skeptical. Decades ago a school friend of mine, when presented with dodgy information used to sing out loud: "And the band played believe it if you like, you like, you like, and the band played believe it if you like!" It seems strangely appropriate where the black art of conjuring up comet gas and comet dust production rates is concerned.

Nevertheless, if you like formulae you may want to see what an *Af(rho)* formula for comets typically looks like. Well, using the *Af(rho)* terms, with ρ substituted for *rho*, the formula for calculating the specific quantity *Af* looks like this:

$$Af = \left(\frac{2\Delta r}{\rho}\right)^2 \frac{Com\ flux}{Sol\ flux}.$$

Obviously the terms can be re-arranged to calculate *AF*ρ if preferred.

The Spanish team members have often reported that their estimates of dust production, based on their own CCD measurements, tie up well with professional data on comets such as 46P/Wirtanen in 2002 and 9P/Tempel in 2005. Many comet magnitudes have been monitored by the group especially the outbursting comet 29P/Schwassmann–Wachmann (see Fig. 14.5). At a BAA comet section meeting in 2005 Mark Kidger admitted that using unfiltered CCDs can produce large errors when comets with high gas output are measured. Where small perihelion distances are involved active comets can exhibit very strong Swan band emissions which confuse the results leading to misleadingly large dust production values. Employing an R band filter can reduce these errors though, although some would argue that a V band filter should be used so that all photometry ties up with the estimates of visual observers. It should be stressed that light from bright comets is often mixed up between broadband scattered sunlight from the dust particles and narrowband emissions from various molecules in the gas. With the zero magnitude C/1996 B2 (Hyakutake) very strong C_2 SWAN band emissions around 514 nm were present and

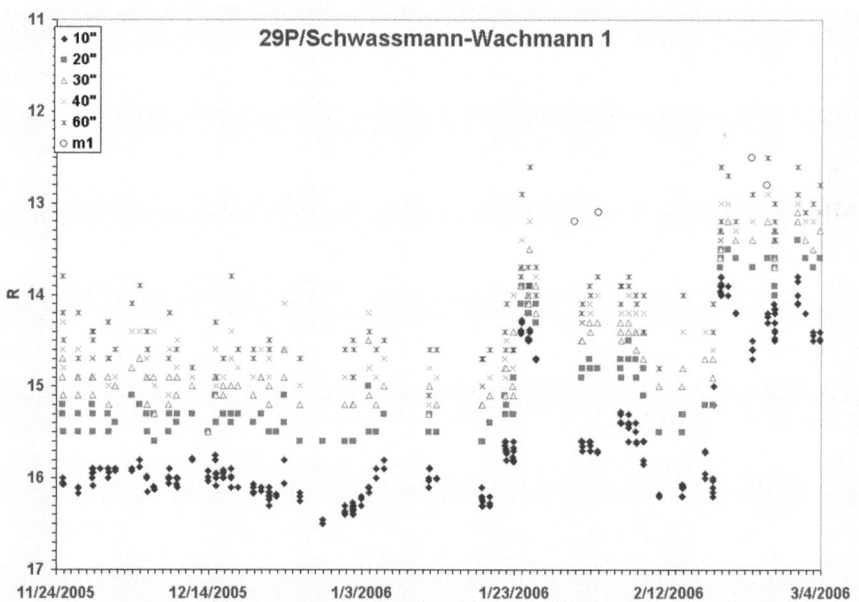

Fig. 14.5. Mark Kidger's Spanish group's magnitude estimates for the outbursting comet 29P/Schwassmann–Wachmann from November 2005 to March 2006. The magnitudes for various photometric apertures are shown. Image: Mark Kidger.

there was strong CN emission in the blue region of the spectrum. To my eye Hyakutake's 90 arc-min wide coma looked positively turquoise! Such strong emission lines can easily cause inaccurate photometry if unfiltered CCDs are used.

The Filtered Approach

The Italian CARA group (Cometary Archive for Amateur Astronomers) uses different photometric techniques to those used by the Spanish observers and their methods are likely to be more accurate with gassy comets, but arguably trickier for the average amateur to undertake, because they use photometric filters. CARA has a website in English which can be found at http://www.cara-project.org/.

Of course, if you make the learning curve too difficult or the cost of equipment too high only a few observers will take up the comet photometry challenge. The Italian CARA group did consider whether it would be possible to use standard RGB photometric filters for comet photometry and concluded that while they were better than an unfiltered system they were still too wide in bandwidth, or not ideal for dust magnitude estimates. To reiterate, the light from a comet consists of two parts, a continuum part due to sunlight scattered and reflected by dust grains and a superimposed emission caused by the coma and tail gas components. Ideally, specific filters are needed to isolate each component and for the continuum (dust) component CARA prefer to use photometric R or I band filters rather than a totally unfiltered system.

Professional astronomers use expensive narrowband interference filters for comet photometry. Almost 30 years ago, in 1981, Michael A'Hearn (yes, him again) proposed a standard cometary photometry filter set designed to isolate primary emissions in the optical region. The gaseous emissions of CN at 387-nm, C_3 at 406-nm, and C_2 at 512-nm were chosen, with bandpasses of five to 12 nm. In addition two regions were chosen to measure the solar-reflected comet dust, called the spectral "continuum", at 365 and 485 nm. Sets of these filters were actually made and distributed to astronomers as part of the 1985/1986 International Halley Watch. A quarter of a century later the CARA group have found it possible to buy useful narrowband filters which cost only $100 or so from suppliers such as Edmund Optics and Andover Corporation. They have found six narrowband (10 nm Full Width Half Maximum) commercial filters with passbands centered on 405, 440, 515, 589, 620 and 647 nm which can be used to isolate (respectively) the following regions of interest: Carbon III (coma), Blue continuum, Carbon II (coma), Sodium (coma and tail),

H_2O (tail) and Red continuum. CARA state that the red 647-nm continuum filter can conveniently be calibrated in the S Vilnius photometric band. However, depending on the brightness of the comets very long (an hour or more) exposures may be necessary. The CARA view is that filters *must* be used for comet photometry as you just never know when a comet's magnitude is going to be distorted by gassy emissions, giving a bright result with an unfiltered detector. Another difference between the CARA group and the Spanish group is that the photometric aperture used by CARA is not fixed at 10 arc-sec but is based on a reference window of 100,000 km in space. This is equivalent to 206 arc-sec at a distance of 100 million kilometers, or 20.6 arc-sec at a distance of a billion kilometers.

However, CARA do not normally use narrowband filters on comets below magnitude 11.5 because the exposure times would simply be impractical. For the faintest comets they use those broadband R or I photometric filters. It is generally accepted that for a periodic comet, 3 AU or further out from the Sun there would be hardly any emission line activity and so the Spanish unfiltered system would be simple to apply and probably accurate too. Mark Kidger has flagged up a potential problem with the Italian method of having a fixed physical photometric aperture in space (in kilometers) which is that in some cases the nearest and furthest measurement points of a comet can vary by a factor of ten if it is passing relatively close to the Earth. So, as I mentioned a few sentences back, one can imagine a fixed physical aperture corresponding to an angular aperture of anywhere from 20 to 200 arc-sec, or bigger. This could cause measurement reduction problems if the observer's sky background was bright and a 3 arc-min wide slab of sky was being sampled. A very accurate sky background measurement would be needed in such cases. Conversely though, if the aperture is fixed in angular terms then different areas of a comet's coma will be sampled as the comet approaches the Earth.

In recent years Roberto Trabatti of the CARA team has developed dedicated software to try to make the measurement process more user-friendly and to standardize the reporting process. CARA now offer two comet photometry software packages, namely one version for *Microsoft Windows* and one for the *Linux* operating system which enables the *Unix* based software used by professionals to be employed. The Linux version is more powerful but special software drivers need to be written to use most of the commercial CCD cameras in amateur hands. These software packages, for *Windows* and *Linux* are respectively called wafrho1.exe and xafrho1.exe (see Fig. 14.6).

Well, that concludes my look at CCD comet photometry. It is a nightmare world, but when all is said and done, any scientific measurement is of value, provided all the details of the measurement are reliably recorded. In many ways it all boils down to whether you prefer the simpler

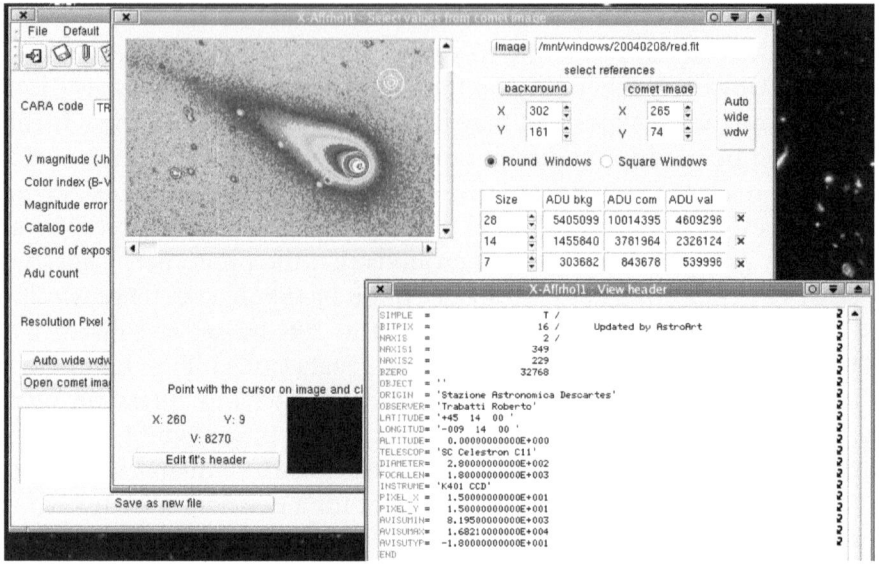

Fig. 14.6. Roberto Trabatti of the CARA team has developed dedicated software for comet photometry for Windows and Linux platforms. This is a screenshot of the Linux xAfrho1.exe software.

Spanish (Kidger) method or the more professional Italian (CARA) method. Comets are horrendous objects to measure the brightness of, and after reading about the CCD pitfalls you may be tempted to make visual magnitude estimates instead! This is a book about imaging comets and hunting for them but I feel I should now include a very small section on visual observing for completeness and comparison.

Visual Photometry

I started off in this hobby as a visual observer but quickly realized that although visual observing was very low cost and relatively hassle-free it was impossible to be a state-of-the art observer if I stuck with the old fashioned way of observing. I do visual observing for relaxation, but where I am submitting data to a scientific body I like my results to be beyond doubt and my photographic efforts of the "1980s and 1990s," followed by my CCD images from the early 1990s to the present date, provide an unbiased record of what my cameras actually recorded. However, as we have seen, CCD photometry of comets is not a precise science unless the most rigorous, and often demoralizing, disciplines are followed. So, the old-fashioned but very affordable way of estimating the magnitude, size and appearance of a comet still has some merit. However, I tend to think its biggest appeal is that the old style observer does not have to learn new tricks or spend money to make a visual observation. The trouble is that visual photometry is just as fraught with problems as CCD photometry. The scatter in reported magnitudes, even amongst experienced observers is proof of that. In addition, comets of 12th or 13th magnitude are sometimes reported as being "glimpsed" when a deep CCD image reveals no trace of the object there. Shades of Percival Lowell's Martian canals perhaps! So, visual photometry is certainly no better than CCD photometry but it is quick and easy to perform, so for those reasons alone I think I had better include a section on visual photometry and observation here.

Visual observers, not surprisingly, observe naturally in the photometric visual band and so see far less into the red than an unfiltered CCD. Some experts regard a CCD filtered to look at a comet in the dominant Swan bands as having a similar response to a visual observer. It is quite common for visual observers to give a brighter magnitude for a comet than one derived from CCD images and it is also common for visual observers to quote a larger physical size for the coma. Some observers with proven exceptional eyesight can sketch very accurate gas tail details in comets. The two British observers that spring to mind to me are the discoverer George Alcock and the planetary observer Paul Doherty; both these legendary figures are sadly no longer with us. One of the most bizarre aspects of visual comet photometry is that it is simply accepted that many comets appear brighter, *relative to the stars*, when viewed through a *smaller* instrument! At first this "smaller means brighter" effect seems totally counter-intuitive, but remember we are talking about a relative comparison here. Of course, the bigger the telescope the fainter you will see, but you may not sense the entire coma at high powers. Mind you, go to a really small aperture and the comet will, quite obviously, totally vanish!

As already mentioned in this book a comet of a specific magnitude will be far harder to see than a star of a specific magnitude because the light from a faint star is focused into a point less than 10 arc-sec across, whereas the heads of comets can often span arc-min. A 400-mm aperture telescope in the hands of a very experienced visual observer may be able to reach stars fainter than magnitude 16 but comets of magnitude 13, or possibly 14, may be barely detectable; the same goes for naked eye stars and comets. In general stars start to be visible from a typical site when they are brighter than magnitude 6.0, whereas for comets they have to be brighter than 4.0 at least. It is all a matter of when the surface brightness of the comet rises sufficiently above the surface brightness of the sky.

It goes almost without saying that the visual observer should adopt a scientific approach to every measurement by keeping a scientific notebook record of every observation made and recording the time in UT (GMT) to an accuracy of at least 0.01 of a day (roughly 15 min precision). A dim red light will enable observations to be entered in the notebook without dazzling and trashing your night vision. Your eyes should be dark adapted for half an hour before commencing serious observations on faint comets. The eye's most sensitive light detectors are called the rods (as opposed to the cones). There is an optimum, ultra-sensitive, rod-packed region of the retina which is roughly eight to 16° away from the eye's center; 12° is a good average value for the best part. This means that to get the most sensitivity out of your retina you have to look to one side of the faint astronomical object you are trying to see; at first this will seem incredibly difficult, but it will improve with practice. This 12° (or so) offset should be arranged so that you appear to place the object nearer to your nose! The reason for this strange requirement is that the eye has a blind spot where the optic nerve leaves the retina and this blind spot appears to be on the other side, away from the nose, due to the inverting property of the lens. The eye is roughly four astronomical magnitudes more sensitive where the rods peak than in the centre of the retina (the bit you are using to read this sentence).

Estimating Visual Magnitudes

Variable star observers use one of two methods to estimate the magnitude of a variable star. These are known as the Fractional method and the Pogson Step method and a modified version of the former method is usually used for estimating comet magnitudes. The Fractional method is very easy to understand. Imagine you have a variable star of roughly magnitude 10. You have already acquired a star chart from, say, the AAVSO (their variable star plotter at http://www.aavso.org/observing/charts/vsp/index.html is a fantastic online resource) and you spot two suitable comparison stars on that chart. One is labeled E and has a magnitude of 9.4. The other is labeled F and has a magnitude of 10.2. Armed with the chart and a dim red light you approach the telescope draw tube, replacing the low power eyepiece you used to find the field, with a high power one. After locating the variable star and the two comparison stars you turn the red light off and stare at the field. Imagine that the variable star appears to be a bit fainter than star E, but a lot brighter than star F. In fact, it is roughly a quarter of the way in brightness from star E to star F (which makes it about magnitude 9.6). Confident of this measurement, you turn your dim red lamp back on and make the following observation in your log book:

$$E\,(1)\,\upsilon\,(3)\,F = 9.6.$$

It is vital to record the fractional estimate in full like this and *not* just the actual value, because occasionally the comparison stars used to estimate magnitudes have their magnitudes revised. If one of the comparison star magnitudes is changed and you know which stars you used to make the estimate, then all is well and good. You can go back a few weeks later and revise the estimate if needed. Of course, if the variable star had been indistinguishable in magnitude from star E, you could simply record a direct comparison magnitude in the form: Variable star = Star $E = 9.4$. What do you do if you simply cannot see the star, but can see many of the comparison stars? The notation for this is simply the "<" symbol. So, for example, <F can be recorded at the telescope, translating to <10.2 in the reduced observation. In other words, the star was fainter than the faintest comparison star you could see on that night. A negative observation like this can still be of value.

The Pogson step method is, quite simply, just using a single comparison star and estimating how many tenths of a magnitude your variable is above or below it in magnitude.

Defocusing

The basic technique is just the same for comet magnitude estimation except you have to defocus the comparison stars until they look the same size (from memory) as the focused fuzzy comet! "Nightmare!" I hear you cry! Yep, it's not exactly precise is it? Nevertheless that is the way visual comet observers estimate cometary magnitudes. This basic defocusing technique is known as the Sidgwick (S) method. Suddenly the whole CCD photometry thing seems a lot more attractive! There is another twist as well and I touched on it earlier. It is invariably recommended that you should always use the lowest magnification and smallest aperture telescope (or binoculars) that will show the comet for making the magnitude estimate. The reason for this is that the smaller the head of the comet appears the more star-like it will seem, and the more the faint outer coma will shrink into the head and seem to contribute to the overall magnitude. However, there is no reason at all why a much larger instrument cannot be used for *studying* the head, but not for actually making the magnitude estimate. Another advantage of using the smallest aperture and magnification is simply that there is more chance of getting the comparison stars in the field of view. Bear in mind that very low powers of less than about three or four times per inch of aperture will produce an eyepiece exit pupil too wide to fit into your dark adapted eye. So, for example, for 80-mm aperture binoculars a magnification of 15× is better than 10×. At 15× the beam of light will be 80/15 mm wide; that's 5.3 mm, which should fit into most dark adapted pupils.

I have not yet finished with defocusing stars though, because there are sub-twists to the plot. I have already mentioned the basic defocusing technique for magnitude estimation called the Sidgwick (S) method but there are three modifications to this method. In the Morris (M) method the comet itself is slightly defocused until it appears to have a uniform surface brightness. The stars are then totally defocused until they have the same diameter (from memory) as the slightly defocused comet. In the Bobrovnikoff (B) method you defocus the comparison stars and comet simultaneously until the bloated stars can be directly compared with the fuzzy comet. At least the Bobrovnikoff method relies less on memory! Finally, there is the Max Beyer (E) method. In this method you defocus to almost infinity and see whether the comet or comparison stars disappear first. This method is arguably only of use on the very brightest comets. If a comet is very low down and the comparison stars are higher up, or visa versa, an extra tier of misery is added as, strictly speaking, a correction for atmospheric extinction of comet or star should be applied to the estimate. As a rough guide: at altitudes of 40° and above the atmosphere dims starlight by about 0.3 magnitudes. At 30° the extinction is 0.4; at 20° it is 0.6; at 10° it is 1.2; and at 5° altitude it is 2.1 magnitudes.

Analysis Software

If a keen observer has followed a comet over many months it is more than likely that he or she will want to produce a light curve of the object's performance. Of course, any spreadsheet package, such as *Microsoft Excel*, can be used to produce a graph of magnitude versus time, but there is a very useful package which is specifically designed for plotting comet light curves, namely *Comet for Windows* by Seiichi Yoshida. One slight issue with this software is that *Spybot-Search & Destroy* or *McAfee SiteAdvisor* judges the Comet.exe file to be spyware or adware, but this can, of course, be ignored! The software can be downloaded from Seiichi Yoshida's site at: http://www.aerith.net/project/comet.html.

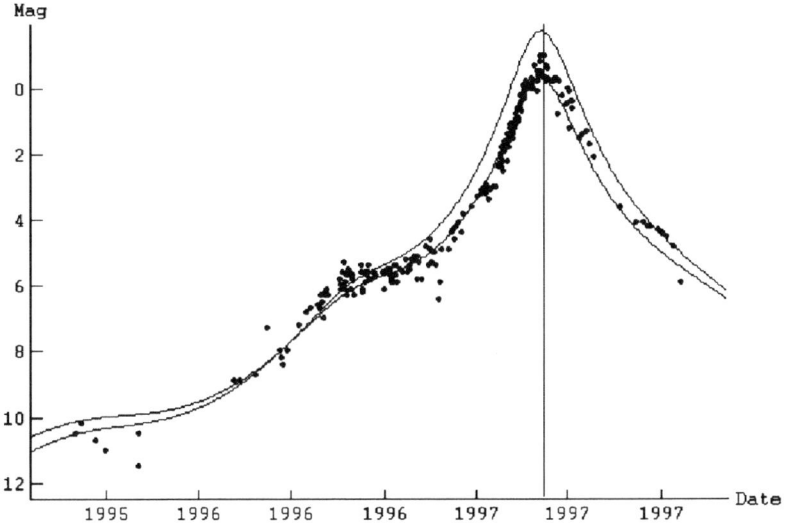

Fig. 14.7. *Comet for Windows* can be used to store magnitude data, plot light curves and compare magnitude law curves to the actual data for each comet. In this example an analysis of comet Hale–Bopp is being plotted.

Now, before you think that *Comet for Windows* is just a graph plotting tool I should stress that it is far more than that. As well as allowing the input of magnitude data for all the comets you might ever observe and the basic plotting of magnitude against time it can, for example, allow the distance corrected magnitude to be plotted against the log of the comet's heliocentric distance. In addition, the points plotted on the graph can be analyzed by the software so as to produce "best fit" magnitude law curves of the form $m = H_0 + 5 \log \Delta + K \log r$ (see Fig. 14.7). You can also see how a magnitude law of your own choosing fits over the curve of the points and use the software to predict how bright a comet might become. All very handy indeed and it is absolutely free as well!

Tails and Degrees of Condensation

Of course, as well as the actual magnitude of the comet the observer will also want to record the diameter of the head (coma) and the length of the tail or tails visible. The direction the tails are pointing should be quoted in position angles from North, through East. Thus, North = 0, East = 90, South = 180 and West = 270. Details of all the instruments used for each observation must be recorded meticulously. As well as the actual diameter of the coma in arc-min (for which you will need to know the field of view of each of your eyepieces well, for each telescope) there is also the: "black art" of estimating the "degree of condensation" of the coma (see Fig. 14.8).

Fig. 14.8. As described in the text the diffuseness of the coma of a comet can be assigned a "degree of condensation" DC value from 0 to 9. In the images above the DC values from top to bottom could be roughly interpreted as DC1, DC4 and DC6.

This somewhat vague parameter is best explained by considering the extremes. A comet whose coma looks like just a floating patch of mist will have a degree of condensation equal to zero, whereas a comet whose coma looks like a star will have a degree of condensation equal to nine. The degrees of condensation as defined by the British Astronomical Association's comet section are:

0. Diffuse coma of uniform brightness.
1. Diffuse coma, with slight brightening towards centre.
2. Diffuse coma, with definite brightening towards centre.
3. Centre of coma much brighter than edges, but still diffuse.
4. Diffuse condensation at centre of coma.
5. Condensation appears as a diffuse spot at the centre of the coma. Described as moderately condensed.
6. Condensation appears as a bright diffuse spot at the centre of the coma.
7. Condensation appears as a fuzzy star that cannot be focused. Described as strongly condensed.
8. Coma has virtually disappeared.
9. Stellar or disc like in appearance.

That concludes my short section on visual comet photometry and observation and in the final chapter I would now like to look at a few of the world's best comet imagers in detail.

CHAPTER FIFTEEN

A Few of the World's Keenest Comet Imagers

There are hundreds of comet imagers throughout the world; some of them take a few comet images per year and others can bag twenty or more comets on a clear night. It would be impossible to list them all and even listing some of the best is always going to be very subjective. Do you rank the best by the quality of their images, the sheer volume of their images, the number of decades spent in the hobby or the amount of science (astrometry and photometry) that they carry out? Other comet imagers (like the tireless NEO and comet astrometrist Peter Birtwhistle) have already been mentioned in previous chapters. Well, all I can do is list some of the very best and hope this gives the reader an idea of what enthusiastic comet imagers across the world are doing. The German comet observer Stefan Beck has compiled a large list of all the comet observers he is aware of and this list can be found at www.cometchaser.de/observers/names.html.

Michael Jäger

Many comet imagers would consider that Michael Jäger, who observes mainly from Stixendorf near Krems, Austria, is the world's finest comet imager. Michael (Fig. 15.1) has been a leading comet photographer since

M. Mobberley, *Hunting and Imaging Comets*, Patrick Moore's Practical Astronomy Series, DOI 10.1007/978-1-4419-6905-7_15, © Springer Science+Business Media, LLC 2011

Fig. 15.1. Michael Jäger is regarded by many as the world's finest comet photographer and imager, and his reputation survived the transition from photography to digital imaging. He is shown here on a visit to the Tivoli Southern Sky Guest Farm in Namibia equipped with a 20-cm f/2.7 ASA astrograph. Image: Michael Jäger.

the 1980s when he used a 200-mm f/1.5 Celestron Schmidt camera and gas hyper-sensitized Kodak Technical Pan film. As well as being one of the few comet imagers to have excelled in both the era of photography and the era of CCD imaging, Michael is also a comet discoverer. As I mentioned in Chap. 4, on the night of October 23–24, 1998 he discovered a fuzzy object on photographs he was taking of comet 52P/Harrington–Abell with his Schmidt camera. The object was magnitude 12.5 and featured a strongly condensed coma 1 arc-min across with a 10 arc-min long fanned tail. The orbital period was soon deduced to be only 14.95 years and the comet was designated P/1998 U3 (Jäger). The comet experienced a very close encounter with Saturn in July 1991. Some of Michael's finest images have been taken from the Austrian Alps. Michael has always preferred fast imaging systems for cometary work and has used both a Starlight Express SXV H9 CCD and, more recently, a Sigma 6303 camera made by Astro-elektronik Fischer (www.nova-ccd.de/index_en.html). This camera uses a KAF-6303E sensor with 3,072×2,048 nine micron pixels and an imaging area of 27.7×18.5 mm. Most of the astrographs Michael has used in recent years have been made by the Austrian company Astro Systeme

Austria (ASA) with optical designs by Phillip Keller. These have included a 30-cm aperture f/3.3 Deltagraph, a 20-cm aperture f/2.8 system and, in recent years, a 25-cm f/3.8 astrograph with a triplet lens field corrector. This most recent system, with the KAF-6303E sensor, delivers a field of view of $1.7 \times 1.1°$ and a resolution of 2 arc-sec per pixel. Michael Jäger's website is at: http://www.cometpieces.at/

Gerald Rhemann

Since the era of large CCD detectors Michael Jäger has often collaborated with his fellow countryman Gerald Rhemann (Fig. 15.2) who is also one

Fig. 15.2. Gerald Rhemann takes spectacular comet and deep sky images from the Austrian Alps, the Canary Islands and the deserts of Namibia. He is shown here with a high quality ASA astrograph. Image: Gerald Rhemann.

of the world's leading deep sky and comet imagers. He also images from the Austrian Alps as well as the Canary Islands and the deserts of Namibia. Gerald works as a sales manager and consultant for the aforementioned Astro Systeme Austria (ASA) Company which produces the high quality astrographs he and Michael Jäger use. As well as imaging comets Gerald specializes in deep, wide-field vistas of the Milky Way using short focal length telescopes and large format CCD cameras. Gerald has a stunning website at www.astrostudio.at/.

Johannes Schedler

Johannes Schedler is based in southern Styria, a south-eastern province of Austria. His home and "Panther" observatory are on a hill at an elevation of 380 m above sea level. The observatory is named Panther because that animal is a heraldic beast of the Styria province. Johannes uses a large and fast 400 mm Cassegrain made by the optical expert Philipp Keller whose telescopes are especially popular with advanced European amateur astronomers. The main Cassegrain focus works at f/10 but a CCD can conveniently be placed at the corrected prime focus at a very fast f/3 ratio where Johannes places his SBIG STL11000M CCD. The correcting lens at the prime focus enables a pin sharp field, 40-mm in diameter, to be illuminated, which matches the 36×24.7-mm dimensions of the STL1000M perfectly. The f/10 Cassegrain focus can also be used at f/6.7 using a reducer/corrector lens. A TEC 140-mm apochromat refractor is also used for imaging from the Panther Observatory. As well as imaging comets Johannes takes some stunning deep sky, planetary, lunar and solar images. His home page can be found at http://panther-observatory.com/.

Bernhard Häusler

Bernhard Häusler (Fig. 15.3) is based in Germany, at Maidbronn, just 7 km north-northeast of the city of Würzburg. He has been imaging comets since 1997 and uses a 30-cm Schmidt–Cassegrain at f/7 together with an SBIG ST10XME CCD. He also uses a Robofocus focusing system and a True Technology filter wheel. Filters used include an IR blocking filter as well as UHC, OIII, Hβ and Swan filters. His comet imaging work is simply referred to as "Bernhard's Comet Project." His cometary astrometry is carried out using Herbert Raab's Astrometrica and photometry

Fig. 15.3. Bernhard Häusler of Maidbronn is one of Germany's top comet imagers. He is shown here with his 30-cm Schmidt–Cassegrain. Image: Bernhard Häusler.

is also undertaken using FOCAS II (see http://astrosurf.com/cometas-obs/_Articulos/Focas/Focas.htm). Bernhard's images are processed using a combination of Astrometrica, CCDSoft, Maxim DL and Adobe Photoshop CS3. Videos and comet animations are produced using Studio Version 9 or Adobe Fireworks CS3. In 2004 Bernhard's telescope, fitted on its giant tripod, fell from the balcony it lives on. A violent squall during a thunderstorm lifted up the entire telescope and hurled it over the balustrade headfirst into a bush 3 m below! Amazingly the bush absorbed the fall and turned the telescope back, tripod down, into its right position just before it finally touched the ground. The tripod deformed in the impact but the telescope still worked afterwards! His home page is at www.amication.de/Bernhards_Comet_Project/.

Erik Bryssinck

Erik Bryssink is one of Belgium's leading amateur astronomers and lives in Kruibeke, roughly 10 km south of Antwerp near the river De Schelde. His main instrument is a powerful 0.4-m f/3.8 Orion Optics AG16

Fig. 15.4. Erik Bryssink's 0.4-m f/3.8 Orion Optics AG16 astrograph mounted on an Italian 10 μm GM2000 Qci mount at his run-off roof observatory in Kruibeke, Belgium. Image: Erik Bryssink.

astrograph mounted on an Italian 10 μm GM2000 Qci mount (see Fig. 15.4). He uses an SBIG ST10-XME CCD camera with a CFW-10 filter wheel. Erik is another contributor to the Italian CARA comet photometry project and he carries out afrho measurements for them. He also carries out cometary astrometry and confirms new NEOs. His observatory is named BRIXIIS which is the oldest known spelling of Eric's surname, dating back more than 700 years to the year 1293. The design is of the run-off roof type and enables full access to the sky, except where neighboring trees block the view. Erik's website can be found at www.astronomie.be/erik.bryssinck/.

Alfons Diepvens

Alfons Diepvens is another comet observer and imager based in Belgium; his observatory is situated at Balens. His instrument for imaging comets is unusual in as much as it is a large aperture long focus high quality refractor, namely a TEC 200 mm f/9 ED apochromat. His observatory sits on the top of his house and is of the sliding roof design. His equatorial mounting is a high quality 10μm GM2000 QCI model manufactured in Italy.

Alfons uses a Canon 40D digital SLR for his comet images which has the benefit that his images are all captured in color. His comet images can be found at http://users.telenet.be/diepvens/cometimages/.

Gustavo Muler

One of the world's most active comet imagers is Gustavo Muler (Fig. 15.5) who lives on Lanzarote, the easternmost island of the Canary Island chain which lies just off the eastern coast of northern Africa. He observes from his Nazaret observatory which has an MPC code of J47. Gustavo's output is extraordinary. On April 22, 2009 he imaged an amazing 23 comets during the night, using his 30-cm Schmidt–Cassegrain. During the course of a year he will typically take as many as two thousand comet images. In 2009 he imaged on 157 nights of the year! He also has an 80 mm f/6 refractor piggybacked on the main instrument for wider field views. Gustavo's main camera is an SBIG ST8XME used with an AO-8 adaptive optics autoguider.

Fig. 15.5. Gustavo Muler of the Nazaret observatory Lanzarote, Canary Islands, is one of the world's most prolific comet imagers, sometimes capturing more than twenty comets per night! Image: Gustavo Muler.

The 30-cm Schmidt–Cassegrain is mounted inside a homemade dome on the roof of his house and an older 20-cm Schmidt–Cassegrain is housed in a run-off roof structure. That earlier instrument is equipped with an SBIG ST7 CCD and an AO-7 adaptive optics autoguider. Remarkably, Gustavo finds time to image other objects in addition to comets. He captures images of supernovae and deep sky objects too. Gustavo's website can be found at http://astrosurf.com/nazaret/index.shtml.

Juan Antonio Henríquez Santana

I have already mentioned Juan Antonio Henríquez Santana (shown in Fig. 15.6) in previous chapters as it was this Tenerife based amateur who first detected the extraordinary 2007 outburst of comet 17P/Holmes.

Fig. 15.6. Juan Antonio Henríquez Santana of Santa Cruz de Tenerife in the Canary Islands. He is shown here with his 20-cm aperture f/9 Vixen VC200L telescope and SBIG ST9XE CCD with which he discovered the spectacular outburst of comet 17P/Holmes and then subsequently alerted the astronomical world. Image: Juan Antonio Henríquez Santana.

He images comets with a portable system from Santa Cruz de Tenerife in the Canary islands using a 20-cm aperture f/9 Vixen VC200L (Visac) telescope and an SBIG ST9XE CCD mounted on a Losmandy G11+Gemini mount. Sometimes a Canon EOS 350D is also used for imaging. At the time of writing Juan had imaged 56 numbered comets and 65 unnumbered comets but that early detection of the 17P/Holmes outburst was his most important observation and one of the most important comet observations ever made by an amateur astronomer. His website can be found at http://atlante.teobaldopower.org/.

Remanzacco Observatory Imagers

The Remanzacco Observatory is situated in the northeastern corner of Italy, not far from the border with Slovenia, approximately 10 km east of the city of Udine. It marks the home of AFAM, the Associazione Friulana di Astronomia e Meteorologia. The observatory is home to an enthusiastic team of amateur astronomers who have a very high reputation in the field of comet imaging and who work closely with the Italian group CARA, which, as we have seen, is an abbreviation for the Cometary Archive for Amateur Astronomers. This group pays special attention to photometric and cometary dust measurement, using photometric filters, so as to obtain the Afrho value used by professionals to study the coma. The Remanzacco Observatory team consists of a number of Italian amateurs including Giovanni Sostero, Ernesto Guido, Luca Donato, Virgilio Gonano, Mario Gonano, Vincenzo Santini and Antonio Lepardo. The most prominent of these are Giovanni Sostero and Ernesto Guido who are frequently the first amateur observers worldwide to obtain images of new comets or novae. Sostero and Guido have also worked with fellow amateur Paul Camilleri and to maximize their observing time they frequently purchase time on the telescopes that are part of the GRAS (Global Rent-a-Scope) facility at Mayhill, New Mexico and Moorock, Australia.

Various instruments are available at the Remanzacco observatory including a 0.45-m f/4.4 Newtonian equipped with a Finger Lakes Instrument 1001E CCD camera. In recent years a new facility has been set up by the Remanzacco team, namely an observatory situated on Mount Matajur, some 20 km east of Remanzacco at an altitude of 1,340 m. This new facility, employing a fast 20-cm f/3 reflector, is above the haze that affects the plains where the Remanzacco observatory is located. The Remanzacco team has made a number of significant contributions to astronomy and during the course of every year usually recovers several periodic comets returning to the inner solar system, confirms 20 or so comet discoveries and follows up with astrometry on 50 or so NEOs. The team has also discovered more than a dozen new asteroids. The Remanzacco blog spot can be found at http://remanzacco.blogspot.com/.

Rolando Ligustri

Another leading Italian comet imager is Rolando Ligustri who is based at Talmassons, roughly 25 km southwest of the Remanzacco observatory in north-eastern Italy, mentioned above. Rolando uses a 35-cm aperture f/5 reflector with an SBIG ST10XME CCD camera for his comet imaging from Talmassons and he also uses the robotic GRAS telescopes in New Mexico via the Internet. He also contributes observations to the Cometary Archive for Amateur Astronomers (CARA) and is a member of his local society Circolo AStrofili Talmassons, abbreviated to CAST (www. castfvg.it/). Rolando Ligustri has a website at: http://picasaweb.google. com/astroligu/CometeDiRolandoLigustriCASTItalia#

Clay Sherrod

Dr P. Clay Sherrod, often better known as Dr Clay (see Fig. 15.7), established his Arkansas Sky Observatory (ASO) in 1971. A total of four IAU coded observatories (H41, H43, H44 and H45) make up the ASO facility and in 2009 alone Clay Sherrod's facility captured more than 2,500 images of 72 comets on the 90 clear nights from Arkansas. The main facility is at H45 on Petit Jean Mountain where the prime ASO instrument, a 0.5-m aperture PlaneWave astrograph, is mounted atop a Mathis Instruments MI-500 equatorial mounting, which uses an Astrophysics GTO servo drive. Dr Clay has been described as an educator and researcher in Earth and physical sciences, astronomy and archeology. Although now retired from formal research and educational astronomy he uses his impressive ASO facilities for private research and outreach programs and the facility operates through the Sherrod family trusts. Comets are not ASO's only objects of study as Dr Clay has taken huge quantities of planetary images too. The ASO has possibly the most comprehensive website of any privately operated observatory and this can be found at http://www. arksky.org/.

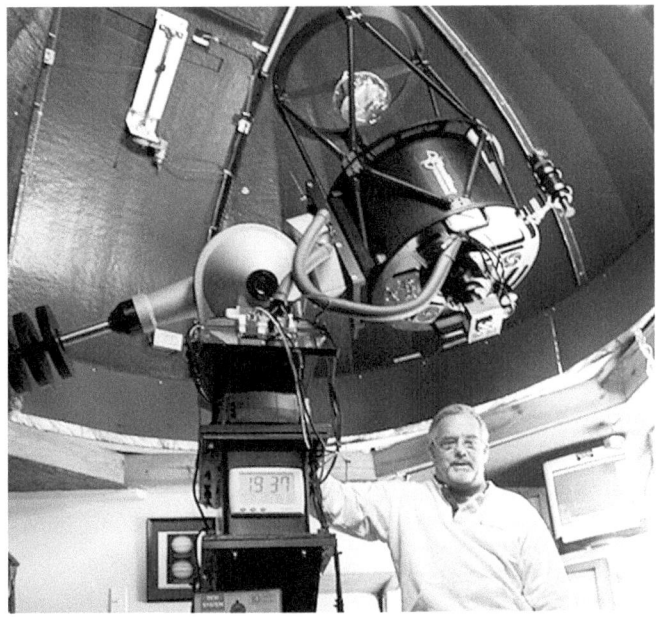

Fig. 15.7. Clay Sherrod of the Arkansas Sky Observatory is one of the world's most active comet imagers and astrometrists. He is pictured here with his 0.5-m PlaneWave CDK (corrected Dall-Kirkham) Cassegrain. Image: Clay Sherrod.

Mike Holloway

Mike Holloway is the moderator of the Yahoo! Comet-Images group and captures comets from his home observatory some twelve miles north of the town of Van Buren in Arkansas, USA (see Fig. 15.8). His telescope is kept in a four by five meter building with a built in 2 m diameter dome made by "Home-Dome." Prior to August 2006 Mike imaged with a Takahashi FSQ-106 refractor, Losmandy G8 mount and Finger Lakes Instruments Maxcam 10ME CCD. Then, on August 4, 2006, disaster struck and after weeks with the temperature at 40°C an incoming cold front generated violent storms resulting in a lightning strike on Mike's dome. The strike trashed two observatory computers, his refractor, his mount and his camera. However, only a few months later he was back up and running with better equipment, namely a TeleVue NP127is apochromat refractor, a Losmandy G-11 mount and an SBIG ST10XMEI CCD.

Fig. 15.8. Mike Holloway's observatory in Van Buren Arkansas had to be re-equipped after a lightning strike in 2006. Since then a TeleVue NP127is apochromat has been his main imaging instrument. Image: Mike Holloway.

Charles Bell

Charles Bell (Fig. 15.9a) of Vicksburg, Mississippi is one of the world's most active comet imagers and lists other interests as photographing the wildlife "critters" in his backyard along with birds, butterflies, bumble bees, possums, raccoons, deer, and local points of interest. He takes his comet images with a 30-cm LX200 GPS Schmidt–Cassegrain and a focal reducer which brings the f-ratio down from f/10 to f/4.7. Charles uses an

Fig. 15.9. (a) Charles Bell of Vicksburg, Mississippi is one of the world's most active comet imagers. Image: Charles Bell. (b) The business end of Charles Bell's 30-cm Schmidt–Cassegrain incorporates a focal reducer, an SBIG ST-8XME CCD, an SBIG AO-7 adaptive optics guider and an SBIG photometric filter wheel. Image: Charles Bell.

SBIG ST-8XME CCD with a built in guide chip along with an SBIG AO-7 adaptive optics (tilt mirror) guider for rapid response autoguiding and a photometric UBVRI filter wheel (see Fig. 15.9b). This arrangement gives him an image scale of about 1.3 arc-sec per pixel and a 33×22 arc-min field of view. Software Bisque's The Sky 6 Pro running on a laptop controls the telescope. Like Gustavo Muler, Charles takes thousands of comet images per year, generally down to seventeenth magnitude.

John Drummond

New Zealand amateur astronomer John Drummond has been interested in astronomy since the age of ten and owns a variety of telescopes including two 410-mm f/4.5 Newtonian reflectors. One is a Dobsonian and the other is an equatorially mounted system. His Possum Observatory rotates in its entirety as the entire structure sits, via eight rollers, atop a 1.8-m steel ring which is fixed to the ground. The observatory is based on a design by the late Bill McLachlan. John lives on New Zealand's North Island, in the town of Gisborne. John's comet images have been taken with a Canon 350D (modified), a Canon 10D and an SBIG STL11000M CCD mainly using the equatorially mounted Newtonian, but bright comets have often been captured with 300-mm focal length lenses. In addition John sometimes uses a 20-cm aperture f/4 Vixen Newtonian.

John's website can be found at www.possumobservatory.co.nz.

Martin Mobberley

I thought I should mention my own comet photography and imaging history within my own book as otherwise it might seem a bit odd! However, I should stress that compared to many other comet enthusiasts mentioned in this book, my scientific contribution is very minor. I am quite a fickle comet observer as I easily get distracted and so I image most types of object in the night sky when there are few good comets around.

My first attempt to see a comet was shortly after I had applied for membership of the British Astronomical Association in 1969, age eleven! The comet in question was called Tago–Sato–Kosaka and had a designation of 1969g, or 1969 IX, but under the modern naming system it becomes C/1969 T1 (Tago–Sato–Kosaka). I was pretty sure I had managed to locate it in my small telescope from the information on the BAA Circular but now, more than 40 years later, I am not 100% convinced that I did. Throughout the 1970s I observed a few comets but cloud and inexperience took their toll on my teenage enthusiasm! However, from 1980 I became a serious observer and searched the skies visually for comets from that year, until 1984, with a 14-in. (35.6-cm) Newtonian at 44×. As there is no comet Mobberley in the database you can see that I was unsuccessful! I took my first comet photograph in 1982. It was of comet Austin 1982g, or, in modern terminology, C/1982 M1 (Austin).

From 1985, with comet Halley returning, photographing comets became my top astronomical priority. I even perfected the time honored practice of offset guided comet photography from 1985 to 1993 and set up darkrooms at two houses for developing and printing the photographs. I started using CCDs for comet imaging from 1992 and my last film based photographs of a comet were for comet Hale–Bopp in 1997. Over the years I have owned a number of large telescopes, including the 14-in. Newtonian and a bigger 19.3-in. Newtonian, as well as a fast 160-mm aperture f/3.6 Takahashi astrograph. These days my main instrument is a Celestron 14 Schmidt–Cassegrain (used at f/7.7) on an excellent Paramount ME equatorial mounting (see Fig. 15.10); this system is coupled to an SBIG ST9XE CCD giving a 13 arc-min field of view. I still use the fast Takahashi 160-mm astrograph with a Canon digital SLR and sometimes I image remotely over the Internet using the Global Rent-a-Scope

Fig. 15.10. The author, with his Celestron 14 (and long dew cap) on its Paramount ME mount. Image: Martin Mobberley.

(GRAS) facilities in New Mexico, USA and Moorock, Australia. In total I have taken more than 500 photographs and images of some 200 comets so far. My all time comet highlight was traveling to Tenerife in March 1996 to see comet Hyakutake and 50° of tail stretch across the sky, at the zenith, as it flew past the Earth. Amazing!

APPENDIX

Comet Resources

Comet Books

Cometography: A Catalog of Comets, by Gary Kronk.

Cometography Volume 1. Ancient – 1799. Cambridge University Press 1999. ISBN-13: 978-0521585040.

Cometography Volume 2. 1800 – 1899. Cambridge University Press 2003. ISBN-13: 978-0521585057.

Cometography Volume 3. 1900 – 1932. Cambridge University Press 2007. ISBN-13: 978-0521585064.

Cometography Volume 4. 1933 – 1959. Cambridge University Press 2008. ISBN-13: 978-0521585071.

Cometography Volume 5. 1960 – 1982. Cambridge University Press 2010. ISBN-13: 978-0521872263.

David Levy's Guide to Observing and Discovering Comets. Cambridge University Press 2003. ISBN-13: 978-0521520515.

The Greatest Comets in History: Broom Stars and Celestial Scimitars by David Seargent. Springer. ISBN-13: 978-0387095127. 2008.

Comets: A Chronological History of Observation, Science, Myth, and Folklore by Donald K. Yeomans. John Wiley & Sons 1991. ISBN-13: 978-0471610113.

Great Comets by Robert Burnham. Cambridge University Press 2000. ISBN-13: 978-0521646000.

383

Comets and How to Observe Them by Richard Schmude. Springer 2010. ISBN-13: 978-1441957894.

Observing Comets by Nick James and Gerald North. Springer 2002. ISBN-13: 978-1852335571.

Comets, Popular Culture, and the Birth of Modern Cosmology by Sara Schechner Genuth. Princeton University Press 1999. ISBN-13: 978-0691009254.

Fire in the Sky: Comets and Meteors, the Decisive Centuries in British Art and Science by Olson and Pasachoff. Cambridge University Press 1999. ISBN-13: 978-0521663595.

Comets II (University of Arizona Space Science Series) by Festou, Keller and Weaver. University of Arizona Press 2004. ISBN-13: 978-0816524501.

Planetarium/Telescope Control Software

Guide 8.0 – http://www.projectpluto.com/
The Sky X – http://www.bisque.com/sc/pages/TheSkyXFamily.aspx
Maxim DL – http://www.cyanogen.com/maxim_main.php

Orbital Elements for Popular Planetarium Software Packages

IAU site – http://www.minorplanetcenter.org/iau/Ephemerides/Comets/SoftwareComets.html
Comet Orbits – http://jcometobs.web.fc2.com/

Comet Astrometry/Photometry/ Twilight Software

Astrometrica – http://www.astrometrica.at/
Comet for Windows – http://www.aerith.net/project/comet.html
Focas II (Spanish) – http://astrosurf.com/cometas-obs/_Articulos/Focas/Focas.htm
CmtWin – http://www.inourfamily.com/sites/cmtwin/
GraphDark – http://www.rfleet.clara.net/graphdark/download.htm

Image Processing Software

AIP4Win – http://www.willbell.com/aip/index.htm
CCDSoft – http://www.bisque.com/sc/shops/store/CCDSoftWin2.aspx

IRIS – http://www.astrosurf.com/buil/us/iris/iris.htm
Maxim DL – http://www.cyanogen.com/maxim_main.php
Astroart – http://www.msb-astroart.com/
Images Plus – http://www.mlunsold.com/
Photoshop – http://www.adobe.com/products/photoshop/
Paint Shop Pro – http://www.corel.com
Registax – http://www.astronomie.be/registax/
Deep Sky Stacker – http://deepskystacker.free.fr/english/

CCD Cameras and DSLRs

SBIG – http://www.sbig.com/
Starlight Xpress – http://www.starlight-xpress.co.uk/
Apogee – http://www.ccd.com/astronomy.html
QSI – http://www.qsimaging.com/
QHY – http://www.qhyccd.com/
Atik – http://www.atik-cameras.com/
Audine project – http://www.astrosurf.com/audine/English/index_en.htm
Canon DSLRs – http://www.canon.com/eos-d/
Nikon – http://imaging.nikon.com/products/imaging/index.htm

Lumicon Swan Band Comet Filters

http://www.lumicon.com/telescope-accessories-list.php?cid=17&cn=Comet+Filters

Optical Tube Assemblies/Astrographs

Orion Optics AG – http://www.orionoptics.co.uk/AG/agrange.html
Takahashi – http://www.takahashiamerica.com/
Astrosysteme Austria – http://www.astrosysteme.at/eng/astrographs.html
Celestron Edge HD – http://www.celestron.com/c3/page.php?PageID=389
RCOS – http://www.rcopticalsystems.com/
Planewave – http://www.planewaveinstruments.com/
Astro Optik – http://www.astrooptik.com/

Coma Correctors

TeleVue Paracorr – http://www.televue.co.uk/paracorr.htm

Telescope Mountings

Paramount ME – http://www.bisque.com/sc/pages/Paramount-ME.aspx
Astrophysics – http://www.astro-physics.com/
Astrotrac (portable) – http://www.astrotrac.com/
Losmandy – http://www.losmandy.com/
Gemini G42 – http://www.astronomy.hu/g42.htm
10micron – http://www.10micron.com/english/homepage-eng.htm
Celestron CGE Pro – http://www.celestron.com/c3/product.php?ProdID=549
Takahashi – http://www.takahashiamerica.com/
Vixen – http://www.vixenoptics.com/mounts.htm
ASA direct drive – http://www.astrosysteme.at/eng/mounts.html
Periodic error data – http://demeautis.christophe.free.fr/ep/pe.htm

Comet Data Websites

Ephemerides – http://www.minorplanetcenter.org/iau/Ephemerides/Comets/
NEO Confirmation Page – http://www.minorplanetcenter.org/iau/NEO/ToConfirm.html
NEO and Comet checker – http://scully.cfa.harvard.edu/~cgi/CheckNEOCMT
IAU Ephemeris Service – http://www.minorplanetcenter.org/iau/MPEph/MPEph.html
Bright comets weekly – http://www.aerith.net/comet/weekly/current.html
Comet/deep sky encounters – http://www.aerith.net/comet/rendezvous/current.html
Comet recovery targets – http://www.aerith.net/comet/recovery.html
Astrosite Groningen – http://www.shopplaza.nl/astro/

Professional Patrol Websites

Catalina Sky Survey – http://www.lpl.arizona.edu/css/
LINEAR – http://www.ll.mit.edu/mission/space/linear/
Siding Spring Survey – http://www.lpl.arizona.edu/css/
Spacewatch – http://spacewatch.lpl.arizona.edu/
LONEOS – http://www.lowell.edu/users/elgb/loneos_disc.html
SOHO sungrazers – http://sungrazer.nrl.navy.mil/
Pan-STARRS – http://pan-starrs.ifa.hawaii.edu/public/
Large Synoptic Survey Telescope – http://www.lsst.org/lsst

Professional Patrol Sky Coverage

http://scully.cfa.harvard.edu/~cgi/SkyCoverage.html

Comet Observer Groups on Yahoo!

Comets mailing list – http://tech.groups.yahoo.com/group/comets-ml/
Comet images – http://tech.groups.yahoo.com/group/Comet-Images/
Comet observing – http://tech.dir.groups.yahoo.com/group/CometObs/
Comet chasing – http://tech.dir.groups.yahoo.com/group/CometChasing/
Spanish comet observers – http://es.groups.yahoo.com/group/Cometas_Obs/

Organizations

BAA Comet Section – http://www.ast.cam.ac.uk/~jds/
The Astronomer Comets page – http://www.theastronomer.org/
comets.html
International Comet Quarterly – http://www.cfa.harvard.edu/iau/icq/
icq.html
Mark Kidger's Spanish page – http://www.observadores-cometas.com/
U.A.I. Sezione Comete (Italy) – http://comete.uai.it/
CARA project (Italy) – http://cara.uai.it/
ALPO – http://alpo-astronomy.org/
VdS-Fachgruppe Kometen – http://kometen.fg-vds.de/fgk_hpe.htm
Astrosurf comet observations – http://astrosurf.com/cometas-obs/
T3 project – http://asteroidi.uai.it/t3.htm
Liada Comets Section – http://cometobservations.freeservers.com/

Comet Enthusiasts' Personal Sites

Michael Jäger – http://www.cometpieces.at/
Rolando Ligustri – http://picasaweb.google.com/astroligu/
CometeDiRolandoLigustriCASTItalia#
Gary Kronk's Cometography – http://cometography.com/
Gerald Rhemann – http://www.astrostudio.at/all.php
Peter Birtwhistle – http://www.birtwhistle.org/
Tsutomu Seki – http://www.comet-web.net/~tsutomu-seki/
Seiichi Yoshida – http://www.aerith.net/
Maik Meyer's pages – http://www.comethunter.de/
Kazuo Kinoshita – http://jcometobs.web.fc2.com/
Gustavo Muler – http://astrosurf.com/nazaret/cometas.shtml
Terry Lovejoy – http://www.pbase.com/terrylovejoy/comet_photos

Stefan Beck – http://www.cometchaser.de/comets.html
Stefan Beck's observer list – http://www.cometchaser.de/observers/country.html
David Levy – http://www.jarnac.org/
Bernhard Häusler – http://www.amication.de/Bernhards_Comet_Project/
Michael Mattiazzo – http://members.westnet.com.au/mmatti/sc.htm
Martin Mobberley – http://martinmobberley.co.uk/
Remanzacco blog – http://remanzacco.blogspot.com/
Crni Vrh Observatory – http://www.observatorij.org/
Rezman Observatory – http://www.rezman-obs.si/
Clay Sherrod (ASO) – http://www.arksky.org/
Johannes Schedler – http://panther-observatory.com/comets.htm
Konrad Horn – http://www.konradhorn.de/
John Drummond – http://www.possumobservatory.co.nz/
Juan Antonio Henríquez Santana – http://atlante.teobaldopower.org/

Index